The Use of Waste Products from the Food Industry to Obtain High Value-Added Products

The Use of Waste Products from the Food Industry to Obtain High Value-Added Products

Editors

Stanisław Kowalski
Dorota Gumul

MDPI

Basel • Beijing • Wuhan • Barcelona • Belgrade • Novi Sad • Cluj • Manchester

Editors
Stanisław Kowalski
University of Agriculture in Krakow
Kraków
Poland

Dorota Gumul
University of Agriculture in Krakow
Kraków
Poland

Editorial Office
MDPI AG
Grosspeteranlage 5
4052 Basel, Switzerland

This is a reprint of articles from the Special Issue published online in the open access journal *Foods* (ISSN 2304-8158) (available at: https://www.mdpi.com/journal/foods/special_issues/EYDBU64HED).

For citation purposes, cite each article independently as indicated on the article page online and as indicated below:

Lastname, A.A.; Lastname, B.B. Article Title. *Journal Name* **Year**, *Volume Number*, Page Range.

ISBN 978-3-7258-1483-1 (Hbk)
ISBN 978-3-7258-1484-8 (PDF)
doi.org/10.3390/books978-3-7258-1484-8

Contents

About the Editors vii

Stanisław Kowalski and Dorota Gumul
The Use of Waste Products from the Food Industry to Obtain High Value-Added Products
Reprinted from: *Foods* **2023**, *12*, 847, doi:10.3390/foods13060847 1

Ibtissam Bouhzam, Rosa Cantero, María Margallo, Rubén Aldaco, Alba Bala, Pere Fullana-i-Palmer and Rita Puig
Extraction of Bioactive Compounds from Spent Coffee Grounds Using Ethanol and Acetone Aqueous Solutions
Reprinted from: *Foods* **2023**, *12*, 4400, doi:10.3390/foods12244400 7

Dorota Gumul, Rafał Ziobro, Jarosław Korus and Magdalena Surma
Pulp from Colored Potatoes (*Solanum tuberosum* L.) as an Ingredient Enriching Dessert Cookies
Reprinted from: *Foods* **2023**, *12*, 3735, doi:10.3390/foods12203735 24

Zuzana Ciesarová, Kristína Kukurová, Viera Jelemenská, Jana Horváthová, Janka Kubincová, Miona Belović and Aleksandra Torbica
Asparaginase Treatment of Sea Buckthorn Berries as an Effective Tool for Acrylamide Reduction in Nutritionally Enriched Wholegrain Wheat, Rye and Triticale Biscuits
Reprinted from: *Foods* **2023**, *12*, 3170, doi:10.3390/foods12173170 39

Sara Lara-Abia, Gloria Lobo, Noelia Pérez-Pascual, Jorge Welti-Chanes and M. Pilar Cano
Improvement in the Stability and Bioaccessibility of Carotenoid and Carotenoid Esters from a Papaya By-Product Using O/W Emulsions
Reprinted from: *Foods* **2023**, *12*, 2654, doi:10.3390/foods12142654 57

Nashi K. Alqahtani, Tareq M. Alnemr, Abdullah K. Alsalem, Mutlaq M. Alotaibi and Maged Mohammed
Experimental Investigation and Modeling for the Influence of Adding Date Press Cake on Drinkable Yogurt Quality
Reprinted from: *Foods* **2023**, *12*, 1219, doi:10.3390/foods12061219 76

Dorota Gumul, Marek Kruczek, Eva Ivanišová, Jacek Słupski and Stanisław Kowalski
Apple Pomace as an Ingredient Enriching Wheat Pasta with Health-Promoting Compounds
Reprinted from: *Foods* **2023**, *12*, 804, doi:10.3390/foods12040804 98

Ibtissam Bouhzam, Rosa Cantero, Mercè Balcells, María Margallo, Rubén Aldaco, Alba Bala, et al.
Environmental and Yield Comparison of Quick Extraction Methods for Caffeine and Chlorogenic Acid from Spent Coffee Grounds
Reprinted from: *Foods* **2023**, *12*, 779, doi:10.3390/foods12040779 114

Nashi K. Alqahtani, Tareq M. Alnemr, Abdullah M. Alqattan, Salah M. Aleid and Hosam M. Habib
Physicochemical and Sensory Properties and Shelf Life of Block-Type Processed Cheeses Fortified with Date Seeds (*Phoenix dactylifera* L.) as a Functional Food
Reprinted from: *Foods* **2023**, *12*, 679, doi:10.3390/foods12030679 133

Joanna Grzelczyk, Joanna Oracz and Ilona Gałazka-Czarnecka
Quality Assessment of Waste from Olive Oil Production and Design of Biodegradable Packaging
Reprinted from: *Foods* **2022**, *11*, 3776, doi:10.3390/foods11233776 151

Liqin Qin, Jinghao Ma, Huafeng Tian, Yanli Ma, Qiuhua Wu, Shuang Cheng and Guangsen Fan
Production of Xylooligosaccharides from *Jiuzao* by Autohydrolysis Coupled with Enzymatic Hydrolysis Using a Thermostable Xylanase
Reprinted from: *Foods* **2022**, *11*, 2663, doi:10.3390/foods11172663 . . . . **167**

About the Editors

Stanisław Kowalski

He is a graduate of the University of Agriculture in Krakow, where he defended his doctoral thesis, and is currently employed as a professor at the university. He has broad research interests that include the following:

- The use of polysaccharide hydrocolloids as substances influencing the texture and rheology of food products;
- Quality characteristics of natural honey, with a particular emphasis on the impact of thermal processes on the quality of honey, particularly on the synthesis of hydroxymethyl furfural and changes in enzymatic activity;
- Supplementation of bread with pseudocereals and other raw materials of plant origin, influencing changes in the quality parameters and nutritional properties of bread;
- Use of edible insects in food technology in the context of the possibility of using this raw material to enrich food products and develop new products, as well as consumer acceptance of food with the addition of edible insects.

Stanisław Kowalski is the author of many scientific articles in the field of food technology published in prestigious international scientific journals.

Dorota Gumul

She is a graduate of the University of Agriculture in Krakow, where she defended her doctoral thesis, and is currently employed as a professor at the university. The main topics of Dorota Gumul's scientific activity include several issues related to the following:

- Examination of the physical, chemical, functional, and morphological properties of starches of various botanical origins, often obtained from alternative sources, with an indication of their application possibilities;
- Searching for various possibilities of modifying starch of various botanical origins and their impact on the physical, chemical, and functional properties of the modification products of this polysaccharide;
- Determining the content of biologically active substances from the group of antioxidants occurring in plant raw materials before and after culinary processing and technological processes;
- Fortification of cereal products, especially gluten-free ones, with raw materials constituting a source of bioactive and health-promoting compounds.

Dorota Gumul is an expert in the field of food technology, the chemical composition of products, food analysis, food chemistry, food and nutrition, and, in particular, research on antioxidants from the group of polyphenols and vitamin E, antioxidant capacity, dietary fibre, and the nutritional value of products. She is also an expert in the extrusion process and its impact on the abovementioned compounds.

foods

MDPI

Editorial

The Use of Waste Products from the Food Industry to Obtain High Value-Added Products

Stanisław Kowalski * and Dorota Gumul

Department of Carbohydrate Technology and Cereal Processing, Faculty of Food Technology, University of Agriculture in Krakow, 122 Balicka Street, 30-149 Krakow, Poland; rrgumul@cyf-kr.edu.pl
* Correspondence: rrkowals@cyf-kr.edu.pl

Globalization and population expansion are driving the evolution of the food industry, offering an expanded array of food choices to cater to increasingly discerning consumers. This includes those opting for alternative dietary approaches, such as gluten-free diets, even in the absence of celiac disease [1–5]. The processing of both plant and animal raw materials generates a large amount of waste. According to FAO 2019, the global level of waste from animal and meat production is around 13% and from fruit and vegetable processes it is around 22%. According to the report "Wastage footprint: Impacts on natural resources–Summary report" [6], fruit processing, packaging, distribution and consumption generate a huge amount of fruit waste worldwide, approximately 1.81, 6.53 and 32.0 million tons, respectively [7]. In fruit and vegetable processing, an important problem is the management or disposal of waste products, i.e., the parts not used in the technological process, which constitute 10 to 35% of the weight of the processed raw material. However, in the juice production process, the mass of pomace depends on the pressing efficiency. During traditional pressing, the proportion of pomace is 20–25% of the initial weight of the raw material [7,8], and the use of enzyme preparations allows the proportion of pomace to be reduced to approximately 12% of the initial fruit weight. In Europe, by-products from fruit processing account for 8% of all food sector waste [7,8].

Unfortunately, discarded solid and liquid materials are rarely used in the production chain and often serve as animal feed products. Much of this waste remains unprocessed, which causes additional disposal costs for treatment plants and increases the biological burden of wastewater [9–12]. However, due to growing environmental concerns, intensive research needs to be carried out so that more food waste can be used to add value to new food products. This will consequently lead to maximum benefits for industry, the environment and consumers. Due to the increased awareness of both producers and consumers regarding ecological issues and sustainable development, there is a growing interest in reusing food industry by-products or waste. The number of scientific articles that have been written and published on this topic over the last 30 years shows an exponential increase, which emphasizes its importance and relevance [10,12–19].

Food industry waste can provide a whole range of substances that can improve the nutritional value and functionality of new products. These substances often contain significant amounts of protein, dietary fiber, fat, vitamins, polyphenols, vitamin C and vitamin E, phytosterols, lignans, etc. [13,20]. Examples of waste products that enhance the health-boosting qualities of new products include fruit and vegetable pomace, flour or post-extraction pulp, molasses and other by-products. These products can also be used as substrates for the production of dyes, vitamins or a whole range of other biologically active substances.

In summary, the possibilities of using the entire range of waste products to obtain products with high added value depend on the ingenuity and creativity of both scientists and food producers. We are therefore constantly facing new challenges but also new opportunities to obtain valuable products from waste. This strategy fits perfectly into the

Citation: Kowalski, S.; Gumul, D. The Use of Waste Products from the Food Industry to Obtain High Value-Added Products. *Foods* **2024**, *13*, 847. https://doi.org/10.3390/foods13060847

Received: 4 March 2024
Accepted: 6 March 2024
Published: 11 March 2024

current trend of zero-waste technology, which is one of the 17 sustainable development goals included in the United Nations resolution [21].

In the first edition of this Special Issue of *Foods*, "The Use of Waste Products from the Food Industry to Obtain High Value-Added Products", we presented some developments in and possibilities of the use of these products, a brief summary of which is presented below (Table 1 and Figure 1).

Table 1. List of articles published in the 1st edition of this Special Issue of *Foods*, "The Use of Waste Products from the Food Industry to Obtain High Value-Added Products".

Article Symbol	Bibliography	By-Product Used
SI 1	Gumul, D.; Ziobro, R.; Korus, J.; Surma, M. Pulp from Colored Potatoes (*Solanum tuberosum* L.) as an Ingredient Enriching Dessert Cookies. *Foods* **2023**, *12*, 3735.	Potato pulp
SI 2	Gumul, D.; Kruczek, M.; Ivanišová, E.; Słupski, J.; Kowalski, S. Apple Pomace as an Ingredient Enriching Wheat Pasta with Health-Promoting Compounds. *Foods* **2023**, *12*, 804.	Apple pomace
SI 3	Lara-Abia, S.; Lobo, G.; Pérez-Pascual, N.; Welti-Chanes, J.; Cano, M.P. Improvement in the Stability and Bioaccessibility of Carotenoid and Carotenoid Esters from a Papaya By-Product Using O/W Emulsions. *Foods* **2023**, *12*, 2654	Papaya pomace
SI 4	Grzelczyk, J.; Oracz, J.; Gałąazka-Czarnecka, I. Quality Assessment of Waste from Olive Oil Production and Design of Biodegradable Packaging. *Foods* **2022**, *11*, 3776.	Olive oil pomace
SI 5	Qin, L.; Ma, J.; Tian, H.; Ma, Y.; Wu, Q.; Cheng, S.; Fan, G. Production of Xylooligosaccharides from Jiuzao by Autohydrolysis Coupled with Enzymatic Hydrolysis Using a Thermostable Xylanase. *Foods* **2022**, *11*, 2663.	Jiuzao (a mixture of grain and rice hull residues after solid-state fermentation)
SI 6	Bouhzam, I.; Cantero, R.; Margallo, M.; Aldaco, R.; Bala, A.; Fullana-i-Palmer, P.; Puig, R. Extraction of Bioactive Compounds from Spent Coffee Grounds Using Ethanol and Acetone Aqueous Solutions. *Foods* **2023**, *12*, 4400.	Spent coffee Grounds
SI 7	Bouhzam, I.; Cantero, R.; Balcells, M.; Margallo, M.; Aldaco, R.; Bala, A.; Fullana-i-Palmer, P.; Puig, R. Environmental and Yield Comparison of Quick Extraction Methods for Caffeine and Chlorogenic Acid from Spent Coffee Grounds. *Foods* **2023**, *12*, 779.	Spent coffee Grounds
SI 8	Alqahtani, N.K.; Alnemr, T.M.; Alqattan, A.M.; Aleid, S.M.; Habib, H.M. Physicochemical and Sensory Properties and Shelf Life of Block-Type Processed Cheeses Fortified with Date Seeds (*Phoenix dactylifera* L.) as a Functional Food. *Foods* **2023**, *12*, 679.	Date fruit seeds
SI 9	Alqahtani, N.K.; Alnemr, T.M.; Alsalem, A.K.; Alotaibi, M.M.; Mohammed, M. Experimental Investigation and Modeling for the Influence of Adding Date Press Cake on Drinkable Yogurt Quality. *Foods* **2023**, *12*, 1219.	Date press cake
SI 10	Ciesarová, Z.; Kukurová, K.; Jelemenská, V.; Horváthová, J.; Kubincová, J.; Belović, M.; Torbica, A. Asparaginase Treatment of Sea Buckthorn Berries as an Effective Tool for Acrylamide Reduction in Nutritionally Enriched Wholegrain Wheat, Rye and Triticale Biscuits. *Foods* **2023**, *12*, 3170.	Sea buckthorn berry pomace

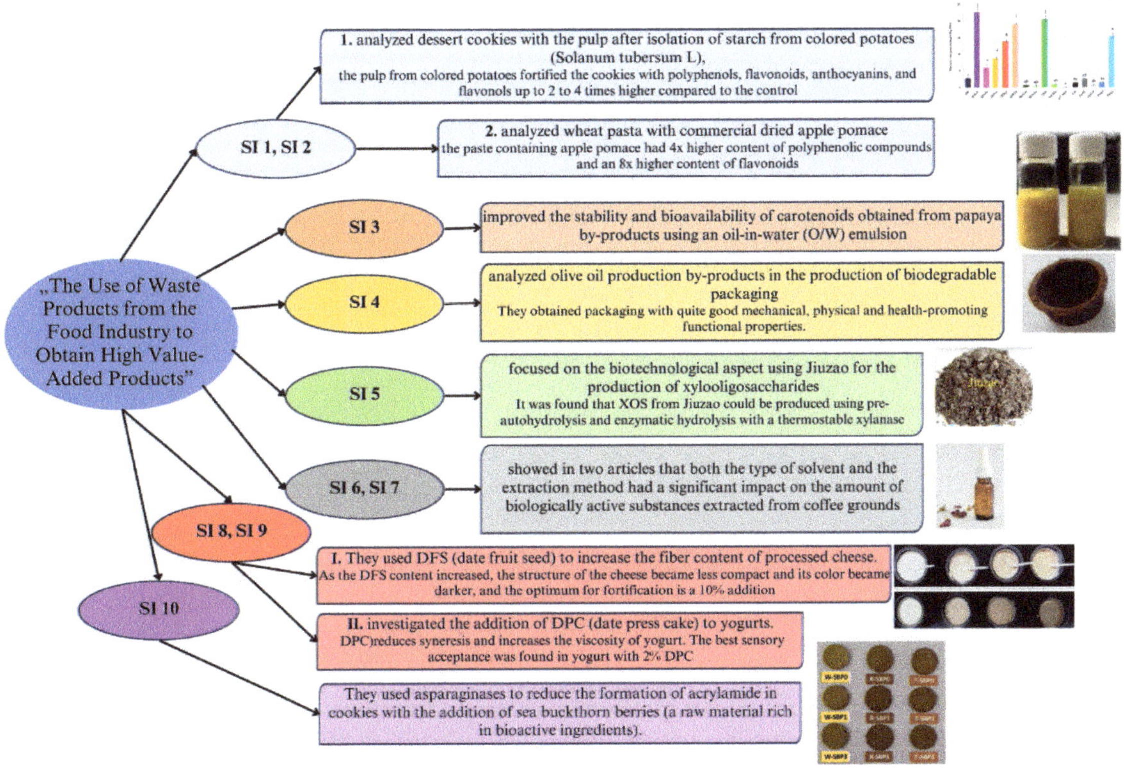

Figure 1. Main directions of research presented in the Special Issue.

Gumul et al. (2023) analyzed the pulp after the isolation of starch from colored potatoes (*Solanum tubersum* L.) as an ingredient to enrich dessert cookies. It was emphasized that the above-mentioned pulp is a very rich source of polyphenols and fiber. It was found that the pulp from colored potatoes fortified the cookies with polyphenols, flavonoids, anthocyanins and flavonols in amounts that were up to two to four times higher compared to the control cookie. The dessert cookies that contained potato pulp showed up to twice the fiber content and 17% higher protein content, while the fat and ash content remained unchanged compared to the control. Moreover, the cookies with potato pulp were characterized by a 30% lower content of hydroxymethylfurfural and an approximately 40% higher acrylamide content, while showing good physical properties of the final product [SI 1]. In a subsequent publication, Gumul et al. (2023) enriched wheat pasta with commercial dried apple pomace. It was found that the paste containing apple pomace had a four times higher content of polyphenolic compounds and an eight times higher content of flavonoids. Due to the fact that apple pomace is characterized by significant amounts of phenolic acids, quercetin and its derivatives, flavonols and dihydrochalcones, especially the unique phloridzin, this study showed that paste enriched with it also has a high content of these bioactive compounds. Moreover, apple-pomace-fortified paste was characterized by a large amount of dietary fiber, lower protein and fat content, and a higher amount of minerals. It was found that 10% addition guaranteed the quality and beneficial health-promoting properties of the pastes [SI 2]. Both of the above publications take into account the global trend of using zero-waste food technology to expand the range of products enriched with health-promoting compounds derived from waste products.

The aim of the study by Lara-Abia et al. (2023) was to improve the stability and bioavailability of carotenoids obtained from papaya by-products using an oil-in-water (O/W) emulsion. The study investigated the impact of varying concentrations of pectin (1%, 2%, and 3%), a high-molecular emulsifier, combined with Tween 20, a low-molecular emulsifier, under different homogenization conditions. The aim was to identify the optimal parameters for creating stable oil-in-water (O/W) emulsions with encapsulated carotenoids. Three oils were used to formulate these O/W emulsions: soybean, sunflower and coconut. The microstructure (confocal and optical microscopy) of the O/W carotenoid emulsions and their behavior during in vitro digestion phases were investigated. Additionally, the bioavailability of select individual encapsulated papaya carotenoids was tested. It was found that O/W carotenoid emulsions from sunflower oil had a smaller average particle size, higher negative ζ potential and higher viscosity than O/W emulsions from soybeans. Particle size reduction in O/W emulsions using the HPH process improved the bioavailability of papaya-encapsulated carotenoids. In these O/W emulsions, depending on the vegetable oil, the carotenoid with the highest bioavailability was lycopene (71–64%), and that with the lowest bioavailability was (all-E)-β-cryptoxanthin laurate (7–4%). Analogous properties were demonstrated by O/W carotenoid emulsions with soybean oil. It was found that the O/W coconut oil microemulsion was not stable. It has been shown that the high degree of unsaturation of fatty acids in sunflower oil and soybean oil can promote the formation of small droplets and increase the elasticity of the interfacial surface in emulsions due to unsaturated bonds. Studies have proven that microencapsulated carotenoids from papaya by-products have better bioavailability compared to non-microencapsulated products due to the protective effect [SI 3].

Grzelczyk et al. (2003) used by-products generated during olive oil production to prepare biodegradable packaging materials. The authors used 70–80% olive oil, flour/groats and lecithins to produce biodegradable products. Then, they baked these dishes for an hour and a half at 180 degrees Celsius and then covered them with beeswax and heated them again. The basic mechanical and physical parameters of these vessels were examined, and a thermal analysis of the final products was performed. The analysis encompassed an examination of the color and functional characteristics of these containers, alongside assessments of their biodegradability. Furthermore, the utilization of waste products with significant health-promoting benefits prompted investigations into antioxidant activity and total polyphenol content. The authors concluded that it is possible to create disposable packaging from natural sources using olive oil by-products, which can easily replace plastic packaging because they have good functional mechanical and physical properties. A storage test of the produced disposable packaging clearly proved that time has no impact on the properties of the product [SI 4].

Qin et al. (2003) analyzed the biotechnological aspect of using Jiuzao for the production of xylooligosaccharides. These authors investigated the preparation of xylooligosaccharides (XOSs) from Jiuzao through autohydrolysis combined with enzymatic hydrolysis using a thermostable xylanase. It was found that XOSs from Jiuzao could be produced using pre-autohydrolysis and enzymatic hydrolysis with a thermostable xylanase. It was shown that treating Jiuzao at a temperature of 181.5 °C for 20 min at a solid-to-liquid ratio of 1:13.6 enabled efficient hydrolysis with thermostable xylanase to produce XOSs. After enzymatic optimization of the hydrolysis conditions, the highest yield of XOSs from Jiuzao was 34.2% at 60 °C and pH 5 with 15 U/mL XynAR for 2 h. This process holds great promise for the practical production of XOSs from Jiuzao [SI 5].

One of the more widely discussed themes in this Special Issue was the use of spent coffee grounds after the coffee brewing process. This issue was described in two articles by Ibtissam Bouhzam and coauthors [SI 6, SI 7]. In their work, they focused on determining the efficiency of the extraction of polyphenolic compounds, especially chlorogenic acid and caffeine, using water–ethanol and water–acetone systems from coffee grounds. In their work, the authors took into account factors such as the method of drying coffee grounds and storage time on the extraction efficiency of individual biologically active compounds.

They also developed quick and effective methods for extracting the above-mentioned compounds using supramolecular solvents and water (mechanical mixing method and ultrasonic extraction support). The results indicated that both the type of solvent and the method of extraction had a significant influence on the amount of extracted biologically active substances.

The next group of articles published as part of this Special Issue were works on the use of by-products from date processing. This topic was studied by Alqahtani et al., and the results were described in two research papers. In the first work [SI 8], the by-products of date fruit seed (DFS) were used to increase the amount of fiber in processed cheese. DFS was used in the range from 0 to 20% and its addition was increased by 5%. The authors observed quality changes in the obtained product, in particular changes in the texture, microstructure, shelf life and sensory characteristics. As the proportion of DFS increased, the structure of the cheese became less compact and the color became darker. The authors suggested that the acceptance of such a fortified product should be achieved up to an additive amount of 10% [SI 8].

In the next work [SI 9], the authors focused on the possibility of using the remains of pressing date syrup. A by-product known as date press cake (DPC) was used to increase the functionality of yogurt. DPC (2, 4 and 6%) was used before fermentation, and the products were tested for changes in pH, acidity, syneresis, water holding capacity, viscosity, and color, among others. The changes in individual parameters were described using appropriate mathematical models. It was found, among other things, that the addition of DPC reduces syneresis and increases the viscosity of yogurt. The best sensory acceptance was found in yogurt with 2% DPC [SI 9].

Ciesarová et al. described the use of asparaginase to reduce the formation of acrylamide in biscuits with the addition of sea buckthorn berries. The sea buckthorn pomace used in this research was a by-product of juice production. Since this pomace is rich in biologically active substances, its reuse is justified. However, due to the high content of the amino acid asparagine in sea buckthorn, this promising material contributes to the undesirable formation of acrylamide. The use of an enzymatic treatment allowed the authors to reduce the asparagine content from 1834 mg/kg to 89 mg/kg and reduce the formation of acrylamide from 35% to 64%, depending on the type of flour from which the biscuits were baked [SI 10].

Author Contributions: Conceptualization, S.K. and D.G.; methodology, S.K. and D.G.; software, S.K. and D.G.; validation, S.K. and D.G.; formal analysis, S.K. and D.G.; investigation, S.K. and D.G.; resources, S.K. and D.G.; data curation, S.K. and D.G.; writing—original draft preparation, S.K. and D.G.; writing—review and editing, S.K. and D.G.; visualization, S.K. and D.G.; supervision, S.K. and D.G.; project administration, S.K. and D.G.; funding acquisition, S.K. and D.G. All authors have read and agreed to the published version of the manuscript.

Funding: This research received no external funding.

Data Availability Statement: No new data were created or analyzed in this study. Data sharing is not applicable to this article.

Acknowledgments: This work was supported by the Ministry of Science and Higher Education of the Republic of Poland.

Conflicts of Interest: The authors declare no conflicts of interest.

References

1. Leonard, M.M.; Sapone, A.; Catassi, C.; Fasano, A. Celiac Disease and Nonceliac Gluten Sensitivity. A Review. *JAMA* **2017**, *318*, 647–656. [CrossRef] [PubMed]
2. Kruczek, B.; Gumul, D.; Olech, E.; Gambuś, H. Diet and the Context of Fruit Industry. *Econ. Environ. Stud.* **2017**, *17*, 389–398. [CrossRef]
3. Gómez, M.; Martinez, M.M. Fruit and vegetable by-products as novel ingredients to improve the nutritional quality of baked goods. *Crit. Rev. Food Sci. Nutr.* **2018**, *58*, 2119–2135. [CrossRef] [PubMed]

4. Padayachee, A.; Day, L.; Howell, K.; Gidley, M.J. Complexity and health functionality of plant cell wall fibers from fruits and vegetables. *Crit. Rev. Food Sci. Nutr.* **2017**, *57*, 59–81. [CrossRef] [PubMed]
5. Makovicky, P.; Makovicky, P.; Caja, F.; Rimarova, K.; Samasca, G.; Vannucci, L. Celiac disease and gluten-free diet: Past, present and future. *Gastroenterol. Hepatol. Bed Bench* **2020**, *13*, 1–7. [PubMed]
6. FAO. *The State of Food and Agriculture 2019. Moving Forward on Food Loss and Waste Reduction*; FAO: Rome, Italy, 2019.
7. Srivastava, N.; Srivastava, M.; Alhazmi, A.; Kausar, T.; Haque, S.; Singh, R.; Ramteke, P.W.; Mishra, P.K.; Tuohy, M.; Leitgeb, M.; et al. Technological advances for improving fungal cellulase production from fruit wastes for bioenergy application: A review. *Environ. Pollut.* **2021**, *287*, 117370. [CrossRef] [PubMed]
8. Kawecka, L.; Galus, S. Wytłoki owocowe—Charakterystyka i możliwości zagospodarowania [Fruit pomace—Characteristics and possibilities of recycling]. *Postępy Tech. Przetwórstwa Spożywczego* **2021**, *1*, 156–167. (In Polish)
9. Sadh, P.K.; Duhan, S.; Duhan, J.S. Agro-industrial Wastes and Their Utilization Using Solid State Fermentation: A Review. *Bioresour. Bioproc.* **2018**, *5*, 1. [CrossRef]
10. Reguengo, L.M.; Salgaço, M.K.; Sivieri, K.; Maróstica, M.R.J. Agro-industrial by-products: Valuable sources of bioactive compounds. *Food Res. Int.* **2022**, *152*, 110871. [CrossRef] [PubMed]
11. Lemes, A.C.; Egea, M.B.; Oliveira Filho, J.G.; Gautério, G.V.; Ribeiro, B.D.; Coelho, M.A.Z. Biological Approaches for Extraction of Bioactive Compounds from Agroindustrial By-products: A Review. *Front. Bioeng. Biotechnol.* **2022**, *9*, 802543. [CrossRef] [PubMed]
12. Santos, D.; da Silva, J.A.L.; Pintado, M. Fruit and Vegetable By-Products' Flours as Ingredients: A Review on Production Process, Health Benefits and Technological Functionalities. *LWT* **2022**, *154*, 112707. [CrossRef]
13. Parveen, H.; Bajpai, A.; Bhatia, S.; Singh, S. Analysis of biscuits enriched with fibre by incorporating carrot and beetroot pomace powder. *Indian. J. Nutr. Diet.* **2017**, *54*, 403. [CrossRef]
14. Kultys, E.; Moczkowska-Wyrwisz, M. Effect of using carrot pomace and beetroot-apple pomace on physicochemical and sensory properties of pasta. *LWT* **2022**, *168*, 113858. [CrossRef]
15. Gumul, D.; Ziobro, R.; Korus, J.; Kruczek, M. Apple pomace as a source of bioactive polyphenol compounds in gluten-free breads. *Antioxidants* **2021**, *10*, 80. [CrossRef] [PubMed]
16. Maner, S.; Sharma, A.K.; Banerjee, K. Wheat flour replacement by wine grape pomace powder positively affects physical, functional and sensory properties of cookies. *Proc. Natl. Acad. Sci. India Sect. B Biol. Sci.* **2015**, *87*, 109–113. [CrossRef]
17. Padalino, L.; Conte, A.; Lecce, L.; Likyova, D.; Sicari, V.; Pellicanò, T.M.; Poiana, M.; Del Nobile, M.A. Functional Pasta with Tomato By-Product as a Source of Antioxidant Compounds and Dietary Fibre. *Czech J. Food Sci.* **2017**, *35*, 48–56. [CrossRef]
18. Xu, J.; Bock, J.E.; Stone, D. Quality and Textural Analysis of Noodles Enriched with Apple Pomace. *J. Food Process. Preserv.* **2020**, *44*, e14579. [CrossRef]
19. Tolve, R.; Pasini, G.; Vignale, F.; Favati, F.; Simonato, B. Effect of Grape Pomace Addition on the Technological, Sensory, and Nutritional Properties of Durum Wheat Pasta. *Foods* **2020**, *9*, 354. [CrossRef] [PubMed]
20. Sagar, N.A.; Pareek, S.; Sharma, S.; Yahia, E.M.; Lobo, M.G. Fruit and vegetable waste: Bioactive compounds, their extraction, and possible utilization. *Compr. Rev. Food Sci. Food Saf.* **2018**, *17*, 512–531. [CrossRef] [PubMed]
21. United Nations. Resolution Adopted by the General Assembly on 25 September 2015, 526 A/RES/70/1, Transforming our World: The 2030 Agenda for Sustainable Development, 527, United Nations. 2015. Available online: https://undocs.org/en/A/RES/70/1 (accessed on 5 March 2023).

foods

MDPI

Article

Extraction of Bioactive Compounds from Spent Coffee Grounds Using Ethanol and Acetone Aqueous Solutions

Ibtissam Bouhzam [1], Rosa Cantero [1], María Margallo [2], Rubén Aldaco [2], Alba Bala [3], Pere Fullana-i-Palmer [3] and Rita Puig [1,*]

[1] Department of Industrial and building Engineering, University of Lleida (UdL), Pla de la Massa, 8, 08700 Igualada, Spain; ibtissam.bouhzam@udl.cat (I.B.); rosa.cantero@udl.cat (R.C.)
[2] Department of Chemical and Biomolecular Engineering, University of Cantabria, Av. de Los Castros s/n, 39005 Santander, Spain; maria.margallo@unican.es (M.M.); ruben.aldaco@unican.es (R.A.)
[3] UNESCO Chair in Life Cycle and Climate Change ESCI-UPF, Pg. Pujades 1, 08003 Barcelona, Spain; alba.bala@esci.upf.edu (A.B.); pere.fullana@esci.upf.edu (P.F.-i.-P.)
* Correspondence: rita.puig@udl.cat

Abstract: Given global coffee consumption, substantial quantities of spent coffee grounds (SCGs) are generated annually as a by-product of brewing coffee. SCG, although rich in bioactive compounds, is nowadays disposed of. The objective of this study is to compare, for the first time and from the same SCG, the efficiency of ethanol–water mixtures and acetone–water mixtures for the recovery of total polyphenols, chlorogenic acid, and caffeine. Acetone at 20% (m/m) was the most convenient solvent to extract all three bioactive compounds simultaneously, yielding 4.37 mg of GAE/g SCG for total polyphenols, chlorogenic acid (0.832 mg 5-CQA/g SCG), and caffeine (1.47 mg/g SCG). Additionally, this study aims to address some challenges associated with the industrial-scale utilization of SCG as a raw material, encompassing factors such as pre-treatment conditions (natural drying and oven drying), storage duration, and the kinetics of the extraction process. No significant difference was observed between the natural drying and oven drying of SCG. In terms of storage duration, it is advisable to process the SCG within less than 3–4 months of storage time. A significant decline of 82% and 70% in chlorogenic acid (5-CQA) and caffeine contents, respectively, was observed after eight months of storage. Furthermore, the kinetic study for the recovery of total polyphenols revealed that the optimal extraction times were 10 min for acetone at 20% and 40 min for water, with a yield increase of 28% and 34%, respectively. What is remarkable from the present study is the approach considered, using the simplest operating conditions (minimal time and solvent-to-solid ratio, and ambient temperature); hence, at an industrial scale, energy and resource consumption and equipment dimensions can be together reduced, leading to a more industrially sustainable extraction process.

Keywords: spent coffee ground; extraction; chlorogenic acid; total polyphenols; caffeine; storage duration; storage conditions; kinetic study

Citation: Bouhzam, I.; Cantero, R.; Margallo, M.; Aldaco, R.; Bala, A.; Fullana-i-Palmer, P.; Puig, R. Extraction of Bioactive Compounds from Spent Coffee Grounds Using Ethanol and Acetone Aqueous Solutions. *Foods* **2023**, *12*, 4400. https://doi.org/10.3390/foods12244400

Academic Editors: Stanisław Kowalski and Dorota Gumul

Received: 9 November 2023
Revised: 1 December 2023
Accepted: 5 December 2023
Published: 7 December 2023

1. Introduction

The loss of biodiversity and unsustainable food waste, driven by a linear growth model, present interconnected global challenges with economic, environmental, and social repercussions amid increasing food demand [1–4].

Therefore, changing our food system and moving towards a more circular economy system is imperative to ensure resource efficiency and waste reduction, rebuild biodiversity, and address climate change issues [2]. In a circular economy, food systems need to be designed to cycle through the economy, where food loss and waste can be converted into a valuable resource [4]. Thus, research is needed to convert food waste into valuable resources.

Instant coffee manufacturing generates significant amounts of waste [5]. Typically, only about 30% of the mass of coffee beans can be extracted into the coffee that we drink, while the remaining 70% ends up as spent coffee grounds (SCG) with no commercial value. The high amount of waste generated highlights the need to develop innovative uses for spent coffee grounds [6–8].

Recent studies have revealed that spent coffee grounds still contain many biological components, such as polyphenols and caffeine, which exhibit important bioactive properties, including antioxidant and anti-inflammatory activities [5].

Plant polyphenols have attracted interest due to their potent antioxidant properties and their ability to protect against cancer development. Plant polyphenols include phenolic acids, flavonoids, and tannins, as well as the less prevalent forms of stilbenes and lignans. Phenolic acids can be divided into the following two classes: derivatives of benzoic acid, such as gallic acid, and derivatives of cinnamic acid, such as coumaric, caffeic, and ferulic acid [9]. The main phenolic acids reported to be present in spent coffee grounds are chlorogenic acids, and they belong to the second class (cinnamic derivatives) [10]. Chlorogenic acids (CGAs) belong to a group of esters that share structural resemblance with quinic acid (QA) and contain one or more cinnamate derivatives such as caffeic, ferulic, and p-coumaric acids [11]. The main groups of CGA are as follows: caffeoylquinic acids (CQA), with three isomers (3-, 4- and 5-CQA); dicaffeoylquinic acids (diCQA), with three isomers (3,4-diCQA; 3,5-diCQA; 4,5-diCQA); feruloylquinic acids (FQA), with three isomers (3-, 4- and 5-FQA); p-coumaroylquinic acids (pCoQA), with three isomers (3-, 4- and 5-pCoQA), and six mixed diesters of caffeoylferuloyl-quinic acids (CFAQ) [12]. Caffeoylquinic acids are the most abundant isomers found in SCG, which are thought to confer an array of health-enhancing advantages [13].

Caffeine is a widely used substance that is consumed in various forms, including coffee, tea, energy drinks, and dietary supplements. The primary source of caffeine is decaffeination. Spent coffee grounds can also be a viable source of caffeine extraction. Extracted caffeine is chemically identical to synthetic caffeine [14]. Therefore, the caffeine extracted from SCG can provide economic and environmental benefits and could be a promising alternative to synthetic caffeine.

Different methods have been developed for the extraction of polyphenols and caffeine from spent coffee grounds, including solid–liquid extraction, ultrasound extraction, microwave-assisted extraction, Soxhlet extraction, hydrothermal treatment, and other methods [5]. Among these techniques, solid–liquid extraction is the most widely used, requiring no specialized equipment [5,15]. Indeed, from the perspective of the sustainable valorization of SCG, avoiding toxic solvents is advised. Some authors recommended the use of polar solvents for the extraction of polyphenols, such as ethanol, methanol, or acetone [13]. On the other hand, caffeine is soluble in acetone, water, methanol, and ethanol [5]. Thus, ethanol and acetone could be used for the extraction of both polyphenols and caffeine.

Water is considered the greenest solvent overall; it has low toxicity, is nonflammable, non-explosive, naturally abundant, inexpensive, and can be readily separated from the reaction mixture [16]. Nevertheless, water alone does not always yield the desired extraction efficiency. On the other hand, some researchers have suggested that the addition of water to organic solvents, such as acetone, methanol, and ethanol, creates a more polar environment that facilitates the extraction of phenolic compounds [17]. Hence, ethanol–water and acetone–water mixtures can be used for the extraction of polyphenols.

Ethanol is a commonly used solvent because it has low toxicity, a low boiling point, and is readily available. These important characteristics make it an environmentally friendly alternative [18]. Furthermore, in the context of large-scale applications, ethanol emerges as a viable bio-solvent that can be generated from the process of alcoholic fermentation utilizing diverse sugar- or starch-containing feedstocks. Notably, the recycling potential inherent to ethanol augments its appeal as an environmentally sustainable option for process development [13].

Ethanol–water mixtures are widely used for the extraction of caffeine and total polyphenols. However, the influence of different proportions of ethanol–water on the extraction yield of total polyphenols, caffeine, and chlorogenic acid (all together) from the same SCG has scarcely been studied.

Additionally, despite numerous studies on the extraction of polyphenols, chlorogenic acid, and caffeine, the use of acetone as a solvent for their recovery has not yet been explored, even though it is considered a preferred solvent according to the solvent selection table of the Pfizer Company [19]. It is accessible, cost-effective, has a low boiling point, and completely dissolves in water.

On the other hand, it is challenging to find studies in the literature comparing various solvents for the simultaneous extraction of total polyphenols, chlorogenic acid, and caffeine from spent coffee grounds (SCG). Many papers tend to focus on introducing novel extraction procedures or optimizing conditions for already established methods. However, if the objective is to efficiently reuse food waste, such as SCG, into a valuable resource that is rich in health-enhancing compounds, it is imperative to compare diverse extraction alternatives using the same SCG, as previously justified by Bouhzam et al. [5].

Therefore, the objective of the present paper is to compare the efficiency of ethanol–water mixtures and acetone–water mixtures for the recovery of total polyphenols, chlorogenic acid, and caffeine from the same SCG. A global aim is to contribute to the circularity of food waste through a practical example. Experiments were performed under the simplest conditions described in the literature, including room temperature, the lowest liquid-to-solid ratio (5.7 mL/g SCG), and a short extraction time (1 min) simultaneously [20]. This study provides new insights into the influence of different solvent mixtures on the extraction yield and explores, for the first time, the potential of acetone–water mixtures as an alternative solvent for the extraction. Additionally, this study aims to address some challenges associated with the industrial-scale utilization of SCG as a raw material, encompassing factors such as storage pre-treatment and duration and the kinetics of the extraction process.

2. Materials and Methods

2.1. Spent Coffee Grounds (SCG)

Wet spent coffee grounds, derived from mixtures of Arabica and Robusta coffee varieties of the brand "Novell gourmet", were supplied by a coffee bar in the province of Barcelona (Spain). The mixture was dried naturally for one week and subsequently stored at room temperature in plastic capsules, covered but not hermetically sealed, for future extractions. Spent coffee grounds were originally obtained from infusions of 100% caffeinated coffee beans.

In Section 3.1.1, where the influence of the drying method was investigated, the SCG used was a blend resulting from the preparation of caffeinated (80% approximately) and decaffeinated (20%) coffee infusions. SCG was collected and separated into two fractions; one was left to dry naturally for 7 days (temperature: 23 °C, and humidity: 50%), while the other was dried in an oven at 50 °C for 24 h. Then, both fractions were stored in plastic capsules at room temperature, covered but not sealed hermetically.

2.2. Chemicals and Reagents

Acetonitrile (99.9%), ethanol (99.9%), acetone (99.8%), glacial acetic acid (99.8%), and Folin–Ciocalteus' phenol reagent were supplied from Scharlab (Barcelona, Spain). Sodium carbonate was purchased from Acros Organics (Geel, Belgium).

The standards used were as follows: 5-chlorogenic acid (5-O-Caffeoylquinic acid, 5-CQA, 98%) caffeine (1,3,7-trimethylxantine), and the HPLC grade were purchased from Sigma-Aldrich (St. Louis, MO, USA). Gallic acid (98%) was supplied from Scharlab (Barcelona, Spain). The chemicals were all of the analytical reagent grade.

A Milli-Q System, equipped with a 0.22-μm filter from Merck Millipore (Burlington, MA, USA), was used to prepare ultrapure water.

2.3. Equipment

The quantification of caffeine (Caf) and chlorogenic acid (5-CQA) was performed using a Waters 2695 high-performance liquid chromatography (HPLC) coupled to a 2998 UV Detector (Milford, MA, USA). A reverse phase column C8 (5 μm particle size, 100 Å pore size, 150 mm length, and 4.6 mm internal diameter) was provided by Scharlab (Barcelona, Spain). The data were processed using the Empower Solutions 2.0 Software (Orlando, FL, USA).

The quantification of the total polyphenols content was conducted using a Perkin Elmer 501S09110511 UV-spectrophotometer (Barcelona, Spain).

A J.P. Selecta ES-05 oven was used to dry SCG (Barcelona, Spain). For sample preparation, a Velp Scientifica F202A0176 vortex shaker (Usmate, MB, Italy) was used together with a J.P. Selecta 7002575 centrifuge and a 3000865 ultrasound bath (Barcelona, Spain).

2.4. Solid/Liquid Extraction

The efficiency of ethanol and acetone–water mixtures for total polyphenols, chlorogenic acid (5-CQA), and caffeine (Caf) extraction was investigated under the simplest conditions (room temperature, 5.7 mL/g SCG solvent/solid ratio, and 1 min extraction time). Different ratios of ethanol/water mixtures (0, 20, 40, 60, 80, and 96%) and acetone/water mixtures (0, 20, and 40%) were tested. A 0.7 g sample of spent coffee grounds (SCG) was mixed with 4 mL of the studied solution in a tube and stirred in a vortex shaker at 3000 rpm for 1 min. The mixture was then centrifuged for 30 min at 4200 rpm and filtered with a syringe filter (0.45 μm) before HPLC analysis. For caffeine quantification, 1:3 dilutions were used for the extracts obtained with 60, 80, and 96% ethanol/water mixtures to better integrate the caffeine peak. All experiments were performed in duplicate.

In Section 3.1.1, the two drying methods (oven and naturally drying) were compared by subjecting the samples to vortex extraction using water or a 40% ethanol/water mixture as the solvents. The concentration of the two analytes in the resulting extracts was measured using high-performance liquid chromatography-ultraviolet (HPLC-UV).

2.5. Analysis of Caffeine and Chlorogenic Acid by HPLC-UV

HPLC-UV analysis was performed using a mobile phase consisting of 92% *v*/*v* water and 8% *v*/*v* acetonitrile, which were both acidified with 0.1% (*v*/*v*) acetic acid in the isocratic mode. The flow rate was 0.6 mL/min for 20 min. The sample injection volume was 20 μL at 36 °C, and the wavelength was set at 278 nm. Chlorogenic acid and caffeine calibration curves were used to quantify both substances in the samples. Both standards were prepared in ultrapure water in the concentration range of 40.4–404 mg/L for chlorogenic acid and 121.2–606 mg/L for caffeine. Chlorogenic acid has different isomers, which can be measured as follows: 3-CQA, 4-CQA, and 5-CQA. In the present study, the standard used was 5-O-caffeoylquinic (5-CQA). Thus, the results show the yield of only this isomer, 5-CQA. The results for both analytes, caffeine, and chlorogenic acid, were the average of two replicates and were expressed as concentrations in mg/L of the extract (ppm) and as the mg of the analyte per g dry SCG (including their coefficients of variation). The results in milligram analyte/g dry SCG were obtained by multiplying the concentration (ppm) with the volume of the extract obtained. The volume of the extract in the test tube was calculated using the cylindrical volume formula ($V = \pi r^2 h$), and the height was measured using a digital caliper. Therefore, it is important to note that the results expressed as mg/g SCG carry a higher uncertainty owing to the volume measurement (see Figure 1).

2.6. Analysis of Total Polyphenols with Folin-Ciocalteau Method

Total polyphenols content (TPC) was determined using the Folin–Ciocalteu method adapted from other studies described in the literature [21–23]. Following a series of preliminary experiments, several key parameters were identified as critical for the process. These include maintaining the pH of the solution within the range of 9–10 after the addition of the Folin reagent 2N, employing a sodium carbonate concentration of 20% m/m, maintaining a specific ratio of Folin to sodium carbonate at 1:3, and allowing for a color development time of 2 h in darkness before conducting UV testing. An aliquot of 0.1 mL of the extract (previously diluted with water 1:4 *v*/*v*) was mixed with a 0.5 mL Folin–Ciocalteu reagent and 1.5 mL of sodium carbonate (20% m/m), all diluted to 10 mL with ultrapure water. The mixture was then incubated in the dark at room temperature for 2 h for color development. Absorbance was then measured at 765 nm. Gallic acid calibration curves were obtained within a concentration range of 50–600 ppm.

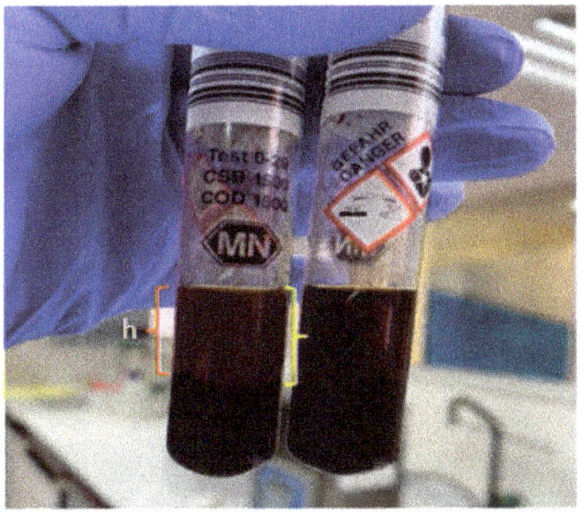

Figure 1. Extraction with acetone at 20% (m/m) (**right**) and 40% (m/m) (**left**): extract (**upper phase**) and SCG (**lower phase**).

The results were the average of four experiments and were reported as concentrations in ppm and as milligrams of gallic acid equivalents per gram of the dried SCG (mg GAE/g), including their coefficients of variation. As mentioned previously, the results expressed as mg/g SCG have higher uncertainty owing to the measurement of the volume.

2.7. *Effect of Extraction Time on Polyphenols Content*

Several experiments were performed under the same conditions but with different extraction times. A sample of 0.7 g of dry SCG was mixed with 4 mL of water or a mixture of acetone and water (20% m/m). Water and acetone 20% were chosen as solvents for the kinetic experiments: the first as a reference and the second because it was the one giving the best results for the three analytes simultaneously (polyphenols, caffeine, and chlorogenic acid) (see the Results and Discussion section).

The mixture was then stirred using a vortex shaker for the following different times: 1, 10, 20, 40, and 60 min for water extraction and 1, 10, 20, and 40 min for acetone 20% extraction. Subsequently, the mixture was centrifuged for 30 min at 4200 rpm, and the obtained extract was used for the determination of polyphenols. The results are the average of four experiments and are reported as concentrations in ppm and milligrams of gallic acid equivalents per gram of dried SCG (mg GAE/g SCG), including their coefficients of variation.

3. Experimental Results and Discussion

3.1. *Storage of Spent Coffee Ground (SCG) after Drying*

3.1.1. Influence of Drying Method

The present study investigated the effects of two different drying methods, including natural drying (at room temperature for 7 days) and oven drying (at 50 °C for 24 h), on the contents of chlorogenic acid (5-CQA) and caffeine in spent coffee grounds.

In these experiments, the SCG used was a blend resulting from the preparation of caffeinated (80% approximately) and decaffeinated (20%) coffee infusions.

As shown in Figure 2, no significant differences (they are within their variation coefficients) were observed in the concentrations of chlorogenic acid and caffeine between the two drying methods for either extraction solvent. Therefore, natural drying or oven drying can be used as viable methods for drying SCG without affecting the concentration of chlorogenic acid and caffeine. Further continued studies presented in this paper were performed using natural dried SCG.

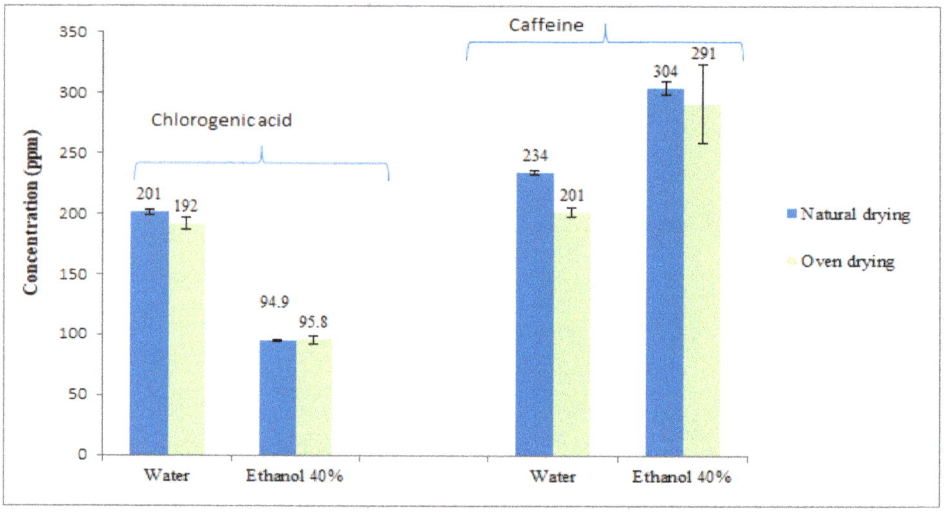

Figure 2. Comparison between natural drying (7 days at room temperature) and oven drying (24 h at 50 °C) SCG for the extraction of chlorogenic acid (5-CQA) and caffeine. The results are the average of duplicate experiments and are presented with their standard deviation.

Figure 2 also illustrates that ethanol 40% is a more efficient solvent than water for extracting caffeine, whereas the opposite is observed in the case of chlorogenic acid. This distinction arises from the varying polarity and chemical characteristics of the two bioactive molecules. Chlorogenic acid possesses an acidic group and five hydroxyls that are capable of forming hydrogen bonds with water, whereas caffeine, with two less polar carbonyl groups, exhibits a preference for a solvent with a higher organic content, such as ethanol at 40%.

3.1.2. Influence of Storage Time

The influence of SCG storage time on the extraction of chlorogenic acid (5-CQA) and caffeine was studied over a period of eight months. The samples were analyzed at the following four time points: initially (time 0), after one month, four months, six months, and eight months of storage. Vortex extraction using water was employed to extract both compounds, which were analyzed through HPLC-UV.

The results presented in Figure 3 reveal a substantial decrease in the concentrations of chlorogenic acid (5-CQA) and caffeine by 82% and 70%, respectively, after an eight-month storage period.

This study suggests that the prolonged storage of SCG may lead to a significant reduction in the concentration of bioactive compounds. Hence, it is recommended to limit the storage time to less than 4 months to maintain initial levels of caffeine and chlorogenic acid as close as possible to the initial ones.

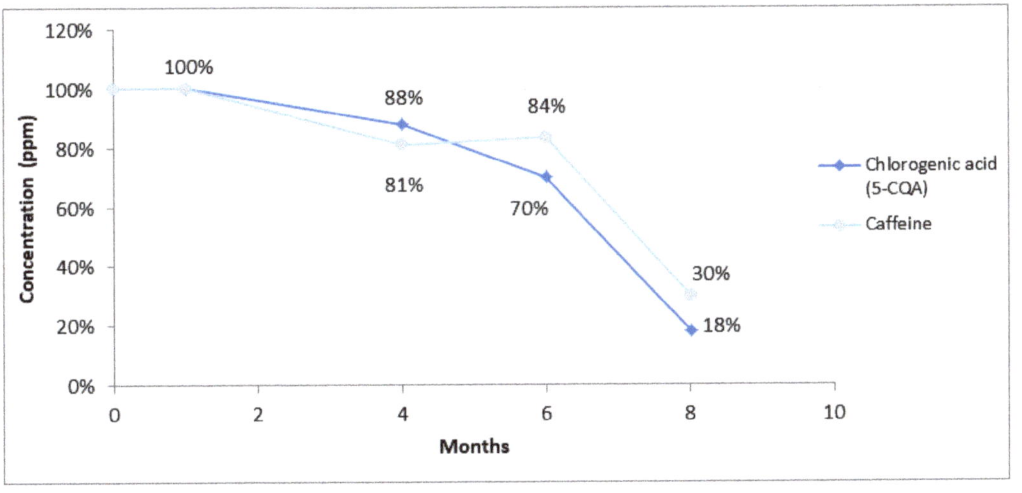

Figure 3. Evolution of chlorogenic acid (5-CQA) and caffeine concentrations in SCG from zero to eight months of storage.

3.2. Extraction of Caffeine and Chlorogenic Acid Using Ethanol/Water Mixtures

The effects of hydroalcoholic solvents with different EtOH/water ratios (0, 20, 40, 60, 80, or 96% m/m) were investigated for chlorogenic acid and caffeine extraction. The chromatographic analysis of the obtained extract (shown in Figure 4) revealed the presence of distinct chlorogenic acid isomers. Spiking experiments were performed using a chlorogenic acid standard (5-CQA, 98%) at a concentration of 400 ppm to identify the peaks corresponding to the standard. The peak area marked in Figure 4 was the one that increased by a factor of 2.53 when the sample was spiked with 4 mL of the chlorogenic acid standard (400 ppm). Therefore, this peak was used for chlorogenic acid (5-CQA) quantification.

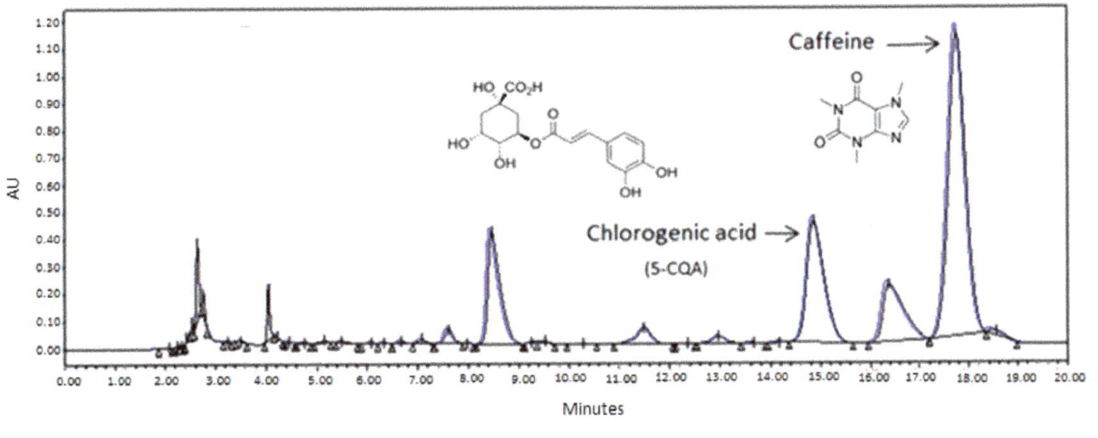

Figure 4. HPLC chromatogram of duplicate solid–liquid extractions using water.

The best result for chlorogenic acid (232 ppm) was observed with water extraction (or 0% ethanol, Figure 5), whereas chlorogenic acid was not detected in the extractions using 96% ethanol as the solvent. It should be noted that the yield of chlorogenic acid decreased progressively as the concentration of ethanol in the solvent mixture increased. The highest caffeine content was obtained with 20–40% ethanol (420–401 ppm, respectively, and, thus, a 5% variation). The caffeine concentration in the extract increased from 325 to 420 ppm, between 0% and 20% ethanol, and then decreased to 24.4 ppm with 96% ethanol (Figure 5). This can be justified by the duality of caffeine, combining both polar and non-polar groups within the same molecule. Initially, the addition of ethanol to water amplifies the organic characteristics of the solvent. Yet, beyond a specific ethanol concentration, the hydrogen bonds between caffeine and water diminished, subsequently decreasing its solubility in the solution.

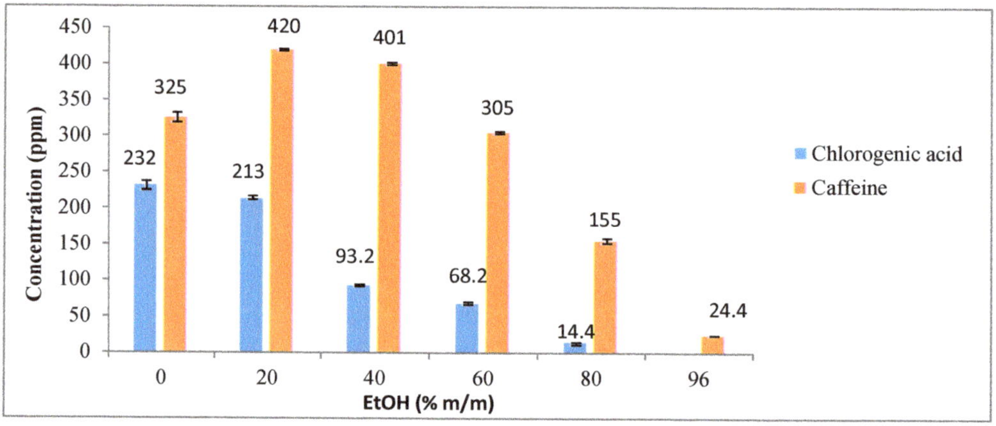

Figure 5. Influence of EtOH/water ratio on chlorogenic acid (5-CQA) and caffeine extraction. The results are the average of duplicate experiments and are presented with their standard deviation. Results in ppm.

Figure 6 just shows the same results but expressed in mg/g of SCG instead of the concentration in the extract solution.

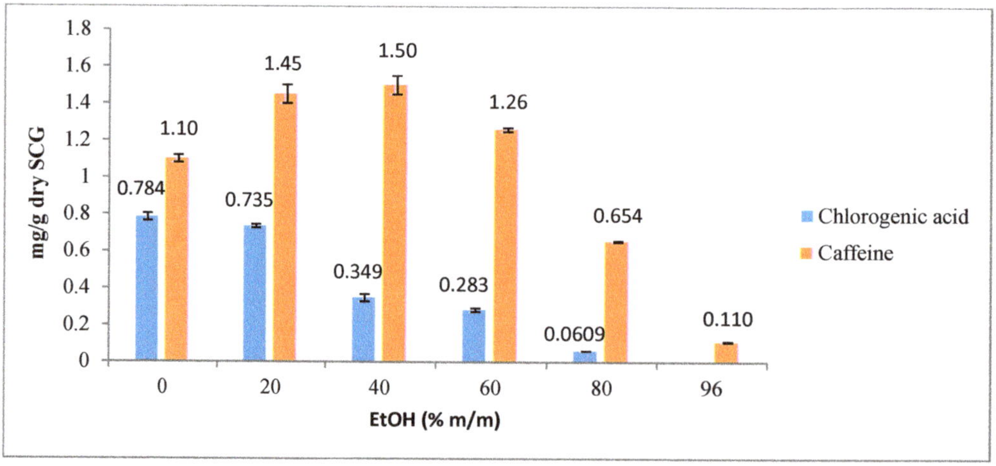

Figure 6. Influence of EtOH/water ratio on chlorogenic acid (5-CQA) and caffeine extraction. The results are the average of duplicate experiments and are presented with their standard deviation. Results in mg/g of dry SCG.

Hence, water is a better solvent to extract chlorogenic acid, while a 20–40% ethanol solvent is preferable to obtain caffeine. This can be explained, as said before, by the nature of chlorogenic acid and caffeine (Figure 4 shows the chemical structures of both molecules).

Chlorogenic acid has five hydroxyl (OH) groups and one acidic group, enabling its solubility in water and facilitating the formation of hydrogen bonds with water molecules. On the other hand, caffeine lacks hydroxyl groups and predominantly contains two carbonyl groups, which have lower polarity compared to the hydroxyl and acidic groups. As a result, while chlorogenic acid is more soluble in water, caffeine demonstrates a higher affinity for mixtures of water with an organic solvent at a specific concentration, making a 20–40% ethanol solvent preferable for its extraction [14,15].

3.3. Extractions Using Acetone/Water Mixtures

The influence of acetone/water mixtures (with 0, 20, and 40% acetone ratios) on chlorogenic acid and caffeine extraction was evaluated.

As mentioned in the previous section, spiking experiments were performed with a chlorogenic acid standard (5-CQA, 98%) to determine the peak corresponding to the standard.

The yields of chlorogenic acid (5-CQA) and caffeine obtained using different acetone/water ratios are shown in Figures 7 and 8, expressed in ppm and mg/g SCG, respectively. As illustrated in the figures, the best result for caffeine extraction was achieved using 20% acetone. However, for chlorogenic acid (5-CQA), the extraction results were comparable between water and 20% acetone solvents. An increase in the acetone concentration beyond 20% resulted in a decrease in both chlorogenic acid and caffeine yields. This can be attributed to a reduction in solubility due to the diminishing of hydrogen bonds that both bioactive compounds can form with water.

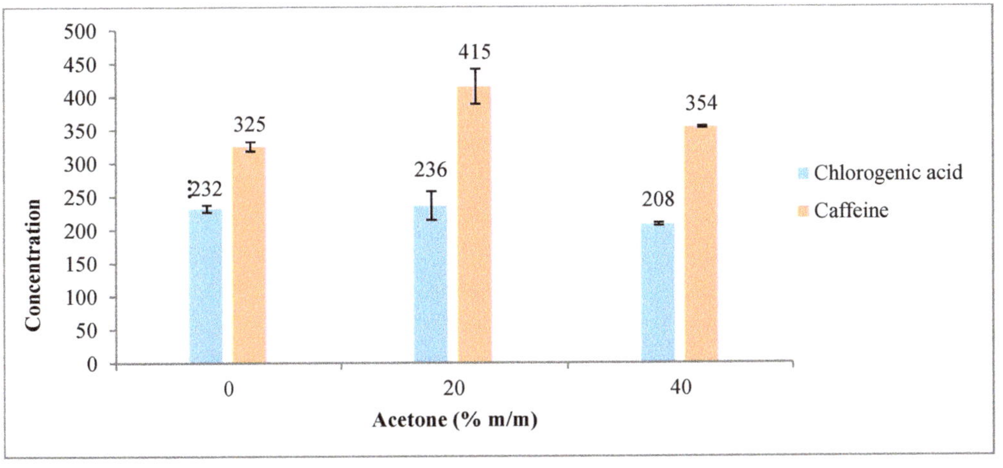

Figure 7. Influence of acetone/water ratio on chlorogenic acid (5-CQA) and caffeine extraction. The results are the average of duplicate experiments and are presented with their standard deviation. Results in ppm.

3.4. Total Polyphenols Content

Water and water mixtures with ethanol (20%, 40%, and 60%) or acetone (20% and 40%) were used as solvents to extract polyphenols from spent coffee grounds (SCG) (Figure 9). The total polyphenols content (TPC) in the extract was measured in terms of the concentration (ppm) or as milligrams of gallic acid equivalents per gram of SCG (mg GAE eq/g SCG). The results (Figure 9) reveal that there were no significant differences in the results between the use of 20% or 40% (m/m) acetone or ethanol, while ethanol at 60% (m/m) resulted in a decrease in total polyphenols. Consequently, to minimize the volume of organic solvent, 20% (m/m) acetone or ethanol is recommended to conduct the experiments.

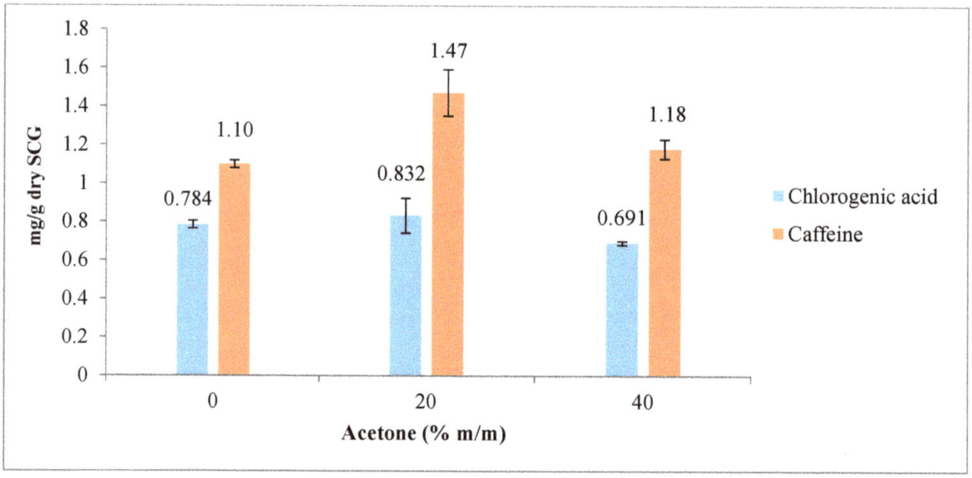

Figure 8. Influence of acetone/water ratio on chlorogenic acid (5-CQA) and caffeine extraction. The results are the average of duplicate experiments and are presented with their standard deviation. Results in mg/g of dry SCG.

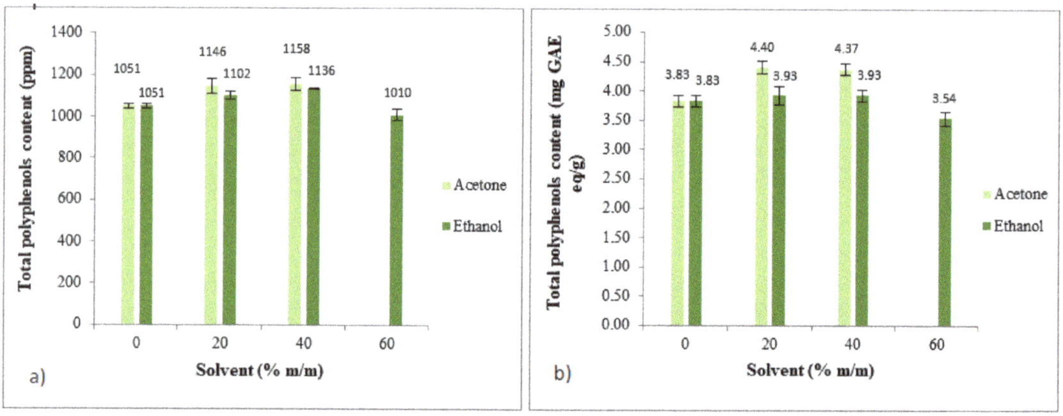

Figure 9. Total polyphenols content (TPC) as a function of different acetone/water and ethanol/water ratios. The results are the average of quadruplicate experiments and are presented with their standard deviation. (**a**) Results in ppm. (**b**) Results in mg GAE eq/g of dry SCG.

3.5. *Comparison of Solvents to Extract Total Polyphenols, Chlorogenic Acid (5-CQA), and Caffeine*

Results from previous sections are included together in Figures 10 and 11 for comparison purposes.

The results revealed that acetone (20% m/m) and ethanol (20% m/m) were more effective than water in extracting phenolic compounds and caffeine (9% and 28% higher yield, respectively). However, in the case of chlorogenic acid (5-CQA), water and acetone (20% m/m) were slightly better (about 11%) than ethanol (20% m/m).

3.6. *Effect of Extraction Time at Room Temperature*

Increasing the extraction time has been identified as a potential factor for enhancing extraction efficiency, as previously mentioned [5]. In this kinetic study, separate applications of acetone (20% m/m) and water were employed as extraction solvents at room temperature.

Figure 12 illustrates the influence of varying extraction times on the total polyphenols content in the SCG extracts.

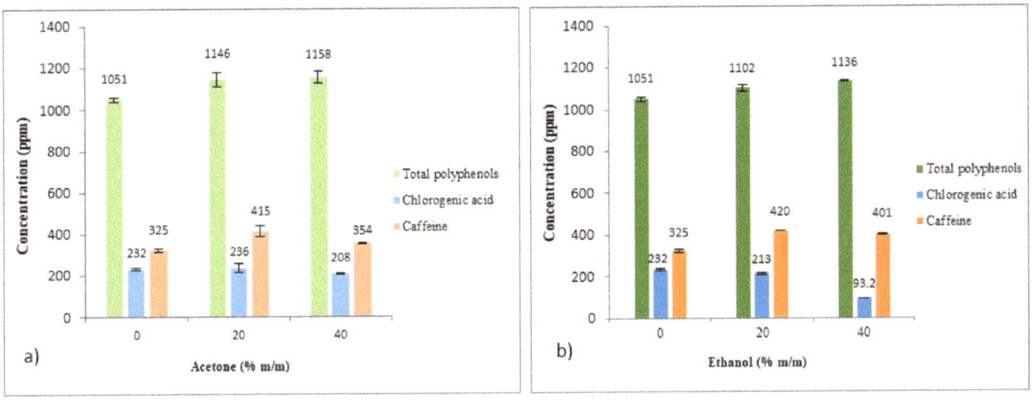

Figure 10. Total polyphenols, chlorogenic acid (5-CQA) and caffeine contents as a function of different acetone/water and ethanol/water ratios. (**a**) Results for acetone–water mixtures (ppm). (**b**) Results for ethanol–water mixtures (ppm).

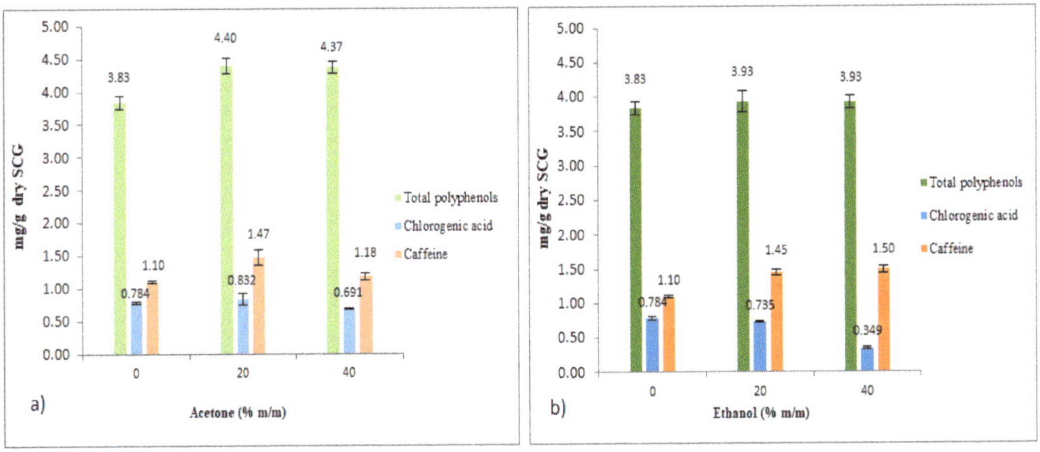

Figure 11. Total polyphenols, chlorogenic acid (5-CQA) and caffeine contents as a function of acetone/water and ethanol/water ratios. (**a**) Results for acetone–water mixtures (mg/g dry SCG). (**b**) Results for ethanol–water mixtures (mg/g dry SCG).

The fastest extraction rate occurred between 0 and 1 min. Beyond this time, the slope of the curve diminished. Hence, after 1 min, a significant portion of the total polyphenols had already been extracted, leading to a subsequent slowdown in the extraction process.

Nevertheless, in the case of water, increasing the extraction time from 1 to 40 min, the total polyphenols content of the extracts increased by +34% (from 1051 to 1408 ppm). For acetone at 20% (m/m), an increase in the total polyphenols content (+28%) was observed over the extraction time range (0–10 min). Thus, the maximal yield of polyphenols was reached faster with Acetone at 20% (within 10 min) than with water (within 40 min). However, in the case of water extraction, increasing the extraction time from 10 to 40 min resulted in an increase of only 13%. This means that 10 min might be considered a sufficient extraction time for both solvents.

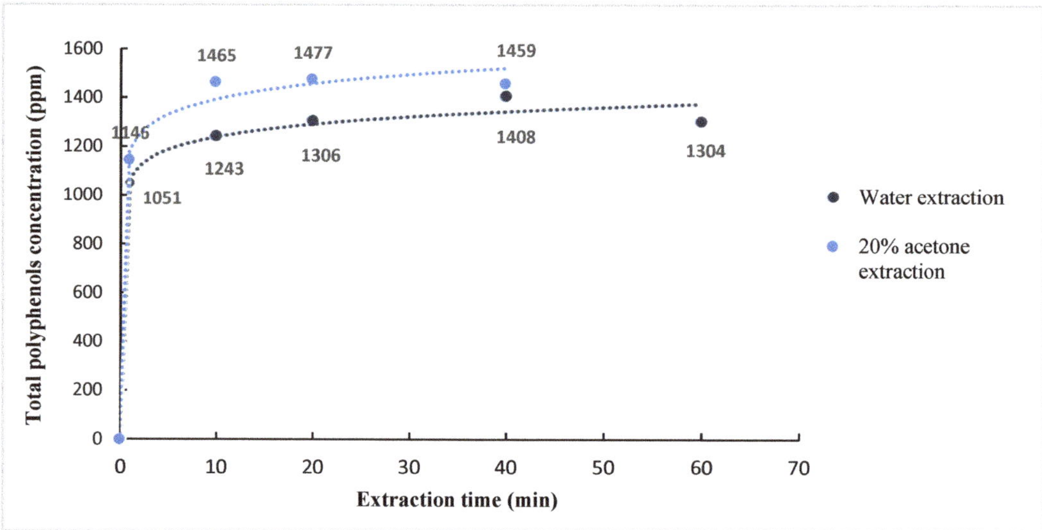

Figure 12. Kinetic curves of total polyphenols at room temperature using water and acetone at 20% (m/m) as the extraction solvents. The results are the average of quadruplicate experiments.

Chlorogenic acid (5-CQA) and caffeine yields were also determined at the optimal extraction times for water and acetone at 20% (40 and 10 min, respectively). These results were obtained using HPLC-UV and are presented along with the total polyphenol yields in Figures 13 and 14. These were the best solvents and extraction times identified in this study.

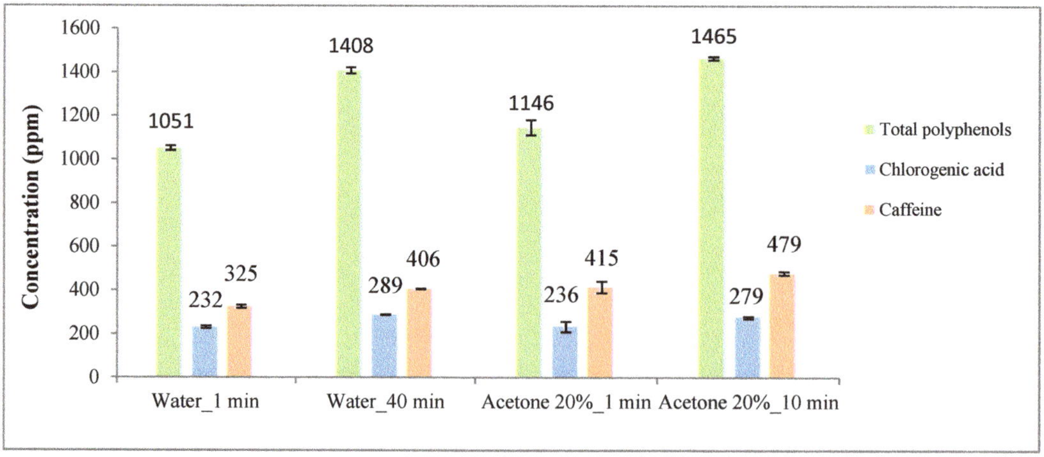

Figure 13. Chlorogenic acid (5-CQA) and caffeine contents in water and acetone at 20% (m/m) at different extraction times. The results for total polyphenols are the average of quadruplicate experiments, and for chlorogenic acid and caffeine, are the average of duplicate experiments presented with their standard deviation. Results in ppm.

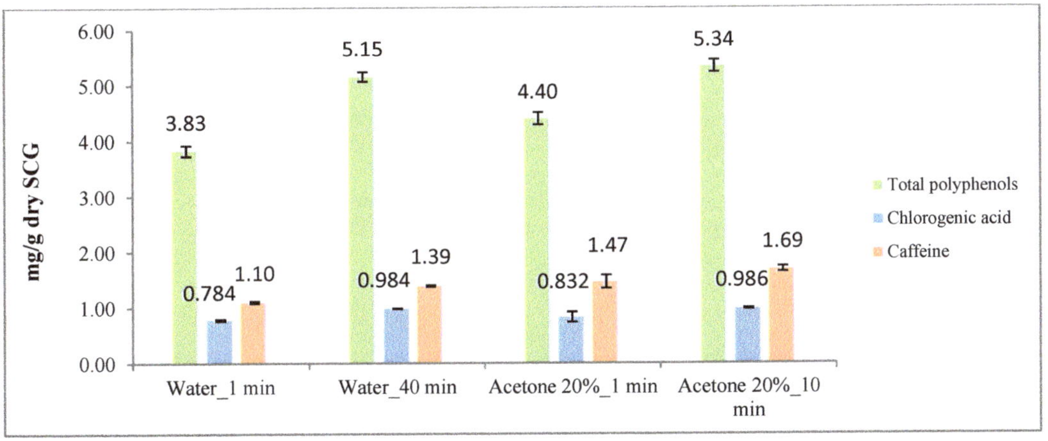

Figure 14. Chlorogenic acid (5-CQA) and caffeine contents in water and acetone at 20% (m/m) and different extraction times. The results for total polyphenols are the average of quadruplicate experiments, and for chlorogenic acid and caffeine are the average of duplicate experiments and are presented with their standard deviation. Results in mg/g.

The results indicate a significant influence of the extraction time and solvent type on the content of these bioactive compounds. Extending the extraction time (from 1 to 40 min for water) resulted in a reasonable yield increase for both chlorogenic acid and caffeine (25% and 25%, respectively). Similarly, using acetone at 20% as the solvent resulted in an 18% increase in chlorogenic acid yield and a 15% increase in caffeine yield when the extraction time was extended from 1 to 10 min.

The results show a more significant increase in the total polyphenols yield over time compared to the increases observed in chlorogenic acid and caffeine (Figure 13), which can be attributed to the comprehensive nature of total polyphenols, including a great diversity of compounds quantified collectively using the Folin method and expressed as mg GAE eq/L. Consequently, these compounds may not present the same behavior in the extraction process as caffeine and chlorogenic acid.

3.7. Critical Discussion of the Present Results Compared with the Literature

First of all, it is noteworthy to acknowledge that results from different studies are not directly comparable because they use SCG from different origins and with different compositions due to inherent variations resulting from the coffee variety, roasting degree, and coffee infusion preparation methodologies. These variables can exert considerable influence on the concentration of bioactive compounds within the original coffee matrix, thereby potentially affecting the extraction outcomes of SCG and limiting direct comparisons across studies, as stated by some authors [5,23–25].

Nevertheless, a literature review was performed to identify all the studies in the literature that have recovered total polyphenols, chlorogenic acid, or caffeine using solid–liquid extraction, and the same solvents were evaluated in this study. A total of 20 papers were identified (some of which are included in Table 1 for discussion purposes).

Several studies have used water and ethanol–water mixtures as extraction solvents. However, only one study employing ethanol 25% quantified all three types of analytes (polyphenols, caffeine, and chlorogenic acid) simultaneously, as conducted in the present study for different ethanol percentages. Among the identified studies, the majority (18 out of 20) analyzed the total polyphenol content. Some of them also quantified caffeine, while only two studies measured the chlorogenic acid content. However, these last two studies did not specify whether they measured only one isomer of chlorogenic acid or included all its isomers in the quantification. The recovery rates reported in the literature vary significantly, ranging from 3.9 to 33.07 mg GAE/g for total polyphenols, from 0.02 to 4.8 mg/g for chlorogenic acid, and from 0.15 to 11.50 mg/g for caffeine.

On the other hand, the use of acetone/water mixtures for extracting these bioactive substances from SCG has not been previously reported in the literature. It is worth mentioning that only one study by Da Silva Araújo et al. [26] used acetone to solely investigate the use of 100% acetone for total polyphenols (TPC) recovery (see Table 1, n° 4), further highlighting the novelty of the present study.

Table 1 presents the different solid liquid extraction studies found in the literature. Studies 1 to 3 show a comparison of TPC, 5-CQA, and the caffeine contents for different SCG types, highlighting that results from different SCGs are not comparable even using the same extraction conditions. On the other hand, studies 4 to 13 present the influence of ethanol–water proportions on the recovery of TPC or caffeine. Only the best ethanol proportion that yielded better results for TPC or caffeine is presented in the table. Lastly, study 14 illustrates the impact of the liquid-to-solid ratio and temperature variations on TPC.

In the present study, and for ethanol–water mixtures, the best results were obtained using ethanol at 20–40% for TPC and caffeine and ethanol at 0–20% for 5-CQA (see Figures 10 and 11). These optimal ethanol–water proportions are in agreement with two other authors [14,26]. However, other studies reported better TPC results using ethanol at 50% [13,15,27,28] or ethanol at 60% [26,29–32]. Nevertheless, these studies did not measure 5-CQA, with most of them only investigating TPC. These variations in the optimal proportions of ethanol among the different studies highlight the complexity of obtaining a global optimization, which requires considering all the variables that affect the process, such as the type of coffee and its production process, as well as the preparation of the infusion and all the operating conditions for the extraction of bioactive compounds from SCG.

On the other hand, when comparing the extraction conditions (such as extraction time, temperature, and solvent-to-solid ratio) of this study with those reported in the literature, it is evident that the current study used the shortest extraction time (1 min), ambient temperature, and a minimal solvent-to-solid ratio (5.7 mL/g). These specific operating parameters are likely to exert a significant influence on the extraction yield. For example, as reported in [30], an increase in the liquid-to-solid ratio from 10 to 50 mL/g led to a considerable increase in the extraction (40%; from 12.12 to 16.94 mg GAE/g) (Table 1, n° 14) and elevating the temperature from 20 to 60 °C resulted in 38% yield augmentation (Table 1, n° 14).

Moreover, the extraction time is also an important factor in the extraction of bioactive compounds (as demonstrated in Section 3.6).

The present study was designed with the objective of working with a minimal solvent-to-solid ratio and ambient temperature. This choice was driven by our interest in obtaining highly concentrated analytes while minimizing the need for extensive solvent removal when using them to produce a new product (as the evaporation process consumes significant energy resources). Simultaneously, we aimed to minimize energy consumption associated with heating processes. The reduction in solvent use not only aligns with energy reduction but also with resource conservation by decreasing solvent use and subsequent industrial equipment dimensions to perform the process at an industrial scale. This approach can contribute to obtaining a more industrially sustainable process. Hence, it might be worth enhancing the present results by conducting the extraction at higher temperatures and a solvent/solid ratio if the gain in terms of yield or efficiency is significant enough to offset the investment in energy and the material resources required. Therefore, it is advisable to conduct further research to evaluate, considering both environmental and economic factors, whether the augmented yield justifies the subsequent increase in energy and resource consumption. This analysis should involve a multi-objective optimization approach.

Table 1. Examples of solid/liquid extraction methods reported in the literature.

Study n°	Year of Publication	Solvent	t (min)	T (°C)	(mL/g) Solvent-to-Solid Ratio	TPC (mg GAE/g)	Caffeine (mg/g)	Chlorogenic Acid (mg/g)	Reference
1	2019	Ethanol 25%	15	60	83	18 * 26 * 9 *	1.9 * 6 * 0.3 *	0.02 * 4.8 * 0.8 *	[23]
2	2011	Ethanol 70%	120	50	40	19.98 * 17.09 *	- -	-	[24]
3	2013	Ethanol 60%	30	60	50	28.26 * 23.9 *	11.50 * 5.99 *	-	[25]
		Water	30	60	50	19.62 * 17.43 *	11.23 * 6.00 *	-	
4	2021	Ethanol 15%	90	60	30	8.6	-	-	[26]
		Acetone 100%	90	65	30	3	-	-	
5	2021	Ethanol 20%	15	Room temperature	25	-	4.1	-	[14]
6	2023	Ethanol 50%	30	50	40	9	-	-	[13]
7	2023	Ethanol 50%	69	65	28	17.79	4.37	-	[15]
8	2023	Ethanol 60%	90	60	40	33.07	-		[29]
9	2013	Ethanol 60%	105	40	30	14.06	-	-	[30]
10	2011	Ethanol 60%	60	25	15	11.50	-	-	[31]
11	2015	Ethanol 50%	90	60	30	23.79	-	-	[27]
12	2019	Ethanol 50%	24 h	4	10	8.4	-	-	[28]
13	2022	Ethanol 66%	-	60	50	22.02	-	-	[32]
14	2013	Ethanol 60%	180 180 180	60 60 20	10 50 50	12.12 16.94 12.28	-	-	[30]

* Results for different SCG types.

4. Conclusions

The present study aims to promote circularity in addressing popular food waste, especially spent coffee grounds, by extracting valuable bioactive compounds with different solvents (total polyphenols, chlorogenic acid, and caffeine) that present different applications in the pharmaceutical and food industries.

The paper investigates water–acetone mixtures as an alternative solvent for this extraction. Additionally, other aspects, such as storage conditions and the kinetics of the extraction process, have also been studied. SCG needs to be dried and stored for less than 3–4 months before extraction to prevent analyte degradation. The kinetics of the extraction at room temperature show that about 70% of polyphenols can be extracted within just 1 min, using either water or acetone at 20% as solvent.

Acetone at a concentration of 20% was the most convenient solvent for the extraction of total polyphenols (1146 ppm or 4.40 mg GAE/g SCG), chlorogenic acid (236 ppm or 0.832 mg 5-CQA/g SCG), and caffeine (415 ppm or 1.47 mg caf/g SCG) simultaneously. These results were obtained using the simplest conditions, including room temperature, the lowest liquid-to-solid ratio (5.7 mL/g SCG), and a short extraction time (1 min) simultaneously. Therefore, they can be enhanced by conducting the extraction at a higher temperature and solvent/solid ratio if the gain in terms of yield or efficiency is significant enough to offset the investment in energy and material resources required.

The present study was conceived with the aim of operating under the most straightforward conditions to achieve highly concentrated analytes in the shortest time and with the lowest energy consumption. This decision was driven by our commitment to reduce environmental and economic costs linked to energy and resource utilization. Decreasing solvent use not only enhances the energy efficiency of analyte concentration but also contributes significantly to resource conservation. By minimizing the use of solvents, there is a consequential reduction in the dimensions of industrial equipment needed for scaling up the process. These insights, often overlooked at the laboratory level, provide a basis for more sustainable industrial processes.

Author Contributions: Conceptualization: R.P.; Methodology: R.C. and R.P.; Formal analysis and investigation: I.B. and R.C.; Resources and data curation: A.B.; Writing—original draft preparation: I.B. and R.C.; Writing—review and editing: I.B., R.P., R.C. and P.F.-i.-P.; Supervision: R.C. and R.P.; Funding acquisition: R.A., M.M. and R.P.; Project administration: R.P.; Conception of the Project's original idea: R.A. and P.F.-i.-P. All authors have read and agreed to the published version of the manuscript.

Funding: The authors wish to acknowledge the financial support from the Spanish Ministry of Science and Innovation to the project KAIROS-BIOCIR (PID2019-104925RB-C32). The first author also appreciates the support (2021FI SDUR 00130) from the Secretariat for Universities and Research of the Ministry of Business and Knowledge of the Government of Catalonia and the European Social Fund.

Data Availability Statement: Data are contained within the article.

Conflicts of Interest: The authors declare no conflict of interest.

References

1. Ellen MacArthur Foundation. Big Food Redesign Challenge: It's Time to Redesign Food for Nature to Thrive. 2021. Available online: https://ellenmacarthurfoundation.org/the-big-food-redesign-challenge/overview (accessed on 22 September 2023).
2. Ellen MacArthur Foundation. Food and the Circular Economy. 2018. Available online: https://ellenmacarthurfoundation.org/food-and-the-circular-economy-deep-dive (accessed on 22 September 2023).
3. Jamaludin, H.; Elmaky, H.S.E.; Sulaiman, S. The future of food waste: Application of circular economy. *Energy Nexus.* **2022**, *7*, 100098. [CrossRef]
4. European Commission. Communication from the Commission to the European Parliament, the Council, the European Economic and Social Committee and the Committee of the Regions. 2014. Available online: https://eur-lex.europa.eu/legal-content/EN/TXT/HTML/?uri=CELEX:52014DC0398&from=ES (accessed on 22 September 2023).
5. Bouhzam, I.; Cantero, R.; Balcells, M.; Margallo, M.; Aldaco, R.; Bala, A.; Fullana-i-Palmer, P.; Puig, R. Environmental and Yield Comparison of Quick Extraction Methods for Caffeine and Chlorogenic Acid from Spent Coffee Grounds. *Foods* **2023**, *12*, 779. [CrossRef] [PubMed]
6. Johnson, K.; Liu, Y.; Lu, M. A Review of Recent Advances in Spent Coffee Grounds Upcycle Technologies and Practices. *Front. Chem. Eng.* **2022**, *4*, 838605. [CrossRef]
7. Leow, Y.; Yew, P.Y.M.; Chee, P.L.; Loh, X.J.; Kai, D. Recycling of spent coffee grounds for useful extracts and green composites. *RSC Adv.* **2021**, *11*, 2682–2692. [CrossRef] [PubMed]
8. Campos-Vega, R.; Loarca-Piña, G.; Vergara-Castañeda, H.A.; Oomah, B.D. Spent coffee grounds: A review on current research and future prospects. *Trends Food Sci. Technol.* **2015**, *45*, 24–36. [CrossRef]
9. Dai, J.; Mumper, R.J. Plant Phenolics: Extraction, Analysis and Their Antioxidant and Anticancer Properties. *Molecules* **2010**, *15*, 7313–7352. [CrossRef]
10. Bravo, J.; Juániz, I.; Monente, C.; Caemmerer, B.; Kroh, L.W.; De Peña, M.P.; Cid, C. Evaluation of spent coffee obtained from the most common coffeemakers as a source of hydrophilic bioactive compounds. *J. Agric. Food Chem.* **2012**, *60*, 12565–12573. [CrossRef] [PubMed]
11. Narita, Y.; Inouye, K. Chlorogenic Acids from Coffee. In *Coffee in Health and Disease Prevention*; Academic Press: Cambridge, MA, USA, 2015; pp. 189–199. [CrossRef]
12. Farah, A.; Donangelo, C.M. Phenolic compounds in coffee. *Braz. J. Plant Physiol.* **2006**, *18*, 23–36. [CrossRef]
13. Beaudor, M.; Vauchel, P.; Pradal, D.; Aljawish, A.; Phalip, V. Comparing the efficiency of extracting antioxidant polyphenols from spent coffee grounds using an innovative ultrasound-assisted extraction equipment versus conventional method. *Chem. Eng. Process.-Process Intensif.* **2023**, *188*, 109358. [CrossRef]
14. Vandeponseele, A.; Draye, M.; Piot, C.; Chatel, G. Study of Influential Parameters of the Caffeine Extraction from Spent Coffee Grounds: From Brewing Coffee Method to the Waste Treatment Conditions. *Clean Technol.* **2021**, *3*, 335–350. [CrossRef]
15. Chatzimitakos, T.; Athanasiadis, V.; Kotsou, K.; Palaiogiannis, D.; Bozinou, E.; Lalas, S.I. Optimized Isolation Procedure for the Extraction of Bioactive Compounds from Spent Coffee Grounds. *Appl. Sci.* **2023**, *13*, 2819. [CrossRef]
16. Zhou, F.; Hearne, Z.; Li, C.J. Water—The greenest solvent overall. *Curr. Opin. Green Sustain. Chem.* **2019**, *18*, 118–123. [CrossRef]
17. Meneses, N.G.T.; Martins, S.; Teixeira, J.A.; Mussatto, S.I. Influence of extraction solvents on the recovery of antioxidant phenolic compounds from brewer's spent grains. *Sep. Purif. Technol.* **2013**, *108*, 152–158. [CrossRef]
18. Capello, C.; Fischer, U.; Hungerbühler, K. What is a green solvent? A comprehensive framework for the environmental assessment of solvents. *Green Chem.* **2007**, *9*, 927–934. [CrossRef]
19. Alfonsi, K.; Colberg, J.; Dunn, P.J.; Fevig, T.; Jennings, S.; Johnson, T.A.; Kleine, H.P.; Knight, C.; Nagy, M.A.; Perry, D.A.; et al. Green chemistry tools to influence a medicinal chemistry and research chemistry based organisation. *Green Chem.* **2008**, *10*, 31–36. [CrossRef]
20. Torres-Valenzuela, L.S.; Ballesteros-Gómez, A.; Sanin, A.; Rubio, S. Valorization of spent coffee grounds by supramolecular solvent extraction. *Sep. Purif. Technol.* **2019**, *228*, 115759. [CrossRef]

21. Boyadzhieva, S.; Angelov, G.; Georgieva, S.; Yankov, D. Characterization of polyphenol content and antioxidant capacity of spent coffee grounds. *Bulg. Chem. Commun.* **2018**, *50*, 85–89.
22. Vega, A.; De León, J.A.; Reyes, S.M. Determinación del contenido de polifenoles totales, flavonoides y actividad antioxidante de 34 cafés comerciales de Panamá. *Inf. Tecnol.* **2017**, *28*, 29–38. [CrossRef]
23. Ramón-Gonçalves, M.; Gómez-Mejía, E.; Rosales-Conrado, N.; León-González, M.E.; Madrid, Y. Extraction, identification and quantification of polyphenols from spent coffee grounds by chromatographic methods and chemometric analyses. *Waste Manag.* **2019**, *96*, 15–24. [CrossRef]
24. Zuorro, A.; Lavecchia, R. Spent coffee grounds as a valuable source of phenolic compounds and bioenergy. *J. Clean. Prod.* **2012**, *34*, 49–56. [CrossRef]
25. Panusa, A.; Zuorro, A.; Lavecchia, R.; Marrosu, G.; Petrucci, R. Recovery of Natural Antioxidants from Spent Coffee Grounds. *J. Agric. Food Chem.* **2013**, *61*, 4162–4168. [CrossRef] [PubMed]
26. Da Silva Araújo, C.; Vimercati, W.C.; MacEdo, L.L.; Pimenta, C.J. Effect of solvent, method, time and temperature of extraction on the recovery of phenolic compounds and antioxidants from spent coffee grounds. *Int. J. Food Eng.* **2022**, *18*, 325–336. [CrossRef]
27. Zuorro, A. Optimization of polyphenol recovery from espresso coffee residues using factorial design and response surface methodology. *Sep. Purif. Technol.* **2015**, *152*, 64–69. [CrossRef]
28. Saada, M.; Kasmi, M.; Ksouri, R. Towards Coffee Processing by-Products Valorization: Phenolic Compounds and Antioxidant Activity of Spent Coffee Grounds and Coffee Silverskin. *Int. J. Biotech Trends Technol. (IJBTT)* **2019**, *9*, 11–15.
29. Brzezińska, R.; Górska, A.; Wirkowska-Wojdyła, M.; Piasecka, I. Response Surface Methodology as a Tool for Optimization of Extraction Process of Bioactive Compounds from Spent Coffee Grounds. *Appl. Sci.* **2023**, *13*, 7634. [CrossRef]
30. Zuorro, A.; Lavecchia, R. Influence of Extraction Conditions on the Recovery of Phenolic Antioxidants from Spent Coffee Grounds. *Am. J. Appl. Sci.* **2013**, *10*, 478–486. [CrossRef]
31. Zuorro, A.; Lavecchia, R. Polyphenols and energy recovery from spent coffee grounds. *Chem. Eng. Trans.* **2011**, *25*, 285–290. [CrossRef]
32. Solomakou, N.; Loukri, A.; Tsafrakidou, P.; Michaelidou, A.M.; Mourtzinos, I.; Goula, A.M. Recovery of phenolic compounds from spent coffee grounds through optimized extraction processes. *Sustain. Chem. Pharm.* **2022**, *25*, 100592. [CrossRef]

foods

MDPI

Article

Pulp from Colored Potatoes (*Solanum tuberosum* L.) as an Ingredient Enriching Dessert Cookies

Dorota Gumul [1,*], Rafał Ziobro [1], Jarosław Korus [1] and Magdalena Surma [2]

1 Department of Carbohydrate Technology and Cereal Processing, Faculty of Food Technology, University of Agriculture in Krakow, Balicka Str. 122, 30-149 Krakow, Poland; rrziobro@cyf-kr.edu.pl (R.Z.); rrkorus@cyf-kr.edu.pl (J.K.)
2 Department of Plant Products Technology and Nutrition Hygiene, Faculty of Food Technology, University of Agriculture in Krakow, Balicka Str. 122, 30-149 Krakow, Poland; magdalena.surma@urk.edu.pl
* Correspondence: rrgumul@cyf-kr.edu.pl

Abstract: Freeze-dried pulp from colored potatoes, obtained after starch isolation, is a rich source of polyphenols. Therefore, it can be used to fortify cookies, contributing to a reduction in industrial waste, aligning with the zero-waste technology. The purpose of this study was to analyze the effects of adding 5% and 10% pulp from two varieties of colored potatoes on the content of polyphenols, antioxidant activity, physical characteristics, nutritional composition, and the levels of hydroxymethylfurfural and acrylamide of the fortified cookies. The findings revealed that colored potato pulp is an outstanding additive for fortifying cookies with polyphenols, flavonoids, anthocyanins, and flavonols (even two to four times in comparison to control). Cookies containing pulp exhibited even two times higher fiber and protein content (up to 17% more), while the fat and ash content remained unchanged compared to control cookies. Furthermore, they contained 30% less HMF and 40% more acrylamide. These cookies also exhibited good physical properties in the final products. The study demonstrated that pulp from the "Magenta Love" potato variety was significantly more effective in enriching cookies with health-promoting compounds and nutrition value compared to pulp from Marleta Blue.

Keywords: antioxidants; acrylamide; HMF nutritional value; physical features; fortified cookies; pulp from red and purple potatoes

Citation: Gumul, D.; Ziobro, R.; Korus, J.; Surma, M. Pulp from Colored Potatoes (*Solanum tuberosum* L.) as an Ingredient Enriching Dessert Cookies. *Foods* **2023**, *12*, 3735. https://doi.org/10.3390/foods12203735

Academic Editor: Gilles Trystram

Received: 25 September 2023
Revised: 7 October 2023
Accepted: 10 October 2023
Published: 11 October 2023

1. Introduction

Processing of raw materials of plant and animal origin contributes to the production of large quantities of organic residues by the food industry. Waste associated with the processing of meat and animal products accounts for 13%, while losses associated with the processing of fruits and vegetables are 22% [1]. These are both solid and liquid materials, which usually have no further use in the production chain. If not properly treated, they can increase the pollution of soil, surface water, and groundwater [2–5].

Starch manufacture is one of the processing methods that involves the occurrence of large quantities of by-products, which may be dangerous for environment. Three types of organic waste are generated during the production of potato starch: potato pulp, potato juice, and juicy water. For every ton of starch produced, there are 1.36 tons of potato pulp, 0.14 cubic meters of potato juice, and 3.66 tons of juice water [6]. Potato juice is a fluid separated from potato pulp containing about 5% dry mass including potato protein of high nutritional value. Potato juice contains 2% nitrogenous compounds, which include protease inhibitors, lectins, phosphorylases, and kinases. The juicy water is produced during the refining of starch milk. It is a tenfold dilution of potato juice. Due to its high content of water and low caloric value, the incineration of solid starch waste does not seem rational. Potato pulp is a by-product obtained after washing out the starch from the disintegrated potato cells with water. The pulp consists mainly of bound starch inside intact cells, minerals (Ca, Cl, Na, P, Mg), and insoluble non-starchy substances including fiber, of which polyphenols are an integral part. Solanine and chaconine, which are toxic glycoalkaloids present in raw potatoes, make up only 0.006% of the potato's fresh weight and are water-soluble and leached out with water during this processing process [7].

Annual production of potato pulp in Europe exceeds one million tons. In northern Japan, about 100,000 tons are produced every year. The pulp is mainly used as fodder or fertilizer. However, these methods are not economical. Still a large part of the waste remains unmanaged, which affects the additional disposal costs incurred by the plant and increases the biological load of wastewater [8]. Potato pulp consists primarily of cell wall fragments containing residual starch. It contains up to 30% pectins, which could be used in the production of jellies and jams [9]. Pectins are water-soluble esters of polygalacturonic acid, partially esterified with methyl alcohol. They are formed from water-insoluble protopectin. In combination with acids and sugar, they may produce stiff gels. We can distinguish between high-methylated and low-methylated pectins [9], which are used for the confectionery and pastry industry and in the production of jellies and jams and beverages. Fermentation of potato pulp could be applied for the production of butanol or acetone. Some attempts have also been made at enzymatic hydrolysis of potato pulp in order to obtain protein preparations [10].

Considering the possible applications of potato pulp, no attention has been paid to polyphenols, which are an integral part of dietary fiber contained in the pulp. Their presence in significant quantities gives a possibility to obtain food products, such as cookies, with improved nutritional and health-promoting properties, maintaining the attractiveness of the final product.

The production of confectionery products is an important branch of the food industry, as the consumption of sponge cakes, cookies, and other confectionery products continues to grow. The global confectionery market is projected to reach $270.5 billion by 2027 (Allied Market Research Report) [11]. Although nutritionists underline that cookies are high in fat and sugar and could be partly responsible for obesity, manufacturers are trying to change this negative perception of confectionery products by introducing health-promoting, nutrient-enriched, reduced-calorie raw materials, or eliminating sugar or replacing it with other sweeteners, and supplying vitamins, fiber, and polyphenols. Efforts are also being made to introduce new products for people with celiac disease or gluten intolerance, who due to the elimination diet could suffer significant oxidative stress and impaired antioxidant enzymes, which normally form a specific antioxidant barrier in the body. Such people are susceptible to oxidant–antioxidant imbalance and DNA damage, and consequently to cancer [12,13], which may partly be prevented by introducing variability of different foods. Such gluten-free products will unquestionably include the cookies with colored potato pulp that are the subject of this manuscript's research.

It should be added that cookies have become a commercially important product due to the practicality of their production and the long shelf life of the finished products. However, customers are looking for new innovative products as the market develops rapidly. The average appearance and quality of the raw materials from which the product is made no longer satisfy the consumers. They want new products with surprising attractive properties and nutritional value, and above all with health-promoting properties. Therefore, producers are looking for new solutions, positively influencing the foodstuffs, preferring not only the pro-health values of products, but also paying attention to harmful substances (acrylamide and HMF), which are formed as products of the Maillard reaction during baking.

The possibilities of enriching cookies with raw materials of high antioxidant potential can be directed towards the introduction of new products on the market, especially for celiacs. On the one hand, the possibility of utilization of the waste product, such as potato pulp, and on the other, the growing awareness of consumers, who look for chemopreventive food ingredients, make it possible to introduce new products on the market. Therefore, this research fits into the current global trend of functional food [14–19].

The aim of this study was to analyze the effect of the level (5 and 10%), origin (two colored flesh cultivars) of potato pulp on the content of polyphenols, phenolic acids, anthocyanins and flavonoids, and the antiradical activity (using the free radical ABTS) in dessert cookies. In addition, the study focused on the physical characteristics of these cookies and their nutritional value. Furthermore, acrylamide and hydroxymethylfurfural (HMF) were determined in the final product.

2. Materials and Methods

2.1. Pulp Extraction and Preparation of Cookies Enriched with Potato Pulp

The batch of potatoes was washed under running water, then a certain part of it was mashed into a homogeneous pulp and transferred in portions to a mill gauze placed on a Buechner funnel, and the starch was washed by pressing the gauze by hand. The leaching process was stopped when the filtrate collected from under the sieve showed a negative reaction with Lugol's iodine. A negative color reaction with iodine allowed the next portion of the pulp to be taken for starch isolation. The pulp was immediately placed in a freezer for about 1 week, 1 day of blast freezing, and freeze-dried. The efficiency of the potato pulp extraction process was 50%. The subjects of the study were dessert cookies (DC) baked under laboratory conditions from corn and potato starches enriched with 5 and 10% of freeze-dried potato pulp. The potato pulp was made from colored flesh potatoes of two varieties, Magenta Love (ML) (red potatoes) and Marleta Blue (MB) (purple potatoes). Potato pulp was obtained as a by-product in laboratory isolation of potato starch, as described above, according to Wischmann et al. [20], subsequently freeze-dried, and included in cookie formulations.

The dough consisted of the ingredients listed in Table 1. The recipe for dessert cookies included corn starch (Bezgluten, Posądza, Poland) and potato starch (Pepees SA, Łomża, Poland). The ingredients, after being weighed according to the recipe (Table 1), were mixed in a Diosna spiral mixer type SP 12 (Osnabrück, Germany) for about 5 min at slow speed until the dough was kneaded. The dough was then rolled out using a confectioners' rolling pin to obtain an even thickness. The cookies were punched out using a single mold with a diameter of 5 cm. Baking was carried out at 230 °C for 12 min. A MIWE CO 2 P608 type oven (Arnstein, Germany) was used for baking. After the cookies cooled to room temperature, they were packed into glass containers (jars) and stored in a storage chamber at a constant temperature of 22 °C, for further analysis.

Table 1. Recipes for dessert cookies with the addition of potato pulp (g).

	Control	DCMB5	DCMB10	DCML5	DCML10
Corn starch	358	340.1	322.2	340.1	322.2
Potato starch	16	16	16	16	16
Icing sugar	180	180	180	180	180
Margarine	96	96	96	96	96
Egg mass	26	26	26	26	26
Milk	64	64	64	64	64
Salt	1.2	1.2	1.2	1.2	1.2
Baking powder	12.8	12.8	12.8	12.8	12.8
Potato Pulp	0	17.9	35.8	17.9	35.8

2.2. Examination of the Physical Characteristics of Cookies Enriched with Potato Pulp

To measure the physical characteristics, 6 pieces of cookie were weighed with an accuracy of 0.1 g, and the average weight of one cookie was calculated. The length of the segment occupied by each cookie was measured in mm, the measurement was repeated in perpendicular axis, and the average diameter of one cookie was calculated. The thickness of the cookies was measured with a ruler by placing all the cookies one on top of the other. The spread ratio was calculated according to the following formula:

$$S = W/T$$

where S—spread ratio, W—diameter of 1 cookie (mm), T—thickness of 1 cookie (mm). Volume of cookies was determined using Volscan 600 (Stable Micro Systems, Surrey, UK) (Supplementary Materials Figure S1) placing all cookies on the rotating base and dividing the result by the number of stacked cookies.

2.3. Chemical Evaluation

The following analyses were performed on each sample:

1. Chemical composition—content of protein, fat, ash—was determined according to AOAC 2006 [21]. Total protein content was carried out using Kjeltec (2200, FOSS, Denmark), according to the AOAC method No 205,950.36. Evaluation of the raw fat content was carried out by using Soxtec Avanti extractor (2055, FOSS, Denmark), following the AOAC method No. 930.05. Total ash content was carried out by using laboratory stove model SM-2002 (Czylok, Poland), following the AOAC method No. 930.05. Available carbohydrates were calculated by difference: 100—fat-protein-ash-water-dietary fiber. Energy value of the cookies was calculated according to EU Directive 1169/2011 [22]: available carbohydrates 4 kcal/g, protein 4 kcal/g, fat 9 kcal/g, fiber 2 kcal/g.
2. Content of non-starch polysaccharides, i.e., total, dietary fiber, by the method 32-07 of AACC [23].
3. Antioxidant constituents, antiradical activity was determined in the ethanol extracts. A total of 0.6 g of the sample was dissolved in 30 mL 80 g/100 g ethanol, shaken in darkness for 120 min (electric shaker: type WB22, Memmert, Schwabach, Germany), and centrifuged (15 min, 4500 rpm. 1050× *g*) in a centrifuge (type MPW-350, MPW MED. Instruments, Warsaw, Poland). The supernatant was decanted and stored at −20 °C for further analyses. Determination of total polyphenols content (TPC) was carried out by spectrophotometric method using Folin–Ciocalteu reagent (with F-C reagent), according to Singleton, Orthofer, and Lamuela-Raventós [24], and the content of flavonoids was evaluated using a spectrophotometrical method, according to El Hariri, Sallé, and Andary [25]. Contents of total phenolic content without F-C reagent, phenolic acids, flavonols, and anthocyanins were evaluated using a spectrophotometrical method, according to Mazza et al. [26] and Oomah et al. [27]. Additionally, antiradical activity was assessed using analytical methods with ABTS (2,2′-azino-bis (3-ethylobenzothiazoline-6-sulphonic acid)-diamonium salt) [28]. Results of antiradical activity were expressed as TEAC (Trolox Equivalent Antioxidant Capacity-mM Trolox/kg dry mass of sample).
4. The content of acrylamide (AA) has been determined. For sample preparation according to Surma, Sadowska-Rociek, Cieślik, and Sznajder-Katarzyńska [29] modified QuEChERS method was applied. AA qualitative and quantitative analysis were performed using HPLC-UV/Vis (according to methods [30]. AA determination was performed using the HPLC-DAD system (VWR HITACHI, LaChrom ELITE, Merck KGaA, Darmstadt, Germany) according to Marconi et al. [30]. The chromatographic separation was performed at room temperature using a mixture of 0.01 M sulfuric acid and water/methanol mixture (water:methanol, 97.5:2.5) in isocratic mode with a C18 reversed-phase column (Lichrospher®® 100, RP-18 end-capped, LiChroCART®® 4 mm ID × 250 mm, 10 μm, Merck KGaA, Darmstadt, Germany). The flow rate was 0.7 mL min^{-1}. AA was determined at 200 nm wavelength. The identification of acrylamide was based on the retention time (4.2 min). The measurements were performed in four replications.

5. In order to determine hydroxymethylfurfural (HMF) content, 10 mL of water was added to 2 g of samples and quantitatively transferred to a 25 mL volumetric flask. After adding 0.25 mL of Carrez solution I and 0.25 mL of Carrez solution II, the volumetric flask was filled to the mark with deionized water. In the case of infusions, 0.07 mL of Carrez solution I and 0.07 mL of Carrez solution II were added to 6.86 mL of the infusion. Before a chromatographic analysis, samples were filtered through a 0.45 µm disc filter. Reagent blank samples were prepared according to the appropriate procedure for all tested analytes. Each sample (real and blank) was prepared in triplicate.
6. HMF qualitative and quantitative analyses were performed using HPLC-UV/Vis system (VWR HITACHI, LaChrom ELITE, Merck KGaA, Darmstadt, Germany) operating under the following conditions: sample volume 20 µL, eluent water/methanol 9:1 (*v*/*v*), flow rate 1 mL/min, wavelength 285 nm, column RP-18 Lichrosphere (250 × 4 mm, 5 µm particle size) (Merck, Germany).

Statistical Analysis

To assess the significance of differences between the averages, the experimental data were subjected to one-way analysis of variance (Duncan's post hoc test), at the confidence level of 0.05, by the use of software Statistica v. 8.0 (Statsoft, Inc., Tulsa, OK, USA). The Pearson correlation coefficients between selected parameters were also calculated to establish the relationships between them.

3. Results and Discussion

3.1. Characteristics of Pulp Derived from Colored Potatoes

The content of different subgroups of polyphenols in the potato pulp is shown in Table 2. It was found that the amount of polyphenols determined with Folin-Ciocalteu reagent was about 80% higher in the pulp derived from Magenta Love as compared to Marleta Blue. Similarly, the amount of polyphenols determined by the method without the Folin-Ciocalteu reagent was 86% higher for Magenta Love potato in comparison to Marleta Blue. Also, the potato pulp derived from variety Magenta Love had 31% more flavonoids than Marleta Blue. In the case of phenolic acids, Magenta Love potato pulp had two times higher content of these components than Marleta Blue samples (Table 2). Phenolic acids are among the most numerous phenolic compounds in the potato tuber. The dominant one is chlorogenic acid and its two isomers: neochlorogenic and cryptochlorogenic. Phenolic acids are referred to as nutraceuticals, i.e., enrichment substances with health-promoting effects [31–33]. They belong to biologically active compounds that influence the course of many biological processes such as adaptive or repair activities of the system into which they are introduced with food. On the other hand, Marleta Blue's potato pulp had six times higher content of flavonols than Magenta Love (Table 2). Taking into account the aforementioned pulps, it was found that the one derived from Marleta Blue potatoes contained 52% less anthocyanins than Magenta Love. Overall, the high content of all subgroups of polyphenols in Magenta Love's potato pulp translated into a high antioxidant potential of this plant material (Table 2).

Taking into account the chemical composition, the analyzed samples did not contain fat, and the product from the red potato variety Magenta Love was characterized by higher protein, ash, and total fiber contents by 48%; 18%, and 17.7%, respectively, in comparison to the variety Marleta Blue (Table 2).

Summarizing, it has been generally demonstrated that colored potato pulp is a source of health-promoting compounds (polyphenols). Such compounds exhibit hypoglycemic, hypocholesterolemic, anticancerogenic effects, and reduce postprandial glycemia and hypertension. They have anti-inflammatory, antiviral, antibacterial, anti-allergenic, and anti-coagulant effects and reduce the risk of diseases such as atherosclerosis and other cardiovascular diseases, cataracts, diabetes, genetic damage, degenerative bone changes, neurodegenerative diseases including Alzheimer's disease [34–39]. Therefore, colored potato residue can be used as a health-promoting ingredient for cookies.

Table 2. Content of polyphenols, flavonoids, phenolic acids, flavonols, and anthocyanins and chemical composition of colored potato pulp.

	Marleta Blue	Magenta Love
Polyphenols		
Total phenolic content—with F-C reagent (mg catechin/100 g d.m.)	188.01 ± 22.84 a	340.46 ± 35.18 b
Total phenolic content—without F-C reagent (mg catechin/100 g d.m.)	197.27 ± 7.19 a	366.54 ± 15.58 b
Flavonoids (mg rutin/100 g d.m.)	113.30 ± 18.53 a	148.75 ± 12.52 b
Phenolic acids (mg gallic acid/100 g d.m)	28.90 ± 0.00 a	61.16 ± 1.69 b
Flavonols (mg quercetin/100 g d.m.)	53.05 ± 0.00 b	8.54 ± 9.81 a
Anthocyanins (mg cyanidin—3-glucoside/100 g d.m.)	23.87 ± 0.92 a	49.27 ± 2.66 b
ABTS (mMTx/kg d.m.)	60.52 ± 0.70 a	77.31 ± 0.83 b
Chemical composition (g/100 g d.m.)		
Protein	4.35 ± 0.03 a	6.44± 0.11 b
Fat	0 ± 0	0 ± 0
Ash	2.01 ± 0 a	2.37 ± 0.07 b
Total dietary fiber	16.71 ± 0.31 a	19.68 ± 0.27 b

Different letters in the rows in the table represent the statistically significant difference of average values ($p = 0.05$).

3.2. Effect of Colored Potato Pulp on the Content of Health-Promoting Compounds in Cookies

The content of different groups of polyphenols in cookies with the participation of colored potato pulp is presented in Table 3. The amount of polyphenols in control cookies was at the level of 22.07 mg of catechin/100 g d. m. This should most probably be explained by the presence of Maillard reaction products, which affect the results of this determination, because according to other authors [40,41] the Folin-Ciocalteu reagent used in this determination not only reacts with polyphenols but also with other compounds such as amino acids, proteins, saccharides, i.e., compounds that can form Maillard reaction products.

Table 3. Content of polyphenols, flavonoids, anthocyanins, and phenolic acids in dessert cookies with the addition of colored potato pulp.

	TPC—with F-C Reagent (mg Catechin /100 g d.m.)	TTPC—without F-C Reagent (mg Catechin /100 g d.m.)	Flavonoids (mg Rutin /100 g d.m.)	Anthocyanins (mg Cyanidin—3-Glucoside /100 g d.m.)	Phenolic Acids (mg Gallic Acid/100g d.m.	ABTS (mMTx /kg d.m.)
Control	22.07 ± 4.14 a	18.22 ± 7.33 ab	14.40 ± 2.97 a	0.95 ± 1.06 a	2.94 ± 0.15 c	15.78 ± 0.85 a
DCMB5	30.46 ± 2.19 b	26.86 ± 3.45 b	17.75 ± 1.77 b	2.60 ± 0.00 b	1.87 ± 0.07 a	22.17 ± 0.10 b
DCMB10	33.74 ± 1.09 b	24.27 ± 1.73 b	23.5 ± 1.77 c	2.60 ± 0.00 b	1.99 ± 0 a	26.96 ± 0.10 c
DCML5	33.01 ± 1.26 b	26.50 ± 0.00 b	36.35 ± 3.61 d	1.00 ± 0.00 a	2.17 ± 0.04 b	25.36 ± 1.05 c
DCML10	47.60 ± 1.67 c	42.56 ± 3.99 c	61.25 ± 2.26 e	4.20 ± 0.00 c	6.05 ± 0.17 d	38.15 ± 0 d

Different letters in the column represent the statistically significant difference of average values ($p = 0.05$).

The presence of polyphenols was also noted in control gluten-free bread in studies by other authors [42,43], which was also explained by the production of Maillard compounds. In the current study, the amount of Maillard compounds is much higher, because they are formed via the exposure of the cookies to high oven temperatures. It was observed that the introduction of colored potato pulp in the cookies resulted in an increase in polyphenols ranging from 38 to 115% compared to the control. The smallest increase in polyphenols in the cookies was observed when 5 and 10% of Marleta Blue potato pulp and 5% of Magenta Love potato pulp were used, and the largest when 10% of Magenta Love preparation was added. Such a large increase results from the fact that the amount of polyphenols in the Magenta Love pulp was significantly higher than in the Marleta Blue (Tables 2 and 3). The determination of polyphenols without Folin–Ciocalteu reagent was also used in this work, giving smaller and more meaningful values compared to those obtained with this reagent. It was noted that the Marleta Blue potato pulp did not contribute to an increase in polyphenols in the cookies regardless of its proportion in the formulation (5 and 10%). Although 5% addition of Magenta Love pulp had no significant influence on polyphenols, its 10% addition caused a 2.5-fold increase in polyphenols in the cookies compared to the control (Table 3). Sudha et al. [44], in a study on the quality of cakes with apple pomace, reported an increase in total polyphenol content from 42.7% to 68.4% in the final products compared to the control. According to the above-mentioned authors, this was related, on the one hand, to the use of apple pomace, from which the polyphenols were derived, and, on the other hand, to the possibility of the formation of products such as reductones. These compounds are formed during the baking process by the oxidation of sugars derived from apple pomace [44]. The addition of 5 to 20% mango peel powder to a cookie recipe resulted in an 8.5-fold increase in polyphenol content [45]. Bertagnolli et al. [46] developed a cookie recipe using flour obtained from guava peels at 30, 50, and 70% substitution level, which resulted in an increase in the polyphenol content of the final product. Kopeć et al. [47] conducted a study on the replacement of wheat flour in sponge cake and shortcrust pastries by buckwheat flour, maize flour, and a mixture of these flours in a ratio of 1:1. The sponge cake and pastries with buckwheat flour had the highest polyphenol content compared to control products containing only wheat flour. In cookies containing dietary fiber isolated from Doum fruit (*Hyphaene thebaica*), the content of polyphenols increased with increasing fiber content. The highest content of these components was determined in cookies with 10% of isolated fiber derived from Doum fruit [48].

The flavonoid content of cookies containing colored potato pulp was 23% higher when 5% Marleta Blue potato pulp was used and four times higher when 10% Magenta Love potato pulp was used, relative to the control (Table 3), with the amount of these health-promoting components increasing accordingly to the level of addition of different types of potato residue. Korus et al. [17] conducted a study on gluten-free cookies using 20%, 40%, and 60% acorn seed flour and hemp flour. Based on the results, it was found that as the amount of hemp and acorn flours increased in the cookies, the flavonoid content increased. The increase in flavonoid content in cookies with acorn flour was five-fold and in cookies with hemp flour 90% compared to control cookies. Maner et al. [49] developed a recipe for cookies enriched with 5, 10, 15, and 20% addition of dried grape pomace. They found that the addition of pomace increased the flavonoid content from 96 to 254% in the final products compared to the control baked goods. Ajibola et al. [50] prepared cookies using dried Moringa oleifera leaves and cocoa powder. They found that 10% participation of Moringa oleifera in the cookie recipe increased the flavonoid content from 0.22 mg/g in the control product to 1.27 mg/g in the final product. Moringa has significant nutritional and health-promoting values, but some studies indicate serious side effects caused by its consumption, so its use in food is not allowed in some countries [51]. Natukunda et al. [52] made a cookie recipe with 2, 4, 6, 8, and 10% tamarind seed powder. The addition increased the flavonoid content from 32 to 153% compared to the control. An increase in flavonoid content as a result of the proportion of purple rice flour was also found by Klunklin and Savage [53]. In the prepared cookies, the flavonoid content was three to eight times higher compared to the control. Van Toan and Vu Quynh Anh [54] used a proportion of 10 to 50% purple sweet potato flour in a cookie recipe. The cookies had a flavonoid content ranging from 5.25 mg quercetin/g to 7.51 mg quercetin/g. The addition of sweet potato flour increased the flavonoid content compared to the control sample of 4.8 mg quercetin/g. When cookies were enriched with acorn flour, Pasqualone and co-workers [55] showed a seven-fold increase in flavonoid content with 30% and a 21-fold increase in the mentioned component with 60% acorn flour addition compared to the control sample.

Dessert cookies with the addition of Marleta Blue pulp contained 36% less phenolic acids at 5% and 32% less phenolic acids at 10% potato pulp compared to control cookies. Also, a 26% decrease in phenolic acids was observed with 5% of Magenta Love potato pulp compared to the unenriched products. It was found that only a 10% addition of Magenta Love preparation caused a two-fold increase in phenolic acids content (Table 3). Analyzing the results, it can be concluded that the low content of phenolic acids in the cookies with Marleta Blue potato pulp at both 5 and 10% addition level may be due to the content of these components in the pulp being lower by half compared to Magenta Love pulp (Tables 2 and 3). The results were significantly influenced by the baking process of the cookies, which is associated with the decarboxylation of phenolic acids. An opposite effect was obtained by enriching the cookies with 60% acorn flour and hemp flour, which caused an approximately six-fold increase in phenolic acids content in the cookies compared to the control product [17].

Cookies with 5 and 10% share of the potato pulp isolated from variety Marleta Blue contained about 174% more anthocyanins compared to the control cookies. The highest anthocyanin content among was found in cookies with 10% of Magenta Love potato pulp (Table 3). Anthocyanins are characterized by thermal instability. The mechanism of anthocyanin degradation is not fully understood, and research suggests that first the glycosidic bonds in the pigment molecule are hydrolyzed and then an unstable aglycone is formed. Elevated temperature also affects the transformation of anthocyanins into colorless chalcones, which undergo oxidation to form high molecular weight brown compounds and dyes. The degree of anthocyanin degradation during thermal treatment depends on the time and temperature of the process. A logarithmic relationship was observed between the destruction of anthocyanins and heating time at constant temperature. The loss of these components also depends on other factors such as pH, the chemical structure of anthocyanins, and the presence in their environment of oxygen, other polyphenolic compounds, proteins, sugars, and their degradation products. Greater glycosidation of anthocyanins and acylation with phenolic acids improves resistance to high temperature [56]. Maner et al. [49] prepared cookies in which part of the wheat flour was replaced by the addition of dried grape pomace. The addition of pomace affected the dark brown color of the products. The color change was influenced by anthocyanins present in grape skin (mainly malvidin 3-O-glucoside and peonidin 3-O-glucoside). The above-mentioned authors found a significant increase in the anthocyanin content (3.512 mg/g) in cookies with pomace addition compared to the control (0.163 mg/g). In a study by Korus et al. [17], the anthocyanin content of cookies with 60% acorn flour was significantly higher than the control cookies. Also in the same study, it was found that cookies with 60% hemp flour showed 75% higher anthocyanin content compared to cookies containing corn flour. Klunklin and Savage [53] made cookies containing between 25 and 100% purple rice flour in their formulation. With the addition of purple rice, a significant increase in anthocyanin content was demonstrated. Wheat cookies (control sample) contained 0.28 mg/kg d.m. of anthocyanins while cookies baked exclusively with rice flour had a content of 51.49 mg/kg d.m.

An increase in total polyphenols and their subgroups resulted in an increase in antioxidant activity in cookies with red potato pulp (Table 3) relative to the control as evidenced by the strong correlation between TPC and ABTS R = 0.921. High antioxidant activity was observed in cookies with 10% Magenta Love potato pulp (38.15 m M Tx/kg d.m.) relative to the control sample (15.78 mM Tx/kg d.m.). An increase in antioxidant activity was also shown in cookies containing grape pomace [49]. The control cookies had an activity of 4.625 mg Tx/g, while the antioxidant activity of the enriched products was 11.651, 29.669, 51.862, and 75.976 (mg Tx/g) at 5, 10, 15, and 20% pomace content, respectively [49]. Pasqualone and co-workers [55] observed an almost 40-fold increase in antioxidant activity in cookies with acorn flour at 60% compared to control cookies. Klunklin and Savage [53] replacing wheat flour with purple rice flour observed an increase in ABTS radical scavenging capacity from 5.49 μmol Tx/g d.m of control cookies to 95.96 μmol Tx/g d.m when wheat flour was completely replaced by purple rice in the final product. In cookies enriched with dietary fiber from Doum fruit, an increase in antioxidant activity was also observed from 2.65 (mM Tx/100 g d.m.) in the control sample to a content of 18.34 (mM Tx/100 g d.m.) with 10% of the said fiber. The increasing antioxidant activity was shown to be adequate for the polyphenol content of the cookies [48].

3.3. Chemical Composition of Cookies with Colored Potato Pulp

Considering the protein content, it was found that the 5% addition of Magenta Love potato pulp did not cause changes in this component compared to the control. The share of 5 and 10% potato pulp isolated from variety Marleta Blue contributed to a small but significant decrease in protein relative to the control. A higher amount of protein was recorded in cookies with a share of 10% potato pulp isolated from variety Magenta Love (Table 4). No fat was determined in the samples, and the amount of ash was identical in each sample analyzed (Table 4). The share of 5% potato pulp isolated from variety Marleta Blue contributed to a two-fold decrease in dietary fiber in relation to the control. The 10% share of potato pulp isolated from variety Marleta Blue contributed to a 60% increase in fiber, and the 10% share of potato pulp isolated from variety Magenta Love contributed to a two-fold increase in dietary fiber in relation to the control (Table 4).

Table 4. Chemical composition of dessert cookies with the addition of colored potato pulp (g/100 g) and their energy value (kcal/100 g).

	Moisture	Protein	Fat	Ash	Total Dietary Fiber	Available Carbohydrates	Energy
Control	3.12 ± 0.05 a	1.18 ± 0.02 b	12.83 ± 0.02 a	1.45 ± 0.02 a	0.49 ± 0.02 b	80.93 ± 1.50 d	444.89 ± 3.20 c
DCMB5	4.15 ± 0.02 b	1.11 ± 0 a	12.85 ± 0.03 a	1.43 ± 0.05 a	0.22± 0 a	80.24 ± 1.12 c	441.49 ± 3.13 c
DCMB10	5.33 ± 0.07 c	1.12 ± 0.01 a	12.84 ± 0.05 a	1.47 ± 0.07 a	0.80 ± 0.07 c	78.44 ± 2.40 a	435.40 ± 2.90 a
DCML5	4.54 ± 0.03 b	1.19 ± 0.04 b	12.82 ± 0.03 a	1.44 ±0.09 a	0.54 ± 0.05 b	79.47 ± 2.12 b	439.10 ± 2.76 b
DCML10	5.10 ± 0.05 c	1.39 ± 0.05 c	12.78 ± 0.04 a	1.50 ± 0.01 a	1.19 ±0.04 d	78.04 ± 1.83 a	435.12 ± 2.55 a

Different letters in the column represent the statistically significant difference of average values ($p = 0.05$).

Considering the available carbohydrates, it was found that the share of pulp from colored potatoes reduced their amount in dessert cookies (in the case of a 10% share of pulp by an average of 3.5%) in relation to the control, also decreasing their energy value in relation to the standard (Table 4).

3.4. Acrylamide and HMF Content in Cookies with Colored Potato Pulp

No differences were noticed in the HMF contents present in the cookies (Table 5). The only statistically significant difference can be seen when comparing the HMF content of the control sample with the other variants regardless of the percentage of pulp and the varieties of potatoes used in the experiment. In each of the variants tested, the HMF content decreased from 32 to 37% (or about 35%) compared to the control sample (Table 5).

Table 5. Acrylamide and HMF in dessert cookies with the addition of colored potato pulp.

	Acrylamide Content (μg/kg)	HMF Content (μg/kg)
Control	65.67 ± 3.25 a	3.1 ± 0.1 b
DCMB5	80.50 ± 7.5 b	2.00 ± 0.008 a
DCMB10	90.11 ± 7.98 d	1.99 ± 0.02 a
DCML5	89.79 ± 1.64 b	1.96 ± 0.1 a
DCML10	95.11 ± 3.44 d	2.22 ± 0.18 a

Different letters in the column represent the statistically significant difference of average values ($p = 0.05$).

In the case of acrylamide, each addition of pulp regardless of its percentage of the variety of potatoes used resulted in a statistically significant increase in its content relative to the control sample. In the case of the Marleta Blue variety, a 5% addition of potato pulp caused a 22% increase in acrylamide content, while a 10% addition caused a 38% increase in acrylamide content relative to the control sample. On the other hand, the addition of Magenta Love potato pulp caused a 37 and 47% increase in acrylamide content for the 5 and 10% additions, respectively. All the results obtained are statistically significantly different from each other if we consider the percentage of the addition of the pulp. Comparing between each other cookies with the same content of 5 and 10% pulp, respectively, higher acrylamide contents were recorded for the Magenta Love variety, by 12 and 11%, respectively (Table 5), which would suggest that the mentioned variety is characterized by a higher content of amino acids (especially asparagine) or reducing sugars. Although there are no limits for acrylamide in food, according to Commission Regulation (EU) 2017/2158 [57], benchmark levels for the presence of acrylamide in foodstuffs have been established. Food was classified in 10 categories, one of which are cookies and wafers for which the benchmark level for the presence of acrylamide has been set at 350 μg/kg. Taking into account the content of acrylamide in the studied samples with or without potato pulp, the detected levels of this component were lower than benchmark level.

In conclusion, it is possible to propose the practical use of pulp from red and purple potatoes for the production of dessert cookies, especially for people with celiac disease. It should be noted that allergies and food intolerances, including celiac disease, according to recent estimates, are the third major threat to human health right after cancer and cardiovascular disease. People with gluten intolerance are forced to exclude products containing gluten, such as most cereal products, from their diet, which can result in deficiencies of many substances valuable for health, such as protein, vitamins (folic acid, B vitamins), minerals (Fe, Ca, Mg, Cu), and dietary fiber [13,58]. Restriction of the above-mentioned components, in turn, induces many diseases in people with celiac disease, such as osteoporosis, esophageal cancer, or infertility, etc. [59]. In addition, recent studies indicate that patients with gluten intolerance have been determined to have significant oxidative stress and impaired antioxidant enzymes, susceptibility to DNA damage and, consequently, cancer [12]. Proposed cakes involving potato pulp were characterized by up to two times higher content of polyphenols and phenolic acids, four times higher content of flavonoids and anthocyanins, two times higher amount of fiber, and 17% higher amount of protein. Another important practical aspect of the use of red and purple potato pulp to enrich dessert cookies with health-promoting compounds such as polyphenols is that these products can be suggested as an alternative for the treatment of type 2 diabetes, since according to Sun and Miao [60] polyphenols have an inhibitory effect on the digestion of starch, and consequently reduce the glycemic index of starchy products. This is because polyphenols inhibit the activity of the amylolytic enzymes (alpha-amylase and gamma-glucosidase) as well as interact directly with starch, forming hardly digestible complexes. The alleviation of postprandial hyperglycemia by polyphenol compounds might be due to both the inhibited starch digestion in vivo and the influenced glucose transport [60].

3.5. Influence of Colored Potato Pulp on Physical Characteristics of Cookies

Table 6 shows the physical characteristics of the dessert cookies with potato pulp isolated from the colored flesh varieties Magenta Love and Marleta Blue. The addition of potato pulp to the traditional cookie recipe resulted in a slight increase in the volume of the final products. In terms of weight, there was little difference from the control products, with the exception of the cookies with 10% of Magenta Love pulp, whose weight was significantly higher than the other samples (Table 6). In the study by Van Toan and Vu Quynh Anh [54] on cookies with colored sweet potato flours, it was found that as the sweet potato flour content increases, the volume of the cookies decreases. This effect may be influenced by the fiber present in the sweet potato flour, which reduces the gas retention capacity of the dough [61,62]. Another factor will be the form of the enrichment additive. It should be noted that the volume of the product is influenced by the form of the additive (whether it is ground or not). The unground form actually breaks the gluten matrix, resulting in a lower volume of the final product [63]. In the case of the study performed in this work, the red potato pulp is ground, and therefore we may be dealing here with an increase in cookie volume (Table 6). Yadav et al. [64] showed a decrease in cookie weight with the addition of banana flour and chickpea flour. The weight of cookies ranged from 8.7 g to 10.1 g with a control cookie weight of 11.4 g. Cookies containing mango seed flour had a 1 to 6% lower weight compared to the control cookies [65]. An increase in the weight of the products was noted for cookies containing rice bran, broken rice, and okara [66]. A similar effect was noted with the contribution of carrot and beetroot powder in the cookie recipe [67]. Srivastava and Singh [68] prepared cookies with 5, 10, and 15% beetroot powder. Their study found no significant differences between the control cookie and the enriched products, and the same was true in the study of this work.

Table 6. Physical properties of dessert cookies enriched with pulp of colored potato varieties.

	Weight (g)	Volume (mL)	Diameter (mm)	Thickness (mm)	Spread Ratio
Control	10.15 ± 0.15 b	56.70 ± 0 a	61.58 ± 0.12 d	11.0 ± 0.10 a	5.59 ± 0.12 c
DCMB5	9.93 ± 0.10 a	58.80 ± 0.15 b	59.16 ± 0.37 c	11.3 ± 0.59 b	4.97 ± 0.20 b
DCMB10	10.18 ±0.20 b	61.00 ± 0.27 c	57.58 ± 0.50 b	12.00 ± 0.12 c	4.80 ± 0.20 b
DCML5	10.27 ± 0.10 b	61.30 ± 0.53 c	59.00 ± 0.20 c	12.5 ± 0.70 c	4.72 ± 0.43 b
DCML10	10.57 ± 0 c	58.00 ± 0.89 ab	56.58 ± 0 a	13.37 ± 0 d	4.23 ± 0.12 a

Different letters in the column represent the statistically significant difference of average values (p = 0.05).

In addition, it was shown that with an increase in the proportion of pulp, the diameter of the cookies decreased in a range of 6.5 to 4% for the Marleta Blue potato variety and 4 to 8% for the Magenta Love cookies compared to the control. The thickness of the cookies increased while the value of the spread ratio decreased as a result of the addition of potato pulp in the final products (Table 6). The spread ratio (width divided by thickness) is considered one of the most important quality parameters for cookies as it correlates with the texture and consistency of the cookie. It is easy to see that the addition of different amounts of flours affects the quality of confectionery [69]. Van Toan and Vu Quynh Anh [54] noted that an increase in the proportion of purple sweet potato flour increased the spread ratio for cookies enriched with the aforementioned flour. The results were directly influenced by the thickness of the cookies decreasing with the addition of purple sweet potato flour, which is directly related to the spread ratio of the products. The diameter value was virtually unchanged regardless of the proportion of sweet potato flour. As a result, cookies made from composite flour had a higher spread ratio than wheat cookies. The result may have been influenced by the lower protein content of the purple potato flour. Makinde and Taibat [70] prepared a cookie recipe with the following flours: maize, almond, and coconut. The contribution of the flours increased the diameter of the cookies from 4 to 8% compared to the control sample of 49.60 mm. The thickness, on the other hand, decreased from approximately 10 to 22% compared to cookies containing only wheat flour. In the final products, the value of the spread ratio increased from 5 to 26% from that characterizing the control cookie. Adeola and Ohizua [71] showed that as sweet potato flour and banana flour increased in the composition of the cookies, the thickness of the cookies increased. These changes were explained by the increasing protein content of the product. An analogous trend was observed in the study of this work because a greater thickness of cookies corresponded to higher protein content (Tables 4 and 6).

4. Conclusions

1. It was found that the Magenta Love potato pulp had a higher content of polyphenols, flavonoids, phenolic acids, and anthocyanins compared to the Marleta Blue variety of potato pulp, which contributed to higher antioxidant activity.
2. It was noted that the Magenta Love potato pulp added at a level of 10% contributed to a significant increase in the total polyphenol content of the dessert cookies compared to the control sample. In addition, an increase in flavonoid content was observed as a

result of the use of colored potato pulp in the cookie recipe, particularly when using the potato variety Magenta Love. It was found that the 10% share of Magenta Love potato pulp had a doubling effect on the phenolic acid content of the final product compared to the control cookies. In other cases, a decrease in phenolic acid content was observed, which could be due to degradation processes of these compounds. It was found that the antioxidant activity of dessert cookies with Magenta Love and Marleta Blue potato pulp was higher in relation to the control cookies, which is adequate to the polyphenol content.

3. Considering nutrients, cookies with 5 and 10% Magenta Love potato pulp had higher amounts of protein and fiber compared to the control and cookies with Marleta Blue potato pulp. Cookies enriched with potato pulp had a 30% lower HMF content relative to the control and a higher acrylamide content, which still fell below the limit for this type of product.
4. It was observed that the addition of pulp isolated from colored potatoes increased the volume of the final products relative to the control products. In contrast, the diameter of the analyzed cookies decreased with an increase in the addition of pulp.
5. In conclusion, the pulp isolated from Magenta Love variety was more beneficial for enriching dessert cookies with health-promoting compounds than the Marleta Blue preparation. It can be suggested that dessert cookies with a 10% share of potato pulp of the Magenta Love potato variety can be recommended on an industrial scale because of the product's health-promoting properties and good physical and nutritional quality. The utilization of by-products in food technology is very important as it is inextricably linked with zero-waste technology. It should be remembered that starch is produced from light-colored potato varieties, which are poorer in phenolic compounds than potatoes with colored flesh. In the future, however, it would be possible to use potatoes with colored flesh for starch production because they are more resistant to diseases due to the very high proportion of anthocyanins in their tubers. Thus, it would be advantageous to use colored potato pulp for the production of food products, especially those labelled gluten free, as it could provide additional value. The gluten-free food sector could be one of the most profitable food industries, not only because of the significant number of people with celiac disease, but also because of the prevailing trend where a significant number of people are switching to a gluten-free diet in search of diets alternative to the traditional one.

Supplementary Materials: The following supporting information can be downloaded at: https://www.mdpi.com/article/10.3390/10.3390/foods12203735/s1, Figure S1: Measuring the volume of a biscuit using a Volscan Profiler laser volume analyzer (Stable Micro Systems, Godalming, UK).

Author Contributions: Conceptualization, D.G.; methodology, D.G., R.Z., J.K. and M.S.; validation, D.G., R.Z., J.K. and M.S.; formal analysis, D.G., R.Z., J.K. and M.S.; investigation, D.G., R.Z., J.K. and M.S.; resources, D.G.; data curation, D.G. and R.Z.; writing—original draft preparation, D.G.; writing—review and editing, D.G., R.Z., J.K. and M.S.; visualization, D.G. and R.Z.; supervision, D.G.; project administration, D.G. and R.Z. All authors have read and agreed to the published version of the manuscript.

Funding: This research was financially supported by the Ministry of Science and Higher Education of Republic of Poland.

Data Availability Statement: The data used to support the findings of this study can be made available by the corresponding author upon request.

Conflicts of Interest: The authors declare no conflict of interest.

References

1. FAO. The State of Food and Agriculture 2019. In *Moving Forward on Food Loss and Waste Reduction*; FAO: Rome, Italy, 2019.
2. Sadh, P.K.; Duhan, S.; Duhan, J.S. Agro-industrial Wastes and Their Utilization Using Solid State Fermentation: A Review. *Bioresour. Bioproc.* **2018**, *5*, 1. [CrossRef]

3. FAOSTAT. *Food and Agriculture Organization of the United Nations Statistical Database*; Statistical Division; FAO: Rome, Italy, 2021. Available online: http://www.fao.org/statistics/en/ (accessed on 24 February 2023).
4. Lemes, A.C.; Egea, M.B.; Oliveira Filho, J.G.; Gautério, G.V.; Ribeiro, B.D.; Coelho, M.A.Z. Biological Approaches for Extraction of Bioactive Compounds from Agroindustrial By-products: A Review. *Front. Bioeng. Biotechnol.* **2022**, *9*, 802543. [CrossRef] [PubMed]
5. Reguengo, L.M.; Salgaço, M.K.; Sivieri, K.; Maróstica, M.R.J. Agro-industrial by-products: Valuable sources of bioactive compounds. *Food Res. Int.* **2022**, *152*, 110871. [CrossRef] [PubMed]
6. Bartek, L.; Sundin, N.; Strid, I.; Andersson, M.; Hansson, P.A.; Eriksson, M. Environmental benefits of circular food systems: The case of upcycled protein recovered using genome edited potato. *J. Clean. Prod.* **2022**, *380*, 20. [CrossRef]
7. Ismail, S.A.; Abdullah, V.S.; Kamel, F.H. Extraction of α-solanine and α-chaconine from green potato tubers and evaluation of its antimicrobial activity. *Plant Arch.* **2022**, *19*, 4009–4014.
8. Sip, A.; Thanh-Blicharz, L.; Siergiej, K.; Lesiecki, M.; Lewandowicz, G. Odpady przemysłu ziemniaczanego jako podłoże do hodowli bakterii mlekowych. *Postępy Nauk. Technol. Przemysłu Rolno-Spożywczego* **2016**, *71*, 65–80.
9. Szarek, D.; Przewodowska, A. Fizykochemiczne metody odzyskiwania białek z soku ziemniaka. *Ziemn. Pol.* **2016**, *26*, 4.
10. Markiewicz, M.; Przewodowska, A.; Przewodowski, W.; Stochła, W. Wykorzystanie chromatografii membranowej do odzyskiwania białek aktywnych biologicznie z odpadów przemysłu skrobiowego. *Rocz. Ochr. Sr.* **2015**, *17*, 1699–1714.
11. Allied Market Research Report. Confectionery Market by Product Type (Hard-Boiled Sweets, Mints, Gums & Jellies, Chocolate, Caramels & Toffees, Medicated Confectionery, Fine Bakery Wares, and Others), Age Group (Children, Adult, and Geriatric), Price Point (Economy, Mid-Range, and Luxury), and Distribution Channel (Supermarket/Hypermarket, Convenience Stores, Pharmaceutical & Drug Stores, Food Services, Duty-Free Outlets, E-Commerce, and Others): Global Opportunity Analysis and Industry Forecast, 2021–2027. Available online: https://www.alliedmarketresearch.com/confectionery-market (accessed on 27 November 2022).
12. Stojiljković, V.; Todorović, A.; Pejić, S.; Kasapović, J.; Saičić, Z.S.; Radlović, N.; Pajović, S.B. Antioxidant Status and LipidPeroxidation in Small Intestinal Mucosa of Children with Celiac Disease. *Clin. Biochem.* **2009**, *42*, 1431–1437. [CrossRef]
13. Caio, G.; Volta, U.; Sapone, A.; Leffler, D.A.; De Giorgio, R.; Catassi, C.; Fasano, A. Celiac disease: A comprehensive current—Review. *BMC Med.* **2019**, *17*, 142. [CrossRef]
14. Filipčev, B.; Lević, L.; Bodroža-Solarov, M.; Mišljenović, N.; Koprivica, G. Quality Characteristics and Antioxidant Properties of Breads Supplemented with Sugar Beet Molasses-Based Ingredients. *Int. J. Food Prop.* **2010**, *13*, 1035–1053. [CrossRef]
15. Altunkaya, A.; Hedegaard, R.V.; Brimer, L.; Gökmen, V.; Skibsted, L.H. Antioxidant capacity versus chemical safety of wheat bread enriched with pomegranate peel powder. *Food Funct.* **2013**, *4*, 722–727. [CrossRef] [PubMed]
16. Świeca, M.; Gawlik-Dziki, U.; Dziki, D.; Baraniak, B.; Czyż, J. The influence of protein-flavonoid interactions on protein digestibility in vitro and the antioxidant quality of breads enriched with onion skin. *Food Chem.* **2013**, *141*, 451–458. [CrossRef] [PubMed]
17. Korus, A.; Gumul, D.; Krystyjan, M.; Juszczak, L.; Korus, J. Evaluation of the quality, nutritional value and antioxidant activity of gluten-free biscuits made from corn-acorn flour or corn-hemp flour composites. *Eur. Food Res. Technol.* **2017**, *243*, 1429–1438. [CrossRef]
18. Zhu, F.; Sun, J. Physicochemical and sensory properties of steamed bread fortified with purple sweet potato flour. *Food Biosci.* **2019**, *30*, 100411. [CrossRef]
19. Martirosyan, D.; Kanya, H.; Nadalet, C. Can functional foods reduce the risk of disease? Advancement of functional food definition and steps to create functional food products. *Funct. Food Health Dis.* **2021**, *11*, 213–221. [CrossRef]
20. Wischmann, B.; Ahmt, T.; Bandsholm, O.; Blennow, A.; Young, N.; Jeppesen, L.; Thomsen, L. Testing properties of potato starch from different scales of isolations—A ringtest. *J. Food Eng.* **2007**, *79*, 970–978. [CrossRef]
21. AOAC. *Official Methods of Analysis*; Association of Official Analytical Chemists: Gaithersburgs, MD, USA, 2006.
22. REGULATION (EU) No 1169/2011 OF THE EUROPEAN PARLIAMENT AND OF THE COUNCILof 25 October 2011 on the Provision of Food Information to Consumers, Amending Regulations (EC) No 1924/2006 and EC) No 1925/2006 of the European Parliament and of the Council, and Repealing Commission Directive 87/250/EEC, Council Directive 90/496/EEC, Commission Directive 1999/10/EC, Directive 2000/13/EC of the European Parliament and of the Council, Commission Directives 2002/67/EC and 2008/5/EC and Commission Regulation (EC) No 608/2004. Available online: https://eur-lex.europa.eu/LexUriServ/LexUriServ.do?uri=OJ:L:2011:304:0018:0063:en:PDF (accessed on 24 February 2023).
23. AACC. *Approved Methods of the American Association of Cereal Chemists*; AACC: Washington, DC, USA, 2000.
24. Singleton, V.L.; Orthofer, R.; Lamuela-Raventós, R.M. Analysis of total phenols and other oxidation substrates and antioxidants by means of Folin-Ciocalteu reagent. In *Methods in Enzymology*; Elsevier: Cham, Switzerland, 1999; pp. 152–178.
25. El Hariri, B.; Sallé, G.; Andary, C. Involvement of flavonoids in the resistance of two poplar cultivars to mistletoe (*Viscum album* L.). *Protoplasma* **1991**, *162*, 20–26. [CrossRef]
26. Mazza, G.; Fukumoto, L.; Delaquis, P.; Girard, B.; Ewert, B. Anthocyanins, phenolics, and color of Cabernet franc, Merlot, and Pinot noir wines from British Columbia. *J. Agric. Food Chem.* **1999**, *47*, 4009–4017. [CrossRef]
27. Oomah, B.D.; Cardador-Martínez, A.; Loarca-Piña, G. Phenolics and antioxidative activities in common beans (*Phaseolus vulgaris* L.). *J. Sci. Food Agric.* **2005**, *85*, 935–942. [CrossRef]

28. Re, R.; Pellegrini, N.; Proteggente, A.; Pannala, A.; Yang, M.; Rice-Evans, C. Antioxidant activity applying an improved ABTS radical cation decolorization assay. *Free Radic. Biol. Med.* **1999**, *26*, 1231–1237. [CrossRef] [PubMed]
29. Surma, M.; Sadowska-Rociek, A.; Cieślik, E.; Sznajder-Katarzyńska, K. Optimization of QuEChERS sample preparation method for acrylamide level determination in coffee and coffee substitutes. *Microchem. J.* **2017**, *131*, 98–102. [CrossRef]
30. Marconi, O.; Bravi, E.; Perretti, G.; Martini, R.; Montanari, L.; Fantozzi, P. Acrylamide risk in food products: The shortbread case study. *Anal. Methods* **2010**, *2*, 1686–1691. [CrossRef]
31. Lachman, J.; Hamouz, K.; Šulc, M.; Orsák, M.; Pivec, V.; Hejtmánková, A.; Dvořák, P.; Čepl, J. Cultivar differences of total anthocyanins and anthocyanidins in red and purple-fleshed potatoes and their relation to antioxidant activity. *Food Chem.* **2009**, *114*, 836–843. [CrossRef]
32. Navarre, D.A.; Pillai, S.S.; Shakya, R.; Holden, M.J. HPLC profiling of phenolics in diverse potato genotypes. *Food Chem.* **2011**, *127*, 34–41. [CrossRef]
33. Kita, A.; Bąkowska-Barczak, A.; Lisińska, G.; Hamouz, K.; Kułakowska, K. Antioxidant activity and quality of red and purple flesh potato chips. *LWT—Food Sci. Technol.* **2015**, *62*, 525–531. [CrossRef]
34. Middleton, E.; Kandaswami, C.; Theoharides, T.C. The effects of plant flavonoids on mammalian cells: Implications for inflammation, heart disease, and cancer. *Pharmacol. Rev.* **2000**, *52*, 673–751. [PubMed]
35. Manach, C.; Mazur, A.; Scalbert, A. Polyphenols and prevention of cardiovascular diseases. *Curr. Opin. Lipidol.* **2005**, *16*, 77–84. [CrossRef]
36. Ramos, S. Effects of dietary flavonoids on apoptotic pathways related to cancer chemoprevention. *J. Nutr. Biochem.* **2007**, *18*, 427–442. [CrossRef]
37. Yokohira, M.; Yamakawa, K.; Saoo, K.; Matsuda, Y.; Hosokawa, K.; Hashimoto, N.; Kuno, T.; Imaida, K. Antioxidant effects of flavonoids used as food additives (purple corn color, enzymatically modified isoquercitrin, and isoquercitrin) on liver carcinogenesis in a rat medium-term bioassay. *J. Food Sci.* **2008**, *73*, C561–C568. [CrossRef]
38. Balasuriya, N.; Rupasinghe, H.V. Antihypertensive properties of flavonoid-rich apple peel extract. *Food Chem.* **2012**, *135*, 2320–2325. [CrossRef]
39. Makarova, E.; Górnaś, P.; Konrade, I.; Tirzite, D.; Cirule, H.; Gulbe, A.; Pugajeva, I.; Seglina, D.; Dambrova, M. Acute anti-hyperglycaemic effects of an unripe apple preparation containing phlorizin in healthy volunteers: A preliminary study. *J. Sci. Food Agric.* **2014**, *95*, 560–568. [CrossRef] [PubMed]
40. Everette, J.D.; Bryant, Q.M.; Green, A.M.; Abbey, Y.A.; Wangila, G.W.; Walker, R.B. Thorough study of reactivity of various compound classes toward the Folin−Ciocalteu reagent. *J. Agric. Food Chem.* **2010**, *58*, 8139–8144. [CrossRef] [PubMed]
41. Gulcin, İ. Antioxidants and antioxidant methods: An updated overview. *Arch. Toxicol.* **2020**, *94*, 651–715. [CrossRef] [PubMed]
42. Korus, J.; Juszczak, L.; Ziobro, R.; Witczak, M.; Grzelak, K.; Sójka, M. Defatted strawberry and blackcurrant seeds as functional ingredients of gluten-free bread. *J. Texture Stud.* **2012**, *43*, 29–39. [CrossRef]
43. Gumul, D.; Ziobro, R.; Korus, J.; Kruczek, M. Apple pomace as a source of bioactive polyphenol compounds in gluten-free breads. *Antioxidants* **2021**, *10*, 807. [CrossRef]
44. Sudha, M.; Baskaran, V.; Leelavathi, K. Apple pomace as a source of dietary fiber and polyphenols and its effect on the rheological characteristics and cake making. *Food Chem.* **2007**, *104*, 686–692. [CrossRef]
45. Ajila, C.; Leelavathi, K.; Rao, U.P. Improvement of dietary fiber content and antioxidant properties in soft dough biscuits with the incorporation of mango peel powder. *J. Cereal Sci.* **2008**, *48*, 319–326. [CrossRef]
46. Bertagnolli, S.M.M.; Silveira, M.L.R.; de Oliveira Fogaça, A.; Umann, L.; Penna, N.G. Bioactive compounds and acceptance of cookies made with Guava peel flour. *Food Sci. Technol.* **2014**, *34*, 303–308. [CrossRef]
47. Kopeć, A.; Syska, A.; Leszczyńska, T.; Piątkowska, E. Skład chemiczny i aktywność antyoksydacyjna biszkoptów i ciastek bezglutenowych. *Właściwości Produktów Surowców Żywnościowych. Wybrane Zagadnienia* **2014**, *153*.
48. Aboshora, W.; Yu, J.; Omar, K.A.; Li, Y.; Hassanin, H.A.M.; Navicha, W.B.; Zhang, L. Preparation of Doum fruit (*Hyphaene thebaica*) dietary fiber supplemented biscuits: Influence on dough characteristics, biscuits quality, nutritional profile and antioxidant properties. *J. Food Sci. Technol.* **2019**, *56*, 1328–1336. [CrossRef]
49. Maner, S.; Sharma, A.K.; Banerjee, K. Wheat flour replacement by wine grape pomace powder positively affects physical, functional and sensory properties of cookies. *Proc. Natl. Acad. Sci. India Sect. B Biol. Sci.* **2015**, *87*, 109–113. [CrossRef]
50. Ajibola, C.; Oyerinde, V.; Adeniyan, O. Physicochemical and antioxidant properties of whole-wheat biscuits incorporated with Moringa oleifera leaves and cocoa powder. *J. Sci. Res. Rep.* **2015**, *7*, 195–206. [CrossRef] [PubMed]
51. Hodas, F.; Rosa, M.; Zorzenon, T.; Milani, P.G. Moringa oleifera potential as afunctional food and a natural food additive: A biochemical approach. *Ann. Acad. Bras. Cienc.* **2021**, *93* (Suppl. S4), e20210571. [CrossRef]
52. Natukunda, S.; Muyonga, J.H.; Mukisa, I.M. Effect of tamarind (*Tamarindus indica* L.) seed on antioxidant activity, phytocompounds, physicochemical characteristics, and sensory acceptability of enriched cookies and mango juice. *Food Sci. Nutr.* **2015**, *4*, 494–507. [CrossRef]
53. Klunklin, W.; Savage, G. Effect of substituting purple rice flour for wheat flour on physicochemical characteristics, in vitro digestibility, and sensory evaluation of biscuits. *J. Food Qual.* **2018**, *2018*, 8052847. [CrossRef]
54. Van Toan, N.; Anh, V.Q. Preparation and improved quality production of flour and the made biscuits from purple sweet potato. *J. Food Nutr.* **2018**, *4*, 1–14.

55. Pasqualone, A.; Makhlouf, F.Z.; Barkat, M.; Difonzo, G.; Summo, C.; Squeo, G.; Caponio, F. Effect of acorn flour on the physico-chemical and sensory properties of biscuits. *Heliyon* **2019**, *5*, e02242. [CrossRef]
56. Ścibisz, I.; Kalisz, S.; Mitek, M. Termiczna degradacja antocyjanow owocow borowki wysokiej. *Zywnosc. Nauka. Technologia. Jakosc* **2010**, *5*, 56–66.
57. COMMISSION REGULATION (EU) 2017/2158 of 20 November 2017 Establishing Mitigation Measures and Benchmark Levels for the Reduction of the Presence of Acrylamide in Food. Available online: https://op.europa.eu/en/publication-detail/-/publication/1f3b45fb-ce6b-11e7-a5d5-01aa75ed71a1/language-en (accessed on 24 February 2023).
58. Saturni, L.; Ferretti, G.; Bacchetti, T. The Gluten-Free Diet: Safety and Nutritional Quality. *Nutrients* **2010**, *2*, 16–34. [CrossRef]
59. Hopper, A.D.; Hadjivassiliou, M.; Butt, S.; Sanders, D.S. Adult Coeliac Disease. *BMJ* **2007**, *335*, 558–562. [CrossRef]
60. Sun, L.; Miao, M. Dietary polyphenols modulate starch digestion and glycaemic level: A review. *Crit. Rev. Food Sci. Nutr.* **2020**, *60*, 541–555. [CrossRef] [PubMed]
61. Hřivna, L.; Zigmundová, V.; Burešová, I.; Maco, R.; Vyhnánek, T.; Trojan, V. Rheological properties of dough and baking quality of products using coloured wheat. *Plant Soil Environ.* **2018**, *64*, 203–208. [CrossRef]
62. Liu, N.; Ma, S.; Wang, Z.; Li, L.; Zheng, X.; Wang, X. Influence of wheat bran dietary fiber on gluten protein structure during dough fermentation. *J. Food Process. Preserv.* **2020**, *45*, e15035. [CrossRef]
63. Wyrwisz, J.; Półtorak, A.; Moczkowska, M.; Żontała, K.; Stelmasiak, A.; Łopacka, J.; Ulanicka, U.; Wierzbicka, A. Wpływ stopnia rozdrobnienia preparatów wysokobłonnikowych na właściwości reologiczne ciasta pszennego. *Inżynieria Rol.* **2013**, *17*, 275–285.
64. Yadav, R.B.; Yadav, B.S.; Dhull, N. Effect of incorporation of plantain and chickpea flours on the quality characteristics of biscuits. *J. Food Sci. Technol.* **2011**, *49*, 207–213. [CrossRef]
65. Awolu, O.O.; Sudha, L.M.; Manohar, B. Influence of defatted mango kernel seed flour addition on the rheological characteristics and cookie making quality of wheat flour. *Food Sci. Nutr.* **2018**, *6*, 2363–2373. [CrossRef]
66. Tavares, B.O.; Silva, E.P.D.; Silva, V.S.N.D.; JÚnior, M.S.S.; Ida, E.I.; Damiani, C. Stability of gluten free sweet biscuit elaborated with rice bran, broken rice and okara. *Food Sci. Technol.* **2016**, *36*, 296–303. [CrossRef]
67. Parveen, H.; Bajpai, A.; Bhatia, S.; Singh, S. Analysis of biscuits enriched with fibre by incorporating carrot and beetroot pomace powder. *Indian J. Nutr. Diet.* **2017**, *54*, 403. [CrossRef]
68. Srivastava, S.; Singh, K. Physical, sensory and nutritional evaluation of biscuits prepared by using beetroot (Beta vulgaris) powder. *Int. J. Innov. Res. Adv. Stud.* **2016**, *3*, 281–283.
69. Jothi, J.; Hashem, S.; Rana, M.; Rahman, M.; Shams-Ud-Din, M. Effect of gluten-free composite flour on physico-chemical and sensory properties of cracker biscuits. *J. Sci. Res.* **2014**, *6*, 521–530. [CrossRef]
70. Makinde, F.M.; Taibat, A.A. Quality characteristics of biscuits produced from composite flours of wheat, corn, almondand coconut. *Ann. Food Sci. Technol.* **2018**, *19*, 116–119.
71. Adeola, A.A.; Ohizua, E.R. Physical, chemical, and sensory properties of biscuits prepared from flour blends of unripe cooking banana, pigeon pea, and sweet potato. *Food Sci. Nutr.* **2018**, *6*, 532–540. [CrossRef] [PubMed]

foods

MDPI

Article

Asparaginase Treatment of Sea Buckthorn Berries as an Effective Tool for Acrylamide Reduction in Nutritionally Enriched Wholegrain Wheat, Rye and Triticale Biscuits

Zuzana Ciesarová [1,*], Kristína Kukurová [1], Viera Jelemenská [1], Jana Horváthová [1], Janka Kubincová [1], Miona Belović [2] and Aleksandra Torbica [2]

[1] National Agricultural and Food Centre, Food Research Institute, Priemyselná 4, 824 75 Bratislava, Slovakia; kristina.kukurova@nppc.sk (K.K.); viera.jelemenska@nppc.sk (V.J.); jana.horvathova@nppc.sk (J.H.); janka.kubincova@nppc.sk (J.K.)

[2] University of Novi Sad, Institute of Food Technology, Bulevar cara Lazara 1, 21000 Novi Sad, Serbia; miona.belovic@fins.uns.ac.rs (M.B.); aleksandra.torbica@fins.uns.ac.rs (A.T.)

* Correspondence: zuzana.ciesarova@nppc.sk; Tel.: +421-911-807-737

Abstract: Sea buckthorn pomace is a by-product of juice production, which is still rich in bioactive compounds. After drying, the pomace can be effectively used as a valuable addition to bakery products supporting their nutritional value. However, due to the high content of the amino acid asparagine in sea buckthorn, this promising material contributes to the undesirable formation of acrylamide. To reduce the risk from this potentially carcinogenic compound, enzymatic treatment of sea buckthorn with asparaginase was applied, which resulted in a substantial reduction of asparagine content from 1834 mg/kg in untreated dried sea buckthorn pomace to 89 mg/kg in enzymatically treated dried sea buckthorn pomace. 10% substitution of wholegrain cereal flour with enzymatically treated sea buckthorn pomace powder in rye and triticale biscuits resulted in a 35% reduction in acrylamide content, in the case of wholegrain wheat biscuits up to a 64% reduction, compared to biscuits with untreated sea buckthorn pomace powder. This study confirmed that treating fruit with asparaginase is an effective way to reduce health risk caused by acrylamide in biscuits enriched with nutritionally valuable fruit pomace.

Keywords: acrylamide; asparaginase; sea buckthorn; wheat; triticale; rye; wholegrain cereals; biscuits

Citation: Ciesarová, Z.; Kukurová, K.; Jelemenská, V.; Horváthová, J.; Kubincová, J.; Belović, M.; Torbica, A. Asparaginase Treatment of Sea Buckthorn Berries as an Effective Tool for Acrylamide Reduction in Nutritionally Enriched Wholegrain Wheat, Rye and Triticale Biscuits. *Foods* **2023**, *12*, 3170. https://doi.org/10.3390/foods12173170

Academic Editor: Shiyi Ou

Received: 14 July 2023
Revised: 15 August 2023
Accepted: 21 August 2023
Published: 23 August 2023

1. Introduction

The current focus of consumers on healthy nutrition has brought into the spotlight many valuable sources of biologically active substances, especially of plant origin, which can be relatively easily incorporated into frequently consumed food products. Fruits or plant parts are usually used in fresh and dried form, respectively, or as extracts of individual valuable compounds [1–3]. The circular economy emphasizes the waste-free use of any resources, especially by-products that can be reused and, moreover, can significantly improve the nutritional profile of common food products. The implementation of selected plant-derived bioactive compounds into various food matrices is well described [4–7]. Among these valuable plants, sea buckthorn has an exceptional position due to its rich and balanced composition of polyphenols, flavonoids, carotenoids, antioxidants, vitamins, fibre, unsaturated fatty acids, essential amino acids, macro and micro elements, etc. [8–11]. Sea buckthorn pomace left over from juice production is still full of valuable bioactive substances, but due to its astringent taste, it is rarely used and consumed separately [9]. For this reason, it is a good option to use it in such matrices that provide dominant sensory attributes to food products to mask the undesirable aspects of sea buckthorn taste and aroma [12]. From this point of view, biscuits are advantageous matrices that are well accepted by consumers, are ready to eat, have a long shelf-life, with huge variabilities in composition [13]. Additionally, used cereal matrices can be valuable in their own right, if nutritionally rich wholegrain ones are used instead of commonly used wheat [14–16].

The necessity of heat treatment—baking—provides many positive aspects: attractive sensory profile, microbial safety and longer shelf life. These properties result from the many specific reactions within the Maillard reaction [17,18]. On the other hand, the formation of undesirable, probably carcinogenic acrylamide can be considered a disadvantage [19–21]. According to the current legislation [22], this process contaminant must be kept below relevant benchmark levels, which are expected to be converted to stricter maximum limits soon. This aspect is an obstacle to the wide-spread use of worthy plant by-products. Besides acrylamide, 5-hydroxymethylfurfural (HMF)—a furanic compound arising from direct dehydration of sugars under acidic conditions—is considered as potentially carcinogenic to humans [23], although no mitigation strategies specifically addressing reduction of HMF content in foods are available. Since the main mechanism of acrylamide formation is known [24–28], although further details are still being investigated, it seems that a good way to break this barrier is to prevent acrylamide formation by implementing some of the mitigation tools offered in Toolbox [29] and verified by many studies [30–33]. The sensory profile of final products, especially cereal-based products, is very sensitive to alterations in food processing. For this reason, the application of enzyme asparaginase is a good choice for flour treatment before baking, as it does not significantly affect the organoleptic properties of the final products. This enzymatic treatment has been successfully used mainly in cereal flour [13,34–36]. However, ingredients other than cereals also have the high potential to form acrylamide, in this case sea buckthorn pomace. The reason is the very high level of the main precursor—amino acid L-asparagine in sea buckthorn berries [9,37]. In order to avoid the presence of acrylamide higher than the required limit attributed to biscuits (350 μg/kg) [22], in this study, a commercial asparaginase was applied to sea buckthorn pomace before drying and subsequent baking of nutritionally enriched wholegrain cereal biscuits. This is a novel approach to prevent acrylamide formation from additional sources besides cereals. Different type of wholegrain flours from wheat, rye and triticale (hybrid of wheat and rye) were used as cereal matrices for the preparation of biscuits. The effect of enzymatically treated and untreated dried sea buckthorn pomace replacing 10% of flours was investigated for the first time to the best of our knowledge. The study followed the hypothesis that the enzymatic treatment of sea buckthorn berries leading to a reduction of asparagine content below 100 mg/kg of dried pomace powder is suitable for keeping acrylamide below the benchmark level (350 μg/kg) even in high-acrylamide rye biscuits with a 10% substitution of flour with no detrimental effect on the quality of the final biscuits.

2. Materials and Methods

2.1. Reagents and Biological Materials

2.1.1. Reagents

The following enzymes, standards and chemicals were used: L-asparaginase (Novozymes, Denmark), amino acids kit of L-alanine, L-arginine hydrochloride, L-asparagine, L-aspartic acid, L-cysteine hydrochloride, L-glutamic acid, L-glutamine, L-glycine, L-histidine hydrochloride monohydrate, L-hydroxyproline, L-isoleucine, L-leucine, L-lysine hydrochloride, L-methionine, L-ornitine, L-phenylalanine, L-proline, L-serine, L-threonine, L-tryptophan, L-tyrosine and L-valine $\geq$ 98–99% (Sigma, St. Louis, MO, USA), acrylamide $\geq$ 99% (Sigma-Aldrich, Steinheim, Germany), internal standards of D3-labelled acrylamide (2,3,3-d_3-2-propenamide $\geq$ 98%) and D3-labelled glutamic acid $\geq$ 99% (Cambridge, Isotope Laboratories, MD, USA), HPLC gradient grade solvents of acetonitrile, methanol and perfluorooctanoic acid $\geq$ 96% (Sigma-Aldrich, Steinheim, Germany), glacial acetic acid, ethyl acetate, potassium hexacyanoferrate trihydrate and zinc sulphate heptahydrate (Merck, Darmstadt, Germany and Lachema, Brno, Czech Republic), sodium hydrogen carbonate (Lach-ner, Neratovice, Czech Republic). Deionized water was prepared with a purification system PUR1TY Select (HP, Oxon, UK).

2.1.2. Biological Material

Biscuits were produced from commercial wholegrain flours from common wheat (*Triticum aestivum*) and rye (*Secale cereale*) provided by Interpak (Kraljevo, Serbia), as well as flour obtained by milling grains of triticale (×*Triticosecale*) variety 'Odisej' provided by the Institute of Field and Vegetable Crops (Novi Sad, Serbia). Reconstituted wholegrain triticale flour was prepared as follows: triticale grain was milled using Quadrumat® Senior roller mill (Brabender, Duisburg, Germany), and the yield of refined flour was in the range of 66–68%. Refined flour was combined with bran in the ratio 70:30 to reconstitute wholegrain triticale flour.

Further ingredients of biscuit recipe: crystal sugar (Sunoko, Novi Sad, Serbia), vanilla sugar (Dr Oetker, Bielefeld, Germany), soybean lecithin (Sojaprotein, Bečaj, Serbia) and table salt (So Produkt, Stara Pazova, Serbia) were bought at a local market. Vegetable fat HF with melting temperature of 36–38 °C (Puratos, Groot—Bijgaarden, Belgium) was used as a shortening.

Fresh sea buckthorn (*Hippophae rhamnoides* L.) berries were provided by PD Tvrdošovce farm (Tvrdošovce, Slovakia) from the crop harvested in 2020. The berries were collected on twigs, frozen at −21 °C, then plucked from the branches.

2.2. Methods

2.2.1. Procedure of Sea Buckthorn Pomace Production

Sea buckthorn berries (5.5 kg) were washed and crushed in a hand mixer (Electro-Pulver EP8, Grâce-Hollogne, Belgium) at 13,000 rpm for 2 min. Mash (1.4 kg) was pressed and filtered manually through gauze to separate juice (1.15 kg) and pomace (0.25 kg). The sea buckthorn pomace (SBP) was dried in a drying machine (Memmert UF260, Schwabach, Germany) at 55 °C for 24 h, then ground (Grindomix GM200, Retsch, Haan, Germany) at 5000 rpm for 15 s to obtain untreated SBP powder. The remaining sea buckthorn mash (4.0 kg) underwent pH adjustment from pH 3.4 to 6.8 by adding sodium hydrogen carbonate with continuous manual stirring. pH-neutral mash (2.0 kg) was pressed and filtered as before to obtain 1.7 kg of pH-neutral juice and 0.3 kg of pH-neutral SBP which was dried using the same drying procedure as untreated SBP. Commercially produced asparaginase Acrylaway®L (Novozymes, Bagsværd, Denmark) provided as a gift by the producer was added to the pH-neutral wet mash (2.0 kg) at an optimized dosage of 5 mL of enzyme (3500 ASN/mL) per kg of mash. Enzyme incubation was performed in a stainless-steel container (Kitchen Aid, Artisan Series, MI, USA) for 60 min at laboratory temperature (20 °C) with low agitation. After the enzymatic treatment of the sea buckthorn mash, the procedure of pressing, filtration and drying was the same. This exact procedure of enzyme treatment of fruit and vegetable in general was submitted for industrial property protection and registered at the Office of Industrial Properties of the Slovak Republic as a utility model No 9572 [38]. Sea buckthorn juice and three samples of dried SBP powders (untreated SBP, pH neutral SBP before enzymatic treatment, and pH neutral SPB after enzymatic treatment) were collected for further characterization of amino acid profile. Two of these, untreated SBP powder and enzymatically treated pH neutral SBP powder, were used for biscuits fortification.

2.2.2. Procedure of Biscuits Production

The procedure of biscuits production was developed by Belović et al. (2020) [39]. The biscuits production procedure started with mixing the fat components (shortening and soya lecithin) with crystal sugar and vanilla sugar in a mixer (Conti, model PL16 5B, Bussolengo, Italy). Then, all the powdered components were added (flour, SBP powder, sodium bicarbonate and salt). When a homogenous mixture was produced, water was added to obtain the desired degree of hydration of the dough. The dough obtained was rolled out to a thickness of 7 mm using a dough sheeter (MAC PAN, model MK 600, Thiene, Italy). The biscuits were shaped using a round mould with a diameter of 45 mm and baked in a pre-heated oven at 180 °C for 14 min (MIWE Michael Wenz GmbH, model Gusto,

Arnstein, Germany). All physical measurements were performed 1 h after baking when biscuits were cooled to room temperature.

2.2.3. Determination of Dimension of Biscuits

Dimensions of biscuits were measured using an analogue caliper with an accuracy of 0.01 mm for eight biscuits from each sample. Biscuit diameter (d) was divided with height (h) to calculate the spread ratio (SR) of biscuits, according to the method AACC 10–50.05 [40].

2.2.4. Determination of Moisture

Moisture was determined using a rapid moisture analyzer (model HE53, Mettler Toledo, Greifensee, Switzerland). Biscuit samples were ground in a kitchen coffee grinder (Gorenje, Velenje, Slovenia) and 3 g of pulverized sample were heated at 105 °C until constant mass was reached.

2.2.5. Determination of Water Activity

The water activity (Aw) was measured in two biscuits from each sample using the Aw meter Testo 205 (Testo AG, West Chester, PA, USA) according to the standard method (ISO, 2017).

2.2.6. Determination of Colour

The colour of biscuits was measured using a Chroma Meter CR-400 (Konica Minolta Co., Ltd., Osaka, Japan), with D65 light source and the observer angle of 2°. Top and bottom surfaces of five biscuits per batch were measured. The results were expressed in CIE L*a*b* colour space, where CIE L* represents lightness (L* = 0—black, L* = 100—white), CIE a* represents red-green hues (+a* = redness, −a* = greenness) and CIE b* represents yellow-blue hues (+b* = yellowness, −b* = blueness).

2.2.7. Determination of Texture

TA.XT Plus Texture Analyser (Stable Micro Systems, Godalming, UK) was used to determine the textural properties of biscuits. Since hardness is the most representative textural property of biscuits, it was determined as the maximum peak force recorded from the force/time curve (break force). The instrument was equipped with a 30 kg load cell and a flat blade (HDP/BS) for biscuit cutting. Instrumental settings of the standard blade cutting test for biscuits were taken from the built-in sample project BIS2_KB included in the texture analysis software Exponent v. 6.1.18.0. (Stable Micro Systems, Godalming, UK).

2.2.8. Determination of Amino Acids

Determination of 22 free amino acids—alanine (Ala), arginine (Arg), asparagine (Asn), aspartic acid (Asp), cysteine (Cys), glutamine (Gln), glutamic acid (Glu), glycine (Gly), histidine (His), hydroxyproline (Hyp), isoleucine (Ile), leucine (Leu), lysine (Lys), methionine (Met), ornithine (Orn), phenylalanine (Phe), proline (Pro), serine (Ser), threonine (Thr), tryptophan (Trp), tyrosine (Tyr) and valine (Val)—was performed by a liquid chromatography-mass spectrometry (LC—MS/MS) apparatus consisting of an Agilent 1200 HPLC system (Agilent Technologies, Santa Clara, CA, USA), a binary pump, an autosampler and a temperature controlled column oven coupled to an Agilent 6410 Triple Quad detector equipped with the electrospray ionisation (ESI) interface. Samples were extracted, cleaned and directly applied directly to the LC-MS/MS analysis without prior derivatisation. The method was adapted from [41] and the exact procedure was as follows: exactly 2.0000× g of homogenised sample was mixed with 20 mL of acetic acid (0.2 mM) extraction solution, vortexed for 1 min and then shaken for 30 min. After centrifugation at 8720× g for 10 min at −5 °C the sample was diluted. 1 mL of the clear supernatant was transferred into a glass tube containing 50 μL of an internal standard solution (0.01 g of Glu-D3 in 100 mL of water) and 9 mL of acetic acid (0.2 mM) and mixed. Samples were injected after filtration through a 0.45 μm pore size nylon syringe filter (Q-Max RR Syringe Filters, Frisenette ApS, Knebel, Denmark). Analytical separation was performed on a Purospher STAR RP-8ec column (150 × 4.6 mm, 2.7 μm; Agilent, USA) using an isocratic mixture of 100 mL of acetonitrile, 500 mL of aqueous solution of PFOA (0.05 mM) and 1 mL of glacial acetic acid at a flow rate of 0.5 mL/min at ambient temperature (25 °C). The ESI mass spectrometry detection was performed in a positive mode with the following optimised parameters: drying gas temperature 320 °C; drying gas flow (N2) 8 L/min; nebulizer pressure 50 psi, capillary voltage 3 kV. Data acquisition was performed using a multiple reaction monitoring (MRM) which monitors only specific mass transitions at specific retention times (Table 1). Calibration curves of individual amino acids were prepared in the range from 0.02 to 4.00 μg/mL; an internal standard Glu-D3 concentration of 0.5 μg/mL was prepared in acidified water (0.1% acetic acid). The analysis time was 31 min. The limits of detection (LOD) were 6–10 μg/L; the limits of quantification (LOQ) were 20–30 μg/L depending on the individual amino acids (Table 1).

Table 1. Optimized MRM parameters and parameters of calibration for LC/MS-MS determination of amino acids.RT—Retention time; PI—Precursor Ion; Q1—Qualifier; Q2—Quantifier; R^2—Coefficient of Determination; LOD—Limit of Detection; LOQ—Limit of Quantification.

Amino Acid	RT (min)	PI	Production Ions		Fragmentor (eV)	Collision Energy (eV)		Dwell (ms)	R^2	LOD (ng/mL)	LOQ (ng/mL)
			Q1	Q2		Q1	Q2				
Ala	4.53	90	44	90	50	6	2	20	0.9990	10	30
Arg	13.64	175	70	175	100	25	5	50	0.9992	10	30
Asn	4.03	133	74	87	50	12	5	20	0.9998	8	25
Asp	3.74	134	74	88	50	12	5	20	0.9996	10	30
Cys	3.92	122	76	59	60	10	25	20	0.9991	10	30
Gln	4.09	147	84	130	50	15	5	20	0.9919	10	30
Glu	4.21	148	84	102	50	15	10	20	0.9994	6	20
Gly	4.47	76	76	30	50	2	6	20	0.9997	11	33
His	11.67	156	110	156	100	15	2	50	0.9996	10	29
Hyp	3.63	132	86	132	50	15	2	20	0.9998	13	39
Ile	7.62	132	86	132	80	5	2	50	0.9997	8	24
Leu	7.11	132	86	132	80	5	2	50	0.9997	11	33
Lys	13.28	147	84	130	80	15	5	50	0.9991	7	23
Met	5.45	150	56	133	80	15	5	50	0.9972	4	12
Orn	12.35	133	116	70	80	5	20	50	0.9997	9	28
Phe	8.21	166	120	103	80	10	30	50	0.9997	10	30
Pro	3.95	116	70	116	50	10	0	5	0.9978	9	29
Ser	4.1	106	60	106	50	5	2	20	0.9987	10	30
Thr	4.18	120	56	74	50	15	8	20	0.9978	8	26
Trp	15.17	205	146	188	100	15	5	50	0.9998	11	34
Tyr	8.14	182	182	136	100	2	10	50	0.9979	10	30
Val	5.32	118	72	118	80	10	2	50	0.9994	3	10

2.2.9. Determination of Acrylamide

The samples were extracted in acetic acid (0.2 mM) water solution and pre-extracted in ethyl acetate to avoid the negative influence of salts in the chromatography system according to the previously used method [18] with a slight modification. The sample preparation was as follows: 1.0000× *g* of a homogenised sample was weighed into a capped 10 mL centrifuge tube, to which 50 μL of internal standard solution (0.0020× *g* of acrylamide-D_3 in 100 mL of water) and 9 mL of acetic acid (0.2 mM) were added. The mixture was vortexed for 30 s and then sonicated for 5 min. Then, 500 μL of Carrez solution I (15 g of ($K_4[Fe(CN)_6].3H_2O$ in 100 mL of water) and 500 μL of Carrez solution II (30 g of ($ZnSO_4.7\,H_2O$) in 100 mL of water) were added and mixed for 1 min. The mixture was then centrifuged in a Sigma 2-16KC centrifuge (Sigma, Osterode am Harz, Germany) at 8720× *g* for 10 min at −5 °C. A volume of 5 mL of the clear supernatant was transferred to a separating funnel; 5 mL of ethyl acetate was added and mixed well. The ethyl acetate layer was removed and the extraction step was repeated twice with 5 mL of ethyl acetate. 3 × 5 mL of ethyl acetate layers were collected and evaporated to dryness in a Heidolph WB 2000 vacuum rotatory evaporator (Heidolph Instruments, Schwabach, Germany) at 35 °C. The residue was dissolved in 1 mL of acetic acid solution (0.2 mM) and filtered through a 0.45 µm pore size nylon syringe filter (Q-Max RR Syringe Filters, Frisenette ApS, Knebel, Denmark). LC-MS/MS analysis was performed on a 1200 series HPLC system (Agilent Technologies, Santa Clara, CA, USA) coupled to an Agilent 6460 Triple Quad detector equipped with an ESI interface. Analytical separation was performed on an Atlantis dC18 column (100 × 3 mm, 3 µm; Waters, Milford, MA, USA) using an isocratic mixture of methanol:acetic acid:deionized water (5:1:500, *v*/*v*/*v*) at a flow rate of 0.4 mL/min at 25 °C. The ESI mass spectrometry detection was performed in positive ESI + mode with drying gas (N2) flow 8 L/min and temperature 350 °C, nebulizer pressure 50 psi, capillary voltage 2.5 kV, sheath gas flow 11 L/min at 250 °C, fragmentor 50 eV, and collision energy 10 eV. Data acquisition was performed using multiple reaction monitoring (MRM) with transition for acrylamide: 72 → 55 and acrylamide-D_3: 75 → 58. The quantification of acrylamide was calculated from a calibration curve of the standard compound in the range from 5 to 200 ng/1 mL. The analysis time was 11 min; the retention time of acrylamide and acrylamide-D_3 was 2.0 min. The LOD of the applied procedure was 10 µg/L; the LOQ was 15 µg/L.

2.2.10. Determination of 5-Hydroxymethylfurfural

5-Hydroxymethylfurfural (HMF) was extracted with a mixture of methanol and water in a ratio of 80:20 (*v*/*v*). Approximately 1.0 g of homogeneous sample was mixed with 10 mL of the extraction mixture and sonicated for 5 min. After filtration through 0.45 µm pore size nylon syringe filters (Frisenette ApS, Knebel, Denmark) samples were ready for injection. The Agilent 1200 HPLC system (Agilent Technologies, Santa Clara, CA, USA) equipped with a quaternary gradient pump at a flow rate of 0.8 mL/min, an autosampler and a photo-diode array detector (DAD) set at 280 nm was used for the separation and quantification of HMF in the samples. The separation was performed on a Zorbax C18 SB column (250 × 4.6 mm, 5 µm; Agilent Technologies, USA). The mobile phase consisted of A (water:acetonitrile:H_3PO_4; 94:5:1, *v*/*v*/*v*) and B (methanol) with the following gradient elution: between 0 and 5 min: 100% A and 0% B; between 5 and 10 min: 90% A and 10% B; between 10 and 15 min: 55% A and 45% B; between 15 and 20 min: 5% A and 95% B; between 20 and 30 min: 5% A and 95% B. The samples (10.0 µL) were injected into the LC-DAD. An external calibration was used with a linearity in the range of 0.05–1.0 µg/mL. The LOD of the applied procedure was 0.5 µg/mL; the LOQ was 1.7 µg/mL. The uncertainty at the concentration level of 1.61 µg/mL was ±0.3 µg/mL.

2.3. *Statistical Analysis*

Statistica 14.0.0.15 software (Tibco Inc., Palo Alto, CA, USA, 2020, https://www.tibco.com/products/data-science (accessed on 6 January 2023) was used for statistical analysis. One-way analysis of variance (ANOVA) and Tukey's honestly significant differences (HSD) test were used to determine the significant differences ($p < 0.05$).

3. Results and Discussion

3.1. *Amino Acid Profiles of Cereal Flours*

Wholegrain wheat, triticale, and rye flours were subjected to full free amino acids profile determination. Each individual amino acid was quantified, then the total amount of amino acids as well as the amount of essential and non-essential amino acids were calculated (Table 2).

Table 2. Amino acid content (mg/kg) in wholegrain cereal flours, sea buckthorn juice and pomaces.

	Wholegrain Flour				Sea Buckthorn Pomace		
Amino Acid (mg/kg)	**Wheat**	**Triticale**	**Rye**	**SBJ**	**SBP1**	**SBP2**	**SBP3**
Asn	447.9 [c]	782.0 [b]	1182.9 [a]	1273.3 [C]	1717.8 [B]	1833.9 [A]	88.5 [D]
Asp	107.9 [b]	92.2 [c]	205.1 [a]	81.2 [B]	94.1 [B]	95.6 [B]	1469.5 [A]
Gln	65.2 [c]	87.6 [b]	121.6 [a]	109.0 [C]	170.8 [B]	209.7 [A]	215.5 [A]
Glu	205.2 [c]	218.4 [b]	373.9 [a]	86.2 [C]	124.2 [B]	150.6 [A]	127.7 [B]
Arg	143.6 [b]	200.6 [a]	146.8 [b]	72.2 [B]	141.9 [A]	113.9 [A]	110.3 [A]
Lys	49.5 [c]	63.7 [a]	58.7 [b]	38.8 [A]	29.6 [B]	30.5 [B]	26.8 [B]
Ala	135.7 [b]	116.8 [c]	173.7 [a]	62.3 [B]	67.4 [AB]	72.6 [A]	63.7 [B]
Phe	26.4 [b]	42.2 [a]	41.6 [a]	86.6 [A]	42.7 [B]	35.6 [B]	32.3 [B]
Pro	36.1 [c]	77.1 [b]	163.3 [a]	39.9 [B]	48.9 [A]	48.9 [A]	42.5 [B]
Trp	90.6 [a]	40.4 [b]	41.5 [b]	22.8 [B]	27.2 [A]	23.6 [B]	23.0 [B]
Ser	28.9 [a]	18.5 [b]	21.3 [b]	39.8 [A]	53.3 [A]	47.2 [A]	52.0 [A]
Val	36.6 [c]	46.4 [b]	82.1 [a]	21.4 [A]	22.2 [A]	22.8 [A]	22.4 [A]
Met	0.9 [c]	1.6 [b]	7.3 [a]	13.3 [A]	11.6 [BC]	11.9 [B]	11.2 [C]
Tyr	nd	nd	nd	41.8 [A]	26.0 [B]	27.6 [B]	24.8 [B]
Ile	40.0 [b]	48.3 [a]	47.7 [a]	38.5 [A]	17.6 [B]	19.8 [B]	19.5 [B]
Leu	13.9 [c]	16.4 [b]	25.9 [a]	9.5 [B]	11.6 [AB]	11.8 [AB]	12.6 [A]
Thr	18.3 [c]	20.9 [b]	25.0 [a]	22.7 [A]	21.3 [A]	23.4 [A]	21.3 [A]
Gly	53.1 [a]	42.1 [b]	44.1 [b]	11.3 [A]	14.3 [A]	11.8 [A]	12.0 [A]
His	17.1 [c]	19.5 [b]	24.1 [a]	31.7 [A]	30.9 [A]	24.6 [B]	25.1 [B]
Orn	3.1 [b]	4.6 [a]	3.5 [b]	10.3 [AB]	14.3 [A]	8.8 [B]	nd
Cys	nd	nd	nd	16.3 [A]	16.7 [A]	nd	15.5 [A]
Hyp	nd	nd	nd	1.1 [A]	1.2 [A]	1.3 [A]	1.0 [A]
Total	1519.9	1939.3	2790.4	2130.1	2705.7	2826.2	2417.3
E-AA	276.0	280.0	330.0	253.6	183.7	179.6	169.1
SemiE-AA	160.7	220.1	170.9	103.9	172.9	138.5	135.4
NonE-AA	1083.0	1439.2	2289.5	1772.5	2349.2	2508.1	2112.2

Values are the average ($n = 4$); nd = not detected; relative standard deviations are below 5% for Asn, Asp, Gln, Glu, Lys, Ala, Pro, Trp, Ser, Val, Ile, Leu, Thr, Gly, His, Cys, Hyp; and below 10% for Arg, Phe, Met, Orn. Different letters (a–c) indicate significant differences ($p < 0.05$) among different cereal flours, and (A–D) suggest significant differences ($p < 0.05$) between different sea buckthorn products. SBJ—fresh sea buckthorn juice; SBP1—residual sea buckthorn pomace after juice pressing without any treatment; SBP2—residual sea buckthorn pomace after juice pressing with neutral pH; SBP3—residual sea buckthorn pomace after juice pressing enzymatically treated at neutral pH; E-AA—essential amino acids; SemiE-AA—semi essential amino acids; NonE-AA—nonessential amino acids.

Asn was identified as a dominant amino acid in all three flours (448; 782; 1183 mg/kg for wheat, triticale, and rye flour, respectively), followed by Glu (205; 218; 374 mg/kg), Arg (144; 201; 147 mg/kg), Ala (136; 117; 174 mg/kg), Asp (108; 92; 205 mg/kg), Trp (91; 40; 42 mg/kg), Gln (65; 88; 122 mg/kg), and Pro (36; 77; 163 mg/kg). For the remaining amino acids, the middle content below 50 mg/kg was observed. The highest total amount of amino acids was in rye flour (2790 mg/kg) followed by triticale flour (1939 mg/kg) and wheat flour (1520 mg/kg). The same order of flours was recognized for total amount of essential amino acids (Lys + Met + Val + Leu + Ile + Phe + Trp + Thr) and for the sum of essential and semi-essential (Arg + His) amino acid amount. This shows that wholegrain rye flour is the richest source of free amino acids, as well as essential amino acids in comparison with wholegrain wheat and wholegrain triticale flours. The content of free amino acids in triticale flour was between wheat and rye flours which is in line with the fact that triticale is a hybrid of wheat and rye. The amino acid profiles of wheat and rye flours are consistent with previously presented data [42–44]. Concentrations of particular amino acids in cereal varieties are affected by genetics, the environment, location, year of harvest including temperature, rainfall, pesticides, diseases and fertilization [42,45,46]. Moreover, amino acid profiles of cereal flours are dependent also on milling process (extraction rate, wet or dry heating etc.) [47]. However, the amino acid profile of triticale flour in published studies [48–52] did not display separately the amounts of Asp/Asn and Glu/Gln, because the analytical methods used did not allow distinguishing carboxylated and amide forms of these amino acids. In the recent studies [17,53], data for free Asn content in triticale varieties have been published, but no information on a full amino acid profile. For this reason, the complete amino acid profile data with Asp/Asn and Glu/Gln amino acids separated are unique.

Since rye typically has a high content of Asn—the main precursor of acrylamide—which is about 2.5 higher in comparison to wheat and 1.5 higher than triticale flours, it belongs to the highest donors of acrylamide in cereal-based foods. Thus, triticale flour appears to be a suitable substitute with higher nutritional value comparing to wheat and lower acrylamide formation capacity comparing to rye. Typical higher content of free Asn was observed in wholegrain flour comparing to refined flour since Asn was concentrated in outer layers of cereal grains [44]. However, the conversion of Gln and Asp to Asn catalysed by asparagine synthetase [54] can contribute to continuously elevated level of Asn in flours during processing enabled enzymatic reactions [18]. For this reason, besides Asn, the content of Gln, Glu and Asp is also important for assessment of the potential of various plant materials to form acrylamide [47]. The supporting effect of glutamine on the yield of acrylamide was described also by other authors [55]. On the other hand, cereal flours were found to contain higher concentrations of such amino acids which can reduce acrylamide level [56]. Amongst others, Pro, Trp, Cys and Gly were identified as the most effective in acrylamide inhibition [57]. Taking these aspects into account, the amounts of aforementioned amino acids were compared in all three wholegrain flours. Free Pro was substantially higher in rye and triticale flours (163.3 and 77.1 mg/kg, respectively) than in wheat one (36.1 mg/kg), contrary to Trp that was higher in wheat flour (90.6 mg/kg) than in triticale and rye (40.4 and 41.5 mg/kg, respectively) (Table 2). The amount of Gly was similar in all three flours, Cys was not detected.

3.2. Amino Acid Profiles of Sea Buckthorn Pomaces

Fresh sea buckthorn juice (SBJ), residual sea buckthorn pomace (SBP) after juice pressing without any treatment (SBP1), as well as SBP with neutral pH (SBP2) and enzymatically treated SBP3 were characterized by their amino acid profile (Table 2). The dominant amino acid in SBJ, SBP1 and SBP2 was Asn (1273; 1718; 1834 mg/kg) followed by Gln (109; 171; 210 mg/kg) which is approximately 10 times lower than Asn. Significant amounts were identified also for Glu (86; 124; 150 mg/kg), Asp (81; 94; 96 mg/kg), Arg (72; 142; 114 mg/kg), and Ala (62; 67; 73 mg/kg). Other amino acids in SBJ, SBP1 and SBP2 were below 50 mg/kg. The total amount of free amino acids varied between 2130 mg/kg in SBJ and 2826 mg/kg in SBP2. The cumulative amount of essential amino acids and semi-essential amino acids was between 11% (SBP2) and 17% (in SBJ) of total amount of amino acids. The highest amounts of total free amino acids (2826 mg/kg), non-essential amino acids (2508 mg/kg) and Asn (1834 mg/kg), were identified in SBP2 which is slightly higher than in SBP1. It can be supposed that free amino acids were released from protein, polypeptides and peptides as a consequence of protein hydrolysis and enzymes activity in a neutral pH environment due to application of sodium hydrogen carbonate for pH adjustment. The dominant position of Asn and a similar amino acid profile of sea buckthorn berries, juices and pomace were recognized also in other studies [11,37,58,59] contrary to the study of Zhang et al. [60] which presented Asp as the amino acid with the highest proportion. This discrepancy can be attributed to the used analytical method used in Zhang's study which did not distinguish between carboxylated amino acids (Asp, Glu) and their amine derivatives (Asn, Gln), similarly to amino acid determination in triticale mentioned earlier. The differences in values of singular amino acid amounts can be related to different varieties, maturity stage, geographical, agronomical, harvesting and processing conditions. It is evident that due to high level of free Asn, sea buckthorn has a great capacity to form acrylamide during heat treatment. Free Asn is located in the whole berry with a higher proportion in flesh than in peel. It is supported by Asn content in SBJ, SPB1 and SBP2 (Table 2), taking into account that SBP consisting of peels and seeds after removing juice contained approximately 60% of water before drying. This means that SBP, although a highly valuable source of bioactive compounds [9,11], is also a large donor of the acrylamide precursor, even when added in small amounts. For this reason, it is fully justified to focus on reducing the content of asparagine in all raw materials used in food processing, not only in cereals. The effectiveness of enzymatic treatment performed on homogenized sea buckthorn mash after pH-adjustment was evident in the Asn and Asp values. After 60 min of enzyme treatment, the striking changes in the mentioned amino acid proportions were observed. The Asn content dropped from 1834 mg/kg to 89 mg/kg with a parallel increase of Asp from 96 mg/kg to 1470 mg/kg. No other significant changes in particular amino acid content before and after enzymatic treatment were noticed (Table 2), neither in Gln and Glu amounts, which pointed to the excellent specificity of the asparaginase. Since Gln has a similar structure to Asn, some asparaginases also have a low activity towards Gln [61]. Because the glutaminase activity can have a serious detrimental effect on human health (liver dysfunction, pancreatitis, leucopoenia) [62], this specific type of asparaginase with known glutaminase activity should not be used for food treatment.

3.3. Characterization of Cereal Biscuits Enriched with SBP Powder

Cereal biscuits are often used as a favourable model system for several reasons. In experiments, they are good matrices for implementations of different ingredients with simple reproducibility of well-defined processing and long shelf-life. As products, they are frequently consumed in countless variations and well-accepted by consumers [13]. In this experiment, cereal biscuits from wholegrain wheat, triticale and rye flours were prepared according to the procedure described in 2.1.4 (as controls), and two kinds of SBP powders—untreated SBP1 and enzymatically treated pH-neutral SBP3—were incorporated into the recipe as a 10% substitution of the respective flour. The impact of SBP presence in cereal biscuits was examined by characterisation of spread ratio, moisture, water activity, colour, texture, acrylamide content, residual free amino acid content and HMF content.

3.3.1. Impact of SBP Powders on Dimensions of Biscuits

Although all biscuits were prepared by the same procedure with the same diameters and heights, the differences in dimension parameters (expressed as a spread ratio value) in final biscuits were observed (Table 3). The higher spread ratio of rye and triticale biscuits in comparison to wheat biscuits can be the consequence of their specific proteins [63]. Doughs prepared from these flours have lower viscosity, resulting in a higher spread ratio. The increase in spread ratio observed in biscuits prepared with SBP1 and SBP3 can be related to the dilution of gluten in biscuit doughs by dietary fibre and subsequent lower viscosity, which was also observed in a paper by Tomić et al. (2016) [64]. Namely, dried SBP was previously shown to be a raw material rich in crude fibre (19.86%) [65]. However, there were no substantial differences between the spread ratios of biscuits prepared with untreated SBP1 and enzymatically treated SBP3. This is in accordance with the results of Gazi et al. (2023) [13], who also observed that the addition of asparaginase in wire-cut cookie dough prepared from wheat flour did not cause a significant difference in the spread ratio. This result is important for the industrial production of biscuits, which can be automated only if the dimensions of biscuits have low variability [66].

Table 3. Characteristics of wholegrain cereal biscuits with sea buckthorn pomace powder addition.

Wholegrain Cereal Biscuits with SBP	Wheat			Triticale			Rye		
	SBP0	SBP1	SBP3	SBP0	SBP1	SBP3	SBP0	SBP1	SBP3
Spread ratio (−)	4.70 ^{bcA}	4.32 ^{cdB}	4.45 ^{bcdAB}	4.05 ^{dC}	4.68 ^{bcB}	5.59 ^{aA}	4.95 ^{bB}	5.54 ^{aA}	5.85 ^{aA}
Weight (g)	8.51 ^{cA}	7.23 ^{eB}	7.39 ^{eB}	8.48 ^{cB}	9.45 ^{aA}	6.70 ^{fC}	8.85 ^{bA}	8.20 ^{dB}	6.78 ^{fC}
Moisture (%)	6.20 ^{aA}	5.09 ^{bB}	3.84 ^{cC}	6.59 ^{aA}	2.78 ^{dC}	4.66 ^{bB}	3.12 ^{dA}	1.22 ^{eB}	0.93 ^{eC}
Aw value	0.475 ^{cA}	0.189 ^{iC}	0.382 ^{fB}	0.539 ^{aA}	0.342 ^{gC}	0.441 ^{dB}	0.513 ^{bA}	0.420 ^{eB}	0.290 ^{hC}

Table 3. *Cont.*

Wholegrain Cereal Biscuits with SBP	Wheat			Triticale			Rye		
	SBP0	SBP1	SBP3	SBP0	SBP1	SBP3	SBP0	SBP1	SBP3
L* (upper)	60.60 aA	54.95 bB	55.31 bB	52.84 bcA	50.73 cdeAB	49.33 defB	52.42 bcdA	47.36 efB	46.01 fB
a* (upper)	5.99 dB	9.40 bcA	8.19 cA	8.93 bcB	10.53 abB	9.35 cA	7.96 cB	11.35 aA	10.14 abA
b* (upper)	24.93 dB	36.33 aA	37.49 aA	24.22 dC	33.38 bA	29.91 cB	25.91 dB	31.60 bcA	30.37 cA
L* (lower)	58.83 aA	52.72 bB	51.32 bcdB	52.51 bA	46.64 eB	48.29 cdeB	52.29 bcA	45.08 eC	47.33 deB
a* (lower)	6.95 fB	10.87 cdA	8.50 efB	9.16 deB	13.16 abA	10.48 cdeB	9.65 cdeC	13.42 aA	11.36 bcB
b* (lower)	26.53 dB	37.10 aA	35.19 abA	25.38 dB	32.46 bcA	31.21 cA	28.00 dB	33.35 bcA	33.30 bcA
Texture Hardness (kg)	7.80 bcA	9.40 bA	7.26 bcA	16.98 aA	17.18 aA	10.45 bB	10.60 bA	8.41 bcA	4.13 cB

Values are the average: spread ratio and weight ($n = 8$); moisture and Aw ($n = 2$); colour and hardness ($n = 5$). Relative standard deviations are below 5% for spread ratio, weight, moisture, Aw, L*, b*; below 10% for a*; below 20% for hardness. Different letters (a–h) indicate significant differences ($p < 0.05$) in a row, and (A–C) suggest significant differences ($p < 0.05$) within the group of cereal biscuits with different sea buckthorn pomaces. SBP0—biscuits with no presence of sea buckthorn pomace powder; SBP1—biscuits with sea buckthorn pomace powder without any treatment; SBP3 biscuits with sea buckthorn pomace powder enzymatically treated at neutral pH.

3.3.2. Impact of SBP Powders on Moisture and Water Activity of Biscuits

Values of moisture varied between 0.9 and 6.6% and water activity (Aw) was in the range of 0.29 to 0.54 (Table 3). Lower moisture content and lower Aw values in biscuits prepared with SBP1 and SBP3 can be related to the lower height of these biscuits. Biscuits with lower heights have consequently higher spread ratio and thus larger surfaces exposed to the heat in the oven, making the evaporation process during baking faster within the same baking time.

3.3.3. Impact of SBP Powders on the Colour of Biscuits

The colour of biscuits is influenced both by the colour of the ingredients used and the baking process. The main colour characteristics from both upper and lower surfaces of final biscuits expressed as a* (redness), b* (yellowness) and L* (lightness) are summarized in Table 3.

It is known that the colour of wholegrain flours is more or less determined by the presence of anthocyanin pigments [67], but flavonoids as a subgroup of anthocyanin pigments are present in sea buckthorn in higher level than in flours (especially from non-coloured grains). Anthocyanins can be found in different chemical forms which depend on the pH of the solution. At pH values higher than 7, the anthocyanins are degraded depending on their substituent groups [68]. Therefore, the increase in red tone (a*) observed after substitution by SBP1 can be related to pH lower than 7, whereas the lower a* values in enzymatically treated SBP3 biscuits compared to SBP1 biscuits can be related to pH values close to 7, which are required for proper asparaginase activity, and subsequent degradation of anthocyanins.

Additionally, brown polymers called melanoidins are formed as Maillard reaction products during baking [17]. Therefore, an increase in red tone (a*) observed after substitution by SBP1 can be also related to the more intensive Maillard reactions as a consequence of the high content of amino acids, especially Asn, in SBP1. Consequently, lower a* values in enzymatically treated SBP3 biscuits compared to SBP1 biscuits can be partly attributed to the action of asparaginase, which reduce the total amount of free amino acids. The substantial increase of yellow tone (b*) in both SBP1 and SBP3 biscuits can be related to the presence of β-carotene and zeaxanthin, the main pigments found in dried SBP [65].

3.3.4. Impact of SBP Powders on Texture of Biscuits

Generally, the texture characteristics of biscuits are strongly dependent on ingredients and processing conditions thus changes in texture are legitimately expected. The hardness of biscuits is influenced by multiple factors, including protein properties, amount of dietary fibre and particle size of flour used for their preparation, as well as by processing conditions.

In our previous study [39], similarly to the presented study (Table 3), cookies prepared from wholegrain triticale flour were shown to be significantly harder than wholegrain wheat and rye cookies, supposedly due to the high content of coarse bran particles. These coarse bran particles were obtained during the milling process by roller mill, which differs from the process used in the production of commercial wholegrain wheat and rye flours. Based on our observations, the substitution of wholegrain flours by SBP1 powder did not significantly influence the hardness of the biscuits. However, the substitution by enzymatically treated SBP3 powder led to a significant decrease in hardness. Although it seems enzymatic treatment had a visible effect on textural properties of biscuits, there is also another aspect which could be taken into consideration. Since enzymatically treated SBP3 had its pH value adjusted to 6.8 to make the pH value closer to asparaginase optimum, contrary to pH value 3.4 measured in untreated SBP1, it can be concluded that this change in pH value can influence the textural properties of biscuits. The significant decrease in the hardness of biscuits with higher concentrations of sodium hydrogen carbonate and potassium hydrogen carbonate in the dough was observed by Chen et al. (2020) [69]. They concluded that the inclusion of high levels of $KHCO_3$ stimulated gluten aggregation via non-redox reactions, i.e., not via SH oxidation. Based on their results, it could be assumed that in a more alkaline environment, more proteins are aggregated via non-covalent bonds, which are weaker and lead to a less stable and less compact protein network. Therefore, they are more easily disaggregated during processing and additional binding of water within the system occurs, resulting in the considerable softening of biscuits. During baking, the water in the protein matrix remains bound longer than it is bound to the dietary fibre, so a biscuit with fewer S-S bonds is softer than one with more S-S bonds.

3.3.5. Impact of SBP Powders on Acrylamide and Residual Free Asparagine in Biscuits

Acrylamide was formed in biscuits prepared from wheat, triticale and rye flours without and with the addition of SBP powders in different amounts. In control biscuits without SBP powder addition (SBP0), acrylamide was detected in levels related to the predisposition of flours used for biscuits preparation. The lowest acrylamide level (27.2 μg/kg DW) was detected in wheat biscuits, followed by triticale biscuits (58.9 μg/kg DW), and the highest acrylamide level was observed in rye biscuits (105.3 μg/kg DW) (Table 4). It is in a good agreement with Asn level in relevant flours which were 448; 782; 1183 mg/kg for wheat, triticale and rye flours, respectively, with a linear correlation between free Asn in dough and acrylamide in biscuits content (R^2 = 0.9970) (Figure 1a).

A 10% substitution of flours with SBP1 powder resulted in a substantial increase of acrylamide level in all biscuits (80.4; 106.0; 485.5 μg/kg DW for wheat, triticale and rye biscuits) which was from 2 to 5 times higher than in the respective control biscuits SBP0 (Table 4).

The increased level of acrylamide in biscuits with SBP1 powder is expected since the higher amount of the main precursor—free Asn—entered the Maillard reaction responsible for acrylamide formation. Untreated SBP1 powder is very high in Asn, thus it is a significant donor of Asn in dough. The amount of available Asn in biscuit dough enriched with SBP1 addition increased by 7% in rye biscuits, by 17% in triticale biscuits and by 30% in wheat biscuits, however, the increase of acrylamide in respected biscuits was substantially higher with exponential correlation between free Asn content in SBP1 biscuit dough and acrylamide (R^2 = 0.9666) (Figure 1a). For this reason, the observations of Borczak et al. [6] declaring the 89% drop of acrylamide after 5% sea buckthorn application in wheat flour cookies, as well as other wild-grown fruits (hawthorns, rowans, wild roses, elderberries, chokeberries), is surprising since many of them are also high in Asn content. Authors attributed this phenomenon to the bioactive molecules present in the tested wild fruits. Actually, in their recipe they replaced only 1% of flour with a lyophilized fruit powder, although they mistakenly declared 5% substitution, which represented only 0.5% in a total dough amount. The effect of additives with antioxidative properties on acrylamide formation is usually dose-dependent [70] and supporting or suppressive effects are ambiguous. A significantly higher tendency of sea buckthorn puree to form acrylamide under thermal conditions was unequivocally demonstrated in a model system [37]. In other studies [70,71], a slight increase of acrylamide level between 10 to 20% was observed in the crumb of bakery products with 5% sea buckthorn powder addition, but 2–5 times higher acrylamide was observed at 10% sea buckthorn powder addition. However, in these studies, no information on Asn content in matrices was presented. Moreover, in these studies, acrylamide was determined by other analytical techniques (HPLC method with a UV-Vis detector [6], capillary electrophoresis system [71,72]) instead of a highly selective and sensitive LC-MS/MS technique which is predominantly used for acrylamide determination [73,74]. On the other hand, the HPLC-UV technique used for acrylamide determination with acceptable parameters was demonstrated in the case of acrylamide detection in olives [75] and deep-fried flour-based foods [76].

Table 4. Acrylamide (μg/kg DW), initial Asn (mg/kg DW), residual Asn (mg/kg DW) and HMF (mg/kg DW) of cereal biscuits with sea buckthorn pomace powder addition.

Cereal Biscuits with SBP	Wheat			Triticale			Rye		
	SBP0	SBP1	SBP3	SBP0	SBP1	SBP3	SBP0	SBP1	SBP3
Acrylamide (μg/kg DW)	27.2 eB	80.4 dA	29.1 eB	58.9 dC	106.0 cA	69.1 dB	105.3 cC	485.5 aA	316.6 bB
Calculated initial Asn (mg/kg DW)	276.9	355.4	254.7	483.5	541.3	440.6	731.3	764.3	663.7
Residual Asn (mg/kg DW)	156.6 cdB	189.3 bcdA	72.9 dC	193.4 bcdAB	256.8 abcA	76.7 dB	342.0 aA	327.1 abA	164.5 cdA
Acrylamide per unit of Asn	0.10	0.23	0.11	0.12	0.20	0.17	0.14	0.64	0.48
HMF (mg/kg DW)	nd	7.72 a	nd	nd	5.25 c	nd	nd	6.58 b	nd

Values are the mean: acrylamide (n = 3); Asn, HMF (n = 2). Relative standard deviations are below 5% for acryalmide and HMF, below 10% for Asn. Different letters (a–e) indicate significant differences ($p < 0.05$) in a row, and (A–C) suggest significant differences ($p < 0.05$) within the group of cereal biscuits with different sea buckthorn pomaces. SBP0—biscuits with no presence of sea buckthorn pomace powder; SBP1—biscuits with sea buckthorn pomace powder without any treatment; SBP3 biscuits with sea buckthorn pomace powder enzymatically treated at neutral pH.

Enzymatic treatment of SBP with asparaginase after previous pH adjustment according to the procedure described in the patent application [38] used in this study, resulted in a substantial decrease of free Asn in SBP from 1834 mg/kg to 89 mg/kg (Table 2) which was lower than the free Asn content in used wholegrain flours. The 10% substitution of flours by SBP3 meant a decrease of free Asn in doughs available for acrylamide formation. Thus, acrylamide level in biscuits with enzymatically treated SBP3 powder was lower (29.1; 69.1; 316.6 μg/kg for wheat, triticale and rye biscuits) comparing to acrylamide in SBP1 biscuits (Table 4). However, despite expectations of similar acrylamide content in SBP3 and SBP0 biscuits, the decrease of free Asn in triticale and rye doughs did not project directly, so acrylamide decreased in the respective biscuits. The correlation of Asn in dough and acrylamide level in biscuits was not linear (e.g., in the case of no SBP addition, but a good exponential correlation was achieved (R^2 = 0.9966), similar to Asn/acrylamide correlation in case of SBP1 dough and biscuits (Figure 1a). It can be supposed that not only free Asn level is responsible for acrylamide formation in complex matrices, but also other factors affected the effectiveness of acrylamide formation. This is evident from a weak linear correlation of free Asn in doughs and final acrylamide content of SBP0, SBP1 and SBP3 biscuits (R^2 = 0.5684) (Figure 1b). This unproportional content of acrylamide can be probably caused by a richer matrix represented by SBP containing antioxidants, flavonoids, carotenes and a high portion of lipids. Some of these substances can support acrylamide formation through other pathways than a Maillard reaction, but also, they can suppress it. Moreover, the opposite reactions may result from the presence of similar factors in flour which inhibit acrylamide formation. Amongst others, the significantly higher concentration of free amino acids Pro, Trp and Gly in flours compared to SBPs are notable which may inhibit acrylamide formation or react with acrylamide and reduce its levels [42,57]. These opposite reactions resulted in less acrylamide formed per unit of Asn at higher presence of these amino acids, notably visible in triticale and rye SPB0 biscuits comparing to SPB3 biscuits (Table 4). The different pH of dough before baking due to the presence of untreated SBP1 powder with pH 3.4 and enzymatically treated SBP3 powder with pH 6.8 can also play the role. This hypothesis with multifactorial impact of matrix on acrylamide formation is supported by determination of residual free Asn in biscuits after baking.

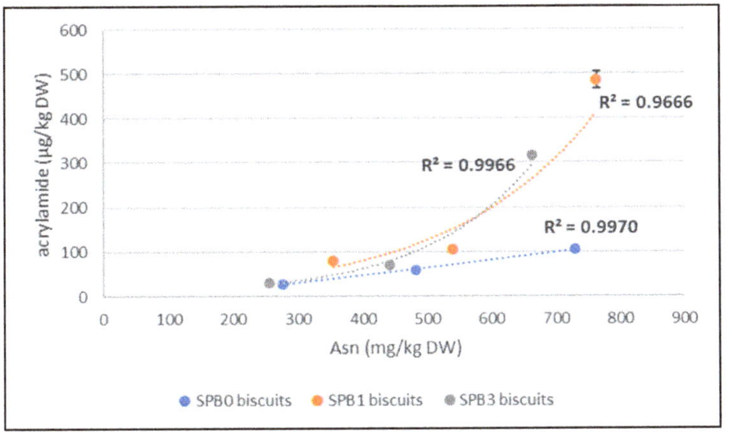

(a)

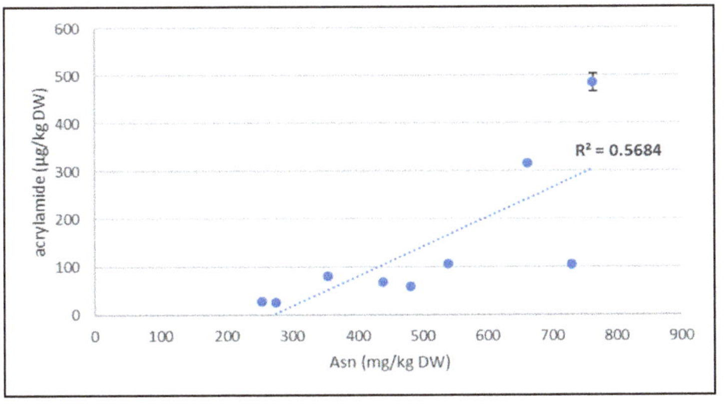

(b)

Figure 1. (**a**,**b**). Correlation of free asparagine (mg/kg DW) in dough and acrylamide (µg/kg DW) in respected biscuits. (**a**)—separate correlations of cereal biscuits SBP0, SBP1 and SBP3; (**b**)—correlation not distinguishing between cereal biscuits and SBP addition.

In control biscuits without addition of SBP powder, the highest level of residual free Asn after baking was observed in rye biscuits (342 mg/kg DW), followed by triticale (193 mg/kg DW) and wheat biscuits (157 mg/kg DW) (Table 4), which was in the same order as free Asn in particular flours before baking. It means that only a partial Asn utilisation was observed and the residual Asn represent 40–56% of the initial free Asn in dough. This was evident by comparison of Asn content calculated from relevant flours and determination of free Asn in biscuits after baking. The similar utilization of Asn was observed in case of the 10% substitution of wholegrain flours by untreated SPB1 powder as shown by determination of residual Asn in SPB1 biscuits. The residual Asn in these biscuits represented 41–50% of the initial Asn content. In biscuits with enzymatically treated SBP3 powders, the utilization of free Asn was higher, since the residual Asn was between 17% and 28% of the initial Asn content. As it was mentioned earlier, the asparaginase treatment was accompanied with pH adjustment to neutral values, thus under these conditions the conversion of Asn to acrylamide was more effective. It is known that the pH value plays an important role in acrylamide formation [77] which decreased in a buffered system [78].

3.3.6. Impact of SBP Powders on HMF of Biscuits

Formation of HMF is supported by lower pH and the presence of acidic matrix [33,78]. In case of untreated SBP1 powder, the pH of SBP before drying is only 3.4 due to a high content of organic and phenolic acids [9,65]. Addition of untreated SBP1 powder to dough of cereal biscuit production resulted in higher HMF content in final biscuits. It can be assumed that HMF formation during baking was promoted by lower pH due to the acidic nature of SBP1. This impact of pH was supported with the observation of HMF detected only in biscuits with untreated SBP1 powder addition (8.13; 5.25 and 6.58 mg/kg DW for wheat, triticale and rye biscuits) (Table 4). In control biscuits without SBP powder addition (SBP0) and in biscuits with enzymatically treated SBP powder addition (SBP3), the HMF was not detected.

4. Conclusions

Following the trend to reuse valuable by-products of plant-based food processing, sea buckthorn rich in exceptionally high levels of health-promoting compounds is worthy of use in commonly well accepted cereal biscuit production. The disadvantage of its high potential to contribute to undesirable acrylamide formation can be eliminated by enzymatic treatment of SBP before its usage in biscuit production. Asparaginase application on SBP under pH neutral conditions decreased free Asn up to 95% from 1834 mg/kg to the final Asn concentration of 89 mg/kg. The 10% substitution of wholegrain cereal flours (wheat, triticale, rye) by enzymatically treated SBP resulted in a substantial reduction of acrylamide in enriched biscuits between 30 and 60% without detrimental effects on quality of final products. Even in high-acrylamide rye biscuits with SBP addition, the enzymatic treatment of SBP brought a decrease of acrylamide from 485.5 µg/kg to 316.6 µg/kg which met the benchmark level established for biscuits and wafers (350 µg/kg). This study confirmed that the enzymatic treatment is a proper tool for acrylamide mitigation not only in cereal based processed foods, but also in fruits and fruit wastes used as nutritionally valuable additives of enriched cereal products.

Author Contributions: Conceptualization, Z.C. and A.T.; methodology, Z.C., K.K., M.B., A.T., V.J., J.H. and J.K.; software, M.B. and A.T.; validation, Z.C., M.B. and A.T.; formal analysis, Z.C., K.K., M.B., A.T., V.J., J.H. and J.K.; investigation, Z.C., K.K., M.B. and A.T.; resources, Z.C. and A.T.; data curation, M.B.; writing—original draft preparation, Z.C., M.B. and A.T.; writing—review and editing, Z.C., K.K., M.B. and A.T.; visualization, Z.C. and M.B.; supervision, Z.C. and A.T.; project administration, Z.C., K.K., M.B. and A.T.; funding acquisition, Z.C., K.K., M.B. and A.T. All authors have read and agreed to the published version of the manuscript.

Funding: This work was supported by the Slovak Research and Development Agency under the contracts No. APVV-17-0212 and SK-SRB-18-0035; the Operational Program Integrated Infrastructure within the project "Demand-driven research for the sustainable and innovative foods", Drive4SIFood, 313011V336, co-financed by the European Regional Development Fund; the Ministry of Science, Technological Development and Innovations of the Republic of Serbia Contract No. 451-03-47/2023-01/200222 The collaboration between international partners was performed in the frame of COST Action 21149 ACRYRED.

Data Availability Statement: The data used to support the findings of this study can be made available by the corresponding author upon request.

Acknowledgments: We thank Mikuláš Šranko (PD Tvrdošovce, Slovakia) for complimentary provided sea buckthorn fresh fruit samples used for experiments. We acknowledge Marek Kunštek for technical assistance at sea buckthorn fruit processing.

Conflicts of Interest: The authors declare no conflict of interest.

References

1. Dimou, C.; Karantonis, H.C.; Skalkos, D.; Koutelidakis, A.E. Valorization of Fruits by-products to Unconventional Sources of Additives, Oil, Biomolecules and Innovative Functional Foods. *Curr. Pharm. Biotechnol.* **2019**, *20*, 776–786. [CrossRef] [PubMed]
2. Gómez, M.; Martinez, M.M. Fruit and vegetable by-products as novel ingredients to improve the nutritional quality of baked goods. *Crit. Rev. Food Sci. Nutr.* **2018**, *58*, 2119–2135. [CrossRef] [PubMed]
3. Trigo, J.P.; Alexandre, E.M.C.; Saraiva, J.A.; Pintado, M.E. High value-added compounds from fruit and vegetable by-products—Characterization, bioactivities, and application in the development of novel food products. *Crit. Rev. Food Sci. Nutr.* **2020**, *60*, 1388–1416. [CrossRef] [PubMed]
4. Quiles, A.; Campbell, G.M.; Struck, S.; Rohm, H.; Hernando, I. Fiber from fruit pomace: A review of applications in cereal-based products. *Food Rev. Int.* **2018**, *34*, 162–181. [CrossRef]
5. Zhang, H.; Troise, A.D.; Sun, S.; Fogliano, V. The water insoluble fraction from red cabbage and black currant pomace reduces the formation of acrylamide, 5-hydroxymethylfurfural and reactive aldehydes in fried potato-based crisps. *LWT* **2023**, *173*, 114238. [CrossRef]
6. Borczak, B.; Sikora, M.; Kapusta-Duch, J.; Fołta, M.; Szewczyk, A.; Zięć, G.; Doskočil, I.; Leszczyńska, T. Antioxidative Properties and Acrylamide Content of Functional Wheat-Flour Cookies Enriched with Wild-Grown Fruits. *Molecules* **2022**, *27*, 5531. [CrossRef]
7. Zhu, F.; Cai, Y.-Z.; Ke, J.; Corke, H. Dietary plant materials reduce acrylamide formation in cookie and starch-based model systems. *J. Sci. Food Agric.* **2011**, *91*, 2477–2483. [CrossRef]
8. Bal, L.M.; Meda, V.; Naik, S.N.; Satya, S. Sea buckthorn berries: A potential source of valuable nutrients for nutraceuticals and cosmoceuticals. *Food Res. Int.* **2011**, *44*, 1718–1727. [CrossRef]
9. Ciesarová, Z.; Murkovic, M.; Cejpek, K.; Kreps, F.; Tobolková, B.; Koplík, R.; Belajová, E.; Kukurová, K.; Daško, L'.; Panovská, Z.; et al. Why is sea buckthorn (*Hippophae rhamnoides* L.) so exceptional? A review. *Food Res. Int.* **2020**, *133*, 109170. [CrossRef]
10. Dienaitė, L.; Pukalskas, A.; Pukalskienė, M.; Pereira, C.V.; Matias, A.A.; Venskutonis, P.R. Phytochemical Composition, Antioxidant and Antiproliferative Activities of Defatted Sea Buckthorn (*Hippophaë rhamnoides* L.) Berry Pomace Fractions Consecutively Recovered by Pressurized Ethanol and Water. *Antioxidants* **2020**, *9*, 274. [CrossRef]
11. Gâtlan, A.-M.; Gutt, G. Sea Buckthorn in Plant Based Diets. An Analytical Approach of Sea Buckthorn Fruits Composition: Nutritional Value, Applications, and Health Benefits. *Int. J. Environ. Res. Public Health* **2021**, *18*, 8986. [CrossRef] [PubMed]
12. Janotková, L.; Potočňáková, M.; Kreps, F.; Krepsová, Z.; Ácsová, A.; Ház, A.; Jablonský, M. Effect of sea buckthorn biomass on oxidation stability and sensory attractiveness of cereal biscuits. *BioResources* **2021**, *16*, 5097–5105. [CrossRef]
13. Gazi, S.; Göncüoğlu Taş, N.; Görgülü, A.; Gökmen, V. Effectiveness of asparaginase on reducing acrylamide formation in bakery products according to their dough type and properties. *Food Chem.* **2023**, *402*, 134224. [CrossRef]
14. Kaur, K.D.; Jha, A.; Sabikhi, L.; Singh, A.K. Significance of coarse cereals in health and nutrition: A review. *J. Food Sci. Technol.* **2014**, *51*, 1429–1441. [CrossRef]
15. McKevith, B. Nutritional aspects of cereals. *Nutr. Bull.* **2004**, *29*, 111–142. [CrossRef]
16. Torbica, A.; Belović, M.; Popović, L.; Čakarević, J.; Jovičić, M.; Pavličević, J. Comparative study of nutritional and technological quality aspects of minor cereals. *J. Food Sci. Technol.* **2021**, *58*, 311–322. [CrossRef] [PubMed]
17. Žilić, S.; Aktağ, I.G.; Dodig, D.; Gökmen, V. Investigations on the formation of Maillard reaction products in sweet cookies made of different cereals. *Food Res. Int.* **2021**, *144*, 110352. [CrossRef] [PubMed]
18. Ciesarová, Z.; Kukurová, K.; Bednáriková, A.; Morales, F.J. Effect of heat treatment and dough formulation on the formation of Maillard reaction products in fine bakery products—Benefits and weak points. *J. Food Nutr. Res.* **2009**, *48*, 20–30.
19. Sung, W.C.; Chen, C.Y. Influence of Cookies Formulation on the Formation of Acrylamide. *J. Food Nutr. Res.* **2017**, *5*, 370–378.
20. Žilić, S.; Dodig, D.; Basić, Z.; Vančetović, J.; Titan, P.; Đurić, N.; Tolimir, N. Free asparagine and sugars profile of cereal species: The potential of cereals for acrylamide formation in foods. *Food Addit. Contam. Part A* **2017**, *34*, 705–713. [CrossRef]
21. Delatour, T.; Stadler, R.H. Two decades of research in dietary acrylamide: What do we know today. *Crit. Rev. Food Sci. Nutr.* **2022**, 1–9. [CrossRef] [PubMed]
22. Commission Regulation (EU). 2017/2158 of 20 November 2017 establishing mitigation measures and benchmark levels for the reduction of the presence of acrylamide in food (Text with EEA relevance.). *J. Eur. Union* **2017**, *60*, 24–44.
23. Capuano, E.; Fogliano, V. Acrylamide and 5-hydroxymethylfurfural (HMF): A review on metabolism, toxicity, occurrence in food and mitigation strategies. *LWT* **2011**, *44*, 793–810. [CrossRef]
24. Mottram, D.S.; Wedzicha, B.L.; Dodson, A.T. Acrylamide is formed in the Maillard reaction. *Nature* **2002**, *419*, 448–449. [CrossRef] [PubMed]
25. Stadler, R.H.; Blank, I.; Varga, N.; Robert, F.; Hau, J.; Guy, P.A.; Robert, M.-C.; Riediker, S. Acrylamide from Maillard reaction products. *Nature* **2002**, *419*, 449–450. [CrossRef] [PubMed]
26. Zyzak, D.V.; Sanders, R.A.; Stojanovic, M.; Tallmadge, D.H.; Eberhart, B.L.; Ewald, D.K.; Gruber, D.C.; Morsch, T.R.; Strothers, M.A.; Rizzi, G.P.; et al. Acrylamide formation mechanism in heated foods. *J. Agric. Food Chem.* **2003**, *51*, 4782–4787. [CrossRef] [PubMed]

27. Blank, I. Current Status of Acrylamide Research in Food: Measurement, Safety Assessment, and Formation. *Ann. N. Y. Acad. Sci.* **2005**, *1043*, 30–40. [CrossRef]
28. Yaylayan, V.A. Acrylamide formation and its impact on the mechanism of the early Maillard reaction. *J. Food Nutr. Res.* **2009**, *48*, 1–7.
29. FoodDrinkEurope. Acrylamide Toolbox. 2019. Available online: https://www.fooddrinkeurope.eu/wp-content/uploads/2021/05/FoodDrinkEurope_Acrylamide_Toolbox_2019.pdf (accessed on 19 May 2023).
30. Schouten, M.A.; Tappi, S.; Rocculi, P.; Romani, S. Mitigation Strategies to Reduce Acrylamide in Cookies: Effect of Formulation. *Food Rev. Int.* **2022**, 1–42. [CrossRef]
31. Shyu, Y.-S.; Hsiao, H.-I.; Fang, J.-Y.; Sung, W.-C. Effects of Dark Brown Sugar Replacing Sucrose and Calcium Carbonate, Chitosan, and Chitooligosaccharide Addition on Acrylamide and 5-Hydroxymethylfurfural Mitigation in Brown Sugar Cookies. *Processes* **2019**, *7*, 360. [CrossRef]
32. Açar, Ö.Ç.; Pollio, M.; Di Monaco, R.; Fogliano, V.; Gökmen, V. Effect of Calcium on Acrylamide Level and Sensory Properties of Cookies. *Food Bioproc. Tech.* **2012**, *5*, 519–526. [CrossRef]
33. Kukurová, K.; Ciesarová, Z.; Mogol, B.A.; Açar, Ö.Ç.; Gökmen, V. Raising agents strongly influence acrylamide and HMF formation in cookies and conditions for asparaginase activity in dough. *Eur. Food Res. Technol.* **2013**, *237*, 1–8. [CrossRef]
34. Hendriksen, H.V.; Kornbrust, B.A.; Østergaard, P.R.; Stringer, M.A. Evaluating the potential for enzymatic acrylamide mitigation in a range of food products using an asparaginase from Aspergillus oryzae. *J. Agric. Food Chem.* **2009**, *57*, 4168–4176. [CrossRef] [PubMed]
35. Xu, F.; Oruna-Concha, M.-J.; Elmore, J.S. The use of asparaginase to reduce acrylamide levels in cooked Food. *Food Chem.* **2016**, *210*, 163–171. [CrossRef] [PubMed]
36. Ciesarova, Z.; Kiss, E.; Boegl, P. Impact of L-asparaginase on acrylamide content in potato products. *J. Food Nutr. Res.* **2006**, *45*, 141–146.
37. Constantin, O.E.; Kukurová, K.; Daško, L'.; Stănciuc, N.; Ciesarová, Z.; Croitoru, C.; Râpeanu, G. Modelling Contaminant Formation during Thermal Processing of Sea Buckthorn Purée. *Molecules* **2019**, *24*, 1571. [CrossRef]
38. Ciesarová, Z.; Kukurová, K.; Jelemenská, V. Method of production of fruit and/or vegetable preparations with reduced potential for acrylamide formation. Utility model No 9572. In *Vestník Úradu Priemyselného Vlastníctva Slovenskej Republiky 2022*; Banská: Bystrica, Slovak Republic, 2022; p. 47.
39. Belović, M.; Torbica, A.; Škrobot, D.; Tomić, J.; Čabarkapa, I.; Živančev, D.; Štatkić, S.; Aćin, V.; Kukurová, K.; Ciesarová, Z. Potential application of triticale cultivar'Odisej'for the production of cookies. *Ratar. I Povrt. Field Veg. Crops Res.* **2020**, *57*, 8–13. [CrossRef]
40. *AACC Approved Methods of Analysis*, 11th ed.; Method 10-50.05 Baking Quality of Cookie Flour; Cereals & Grains Association: St. Paul, MN, USA, 2000. [CrossRef]
41. Kowalski, S.; Kopuncová, M.; Ciesarová, Z.; Kukurová, K. Free amino acids profile of Polish and Slovak honeys based on LC-MS/MS method without the prior derivatisation. *J. Food Sci. Technol.* **2017**, *54*, 3716–3723. [CrossRef]
42. Curtis, T.Y.; Powers, S.J.; Balagiannis, D.; Elmore, J.S.; Mottram, D.S.; Parry, M.A.J.; Rakszegi, M.; Bedö, Z.; Shewry, P.R.; Halford, N.G. Free Amino Acids and Sugars in Rye Grain: Implications for Acrylamide Formation. *J. Agric. Food Chem.* **2010**, *58*, 1959–1969. [CrossRef]
43. Curtis, T.Y.; Muttucumaru, N.; Shewry, P.R.; Parry, M.A.J.; Powers, S.J.; Elmore, J.S.; Mottram, D.S.; Hook, S.; Halford, N.G. Effects of Genotype and Environment on Free Amino Acid Levels in Wheat Grain: Implications for Acrylamide Formation during Processing. *J. Agric. Food Chem.* **2009**, *57*, 1013–1021. [CrossRef]
44. Mustafa, A.; Åman, P.; Andersson, R.; Kamal-Eldin, A. Analysis of free amino acids in cereal products. *Food Chem.* **2007**, *105*, 317–324. [CrossRef]
45. Curtis, T.Y.; Powers, S.J.; Halford, N.G. Effects of Fungicide Treatment on Free Amino Acid Concentration and Acrylamide-Forming Potential in Wheat. *J. Agric. Food Chem.* **2016**, *64*, 9689–9696. [CrossRef] [PubMed]
46. Curtis, T.Y.; Powers, S.J.; Wang, R.; Halford, N.G. Effects of variety, year of cultivation and sulphur supply on the accumulation of free asparagine in the grain of commercial wheat varieties. *Food Chem.* **2018**, *239*, 304–313. [CrossRef] [PubMed]
47. Ciesarová, Z.; Kukurová, K.; Torbica, A.; Belović, M.; Horváthová, J.; Daško, L'.; Jelemenská, V. Acrylamide and 5-hydroxymethylfurfural in thermally treated non-wheat flours and respective breads. *Food Chem.* **2021**, *365*, 130491. [CrossRef] [PubMed]
48. Fernandez-Figares, I.; Marinetto, J.; Royo, C.; Ramos, J.M.; Garcia del Moral, L.F. Amino-Acid Composition and Protein and Carbohydrate Accumulation in the Grain of Triticale Grown under Terminal Water Stress Simulated by a Senescing Agent. *J. Cereal Sci.* **2000**, *32*, 249–258. [CrossRef]
49. Jaskiewicz, B. Chemical composition of winter triticale grain depending on type of tillage in crop rotation. In Proceedings of the 18th International Scientific Conference Engineering for Rural Development, Jelgava, Latvia, 22–24 May 2019; pp. 319–323.
50. Rodehutscord, M.; Rückert, C.; Maurer, H.P.; Schenkel, H.; Schipprack, W.; Bach Knudsen, K.E.; Schollenberger, M.; Laux, M.; Eklund, M.; Siegert, W.; et al. Variation in chemical composition and physical characteristics of cereal grains from different genotypes. *Arch. Anim. Nutr.* **2016**, *70*, 87–107. [CrossRef]
51. Sibian, M.S.; Saxena, D.C.; Riar, C.S. Effect of germination on chemical, functional and nutritional characteristics of wheat, brown rice and triticale: A comparative study. *J. Sci. Food Agric.* **2017**, *97*, 4643–4651. [CrossRef]

52. Mossé, J.; Huet, J.C.; Baudet, J. The amino acid composition of triticale grain as a function of nitrogen content: Comparison with wheat and rye. *J. Cereal Sci.* **1988**, *7*, 49–60. [CrossRef]
53. Žilić, S.; Aktağ, I.G.; Dodig, D.; Filipović, M.; Gökmen, V. Acrylamide formation in biscuits made of different wholegrain flours depending on their free asparagine content and baking conditions. *Food Res. Int.* **2020**, *132*, 109109. [CrossRef]
54. Richards, N.G.; Humkey, R.N.; Li, K.; Meyer, M.E.; de Sintjago, T.C.C. Tunnels and Intermediates in the Glutamine-Dependent Amidotransferases. In *Comprehensive Natural Products II*; Liu, H.-W., Mander, L., Eds.; Elsevier: Amsterdam, The Netherlands, 2010; pp. 161–230.
55. De Vleeschouwer, K.; Van der Plancken, I.; Van Loey, A.; Hendrickx, M.E. Role of precursors on the kinetics of acrylamide formation and elimination under low moisture conditions using a multiresponse approach—Part II: Competitive reactions. *Food Chem.* **2009**, *114*, 535–546. [CrossRef]
56. Claeys, W.L.; de Vleeschouwer, K.; Hendrickx, M.E. Effect of amino acids on acrylamide formation and elimination kinetics. *Biotechnol. Prog.* **2005**, *21*, 1525–1530. [CrossRef]
57. Koutsidis, G.; Simons, S.P.J.; Thong, Y.H.; Haldoupis, Y.; Mojica-Lazaro, J.; Wedzicha, B.L.; Mottram, D.S. Investigations on the effect of amino acids on acrylamide, pyrazines, and Michael addition products in model systems. *J. Agric. Food Chem.* **2009**, *57*, 9011–9015. [CrossRef] [PubMed]
58. Tkacz, K.; Gil-Izquierdo, Á.; Medina, S.; Turkiewicz, I.P.; Domínguez-Perles, R.; Nowicka, P.; Wojdyło, A. Phytoprostanes, phytofurans, tocopherols, tocotrienols, carotenoids and free amino acids and biological potential of sea buckthorn juices. *J. Sci. Food Agric.* **2022**, *102*, 185–197. [CrossRef] [PubMed]
59. Li, Y.; Li, P.; Yang, K.; He, Q.; Wang, Y.; Sun, Y.; He, C.; Xiao, P. Impact of Drying Methods on Phenolic Components and Antioxidant Activity of Sea Buckthorn (*Hippophae rhamnoides* L.) Berries from Different Varieties in China. *Molecules* **2021**, *26*, 7189. [CrossRef]
60. Zhang, W.; Yan, J.; Duo, J.; Ren, B.; Guo, J. Preliminary study of biochemical constitutions of berry of sea buckthorn growing in Shanxi province and their changing trend. In Proceedings of International Symposium on Sea Buckthorn (*H. rhamnoides* L.), Xi'an, China, 19–23 October 1989.
61. Krasotkina, J.; Borisova, A.A.; Gervaziev, Y.V.; Sokolov, N.N. One-step purification and kinetic properties of the recombinant l-asparaginase from Erwinia carotovora. *Biotechnol. Appl. Biochem.* **2004**, *39*, 215–221. [CrossRef]
62. Mahajan, R.V.; Saran, S.; Kameswaran, K.; Kumar, V.; Saxena, R.K. Efficient production of L-asparaginase from Bacillus licheniformis with low-glutaminase activity: Optimization, scale up and acrylamide degradation studies. *Bioresour. Technol.* **2012**, *125*, 11–16. [CrossRef]
63. Blanco Canalis, M.S.; Steffolani, M.E.; León, A.E.; Ribotta, P.D. Effect of different fibers on dough properties and biscuit quality. *J. Sci. Food Agric.* **2017**, *97*, 1607–1615. [CrossRef] [PubMed]
64. Tomić, J.; Belović, M.; Torbica, A.; Pajin, B.; Lončarević, I.; Petrović, J.; Fišteš, A. The influence of addition of dried tomato pomace on the physical and sensory properties of whole grain rye flour cookies. *Food Feed Res.* **2016**, *43*, 145–152. [CrossRef]
65. Nour, V.; Panaite, T.D.; Corbu, A.R.; Ropota, M.; Turcu, R.P. Nutritional and bioactive compounds in dried sea-buckthorn pomace. *Erwerbs-Obstbau* **2021**, *63*, 91–98. [CrossRef]
66. Gani, A.; Jan, R.; Ashwar, B.A.; Ashraf, Z.U.; Shah, A.; Gani, A. Encapsulation of saffron and sea buckthorn bioactives: Its utilization for development of low glycemic baked product for growing diabetic population of the world. *LWT* **2021**, *142*, 111035. [CrossRef]
67. Zykin, P.A.; Andreeva, E.A.; Lykholay, A.N.; Tsvetkova, N.V.; Voylokov, A.V. Anthocyanin Composition and Content in Rye Plants with Different Grain Color. *Molecules* **2018**, *23*, 948. [CrossRef]
68. Castañeda-Ovando, A.; Pacheco-Hernández, M.; Páez-Hernández, M.E.; Rodríguez, J.A.; Galán-Vidal, C.A. Chemical studies of anthocyanins: A review. *Food Chem.* **2009**, *113*, 859–871. [CrossRef]
69. Chen, G.; Hu, R.; Li, Y. Potassium bicarbonate improves dough and cookie characteristics through influencing physicochemical and conformation properties of wheat gluten. *Food Chem. X* **2020**, *5*, 100075. [CrossRef]
70. Ciesarová, Z.; Suhaj, M.; Horváthová, J. Correlation between acrylamide contents and antioxidant capacities of spice extracts in a model potato matrix. *J. Food Nutr. Res.* **2008**, *47*, 1–5.
71. Nilova, L.; Malyutenkova, S.; Kruchina-Bogdanov, I. The impact of plant powders on acrylamide content in bakery products. *Agron. Res.* **2019**, *17*, 1401–1413.
72. Kruchina-Bogdanov, I.V.; Nilova, L.P.; Malyutenkova, S.M.; Naumenko, N.V. Using capillary electrophoresis to determine acrylamide in bakery products. In *International Scientific and Practical Conference "Agro-SMART-Smart Solutions for Agriculture" (Agro-SMART 2018)*; Atlantis Press: Amsterdam, The Netherlands, 2018; pp. 528–533.
73. Wenzl, T.; de la Calle, M.B.; Anklam, E. Analytical methods for the determination of acrylamide in food products: A review. *Food Addit. Contam.* **2003**, *20*, 885–902. [CrossRef]
74. Pan, M.; Liu, K.; Yang, J.; Hong, L.; Xie, X.; Wang, S. Review of Research into the Determination of Acrylamide in Foods. *Foods* **2020**, *9*, 524. [CrossRef] [PubMed]
75. Crawford, L.M.; Wang, S.C. Comparative Study of Four Analytical Methods for the Routine Determination of Acrylamide in Black Ripe Olives. *J. Agric. Food Chem.* **2019**, *67*, 12633–12641. [CrossRef]
76. Wang, H.; Lee, A.W.M.; Shuang, S.; Choi, M.M.F. SPE/HPLC/UV studies on acrylamide in deep-fried flour-based indigenous Chinese foods. *Microchem. J.* **2008**, *89*, 90–97. [CrossRef]

77. De Vleeschouwer, K.; Van der Plancken, I.; Van Loey, A.; Hendrickx, M.E. Impact of pH on the Kinetics of Acrylamide Formation/Elimination Reactions in Model Systems. *J. Agric. Food Chem.* **2006**, *54*, 7847–7855. [CrossRef] [PubMed]
78. Wang, Y.; Hu, H.; McClements, D.J.; Nie, S.; Shen, M.; Li, C.; Huang, Y.; Zhong, Y.; Chen, J.; Zeng, M.; et al. PH and lipid unsaturation impact the formation of acrylamide and 5-hydroxymethylfurfural in model system at frying temperature. *Food Res. Int.* **2019**, *123*, 403–413. [CrossRef] [PubMed]

foods

MDPI

Article

Improvement in the Stability and Bioaccessibility of Carotenoid and Carotenoid Esters from a Papaya By-Product Using O/W Emulsions

Sara Lara-Abia [1,2], Gloria Lobo [3], Noelia Pérez-Pascual [1], Jorge Welti-Chanes [2] and M. Pilar Cano [1,*]

1 Laboratory of Phytochemistry and Plant Food Functionality, Biotechnology and Food Microbiology Department, Institute of Food Science Research (CIAL) (CSIC-UAM), Nicolas Cabrera 9, 28049 Madrid, Spain; sara.lara.abia@gmail.com (S.L.-A.); nellyper2000@gmail.com (N.P.-P.)
2 School of Sciences and Engineering, Tecnologico de Monterrey (ITESM), Monterrey 64849, NL, Mexico; jwelti@tec.mx
3 Department of Crop Production in Tropical and Subtropical Areas, Instituto Canario de Investigaciones Agrarias (ICIA), 38297 Tenerife, Spain; globo@icia.es
* Correspondence: mpilar.cano@csic.es; Tel.: +34-910019737

Abstract: The aim of the present work was to improve the stability and bioaccessibility of carotenoids from green oil extracts obtained from papaya by-products using oil-in-water (O/W) emulsions. The effects of different concentrations of pectin (1%, 2%, and 3%), a high-molecular-size emulsifier, together with Tween 20, a low-molecular-size emulsifier, high-speed homogenization conditions (time: 2, 3, 4, and 5 min; rpm: 9500, 12,000, 14,000, and 16,000 rpm), and high-pressure homogenization (HPH) (100 MPa for five cycles) were evaluated to determine the optimal conditions for obtaining O/W stable emulsions with encapsulated carotenoids. Soybean, sunflower, and coconut oils were used to formulate these O/W emulsions. The bioaccessibility of the main individual encapsulated papaya carotenoids was evaluated using the INFOGEST digestion methodology. In addition, the microstructures (confocal and optical microscopy) of the O/W carotenoid emulsions and their behavior during in vitro digestion phases were studied. Sunflower O/W carotenoid emulsions showed smaller mean particle size, higher negative ζ-potential, and higher viscosity than soybean O/W emulsions. Particle size reduction in the O/W emulsions using the HPH process improved the bioaccessibility of papaya encapsulated carotenoids. In these O/W emulsions, depending on the vegetable oil, lycopene was the carotenoid with the highest bioaccessibility (71–64%), followed by (*all-E*)-β-carotene (18%), (*all-E*)-β-cryptoxanthin (15%), and (*all-E*)-β-cryptoxanthin laurate (7–4%). These results highlight the potential of using green carotenoid papaya extracts to formulate O/W emulsions to enhance carotenoid bioactivity by efficiently preventing degradation and increasing in vitro bioaccessibility.

Keywords: *Carica papaya* L.; by-products; carotenoids; O/W emulsions; encapsulation; vegetable oils; stability; bioaccessibility; emulsion microstructure

Citation: Lara-Abia, S.; Lobo, G.; Pérez-Pascual, N.; Welti-Chanes, J.; Cano, M.P. Improvement in the Stability and Bioaccessibility of Carotenoid and Carotenoid Esters from a Papaya By-Product Using O/W Emulsions. *Foods* **2023**, *12*, 2654. https://doi.org/10.3390/foods12142654

Academic Editors: Stanisław Kowalski, Dorota Gumul and Wenjun Wang

Received: 22 May 2023
Revised: 8 June 2023
Accepted: 22 June 2023
Published: 10 July 2023

1. Introduction

In the last few years, a number of studies have proved that consuming carotenoids may provide health benefits, such as reducing the incidence of certain types of cancers, reducing the risk of cardiovascular disease, and improving eye health. Papaya fruits are rich sources of bioactive compounds such as carotenoids. Several studies have characterized their carotenoid profile, reporting that the most abundant carotenoids are β-carotene, β-cryptoxanthin and some of its isomers (such as laurate and caprate), and lycopene [1,2]. The Sweet Mary papaya variety is a hybrid cultivated in Tenerife, Canary Islands (Spain), and was recently introduced onto the market. Its physical–chemical properties (° Brix, color, shape, weight, etc.) and sensorial characteristics have made the Sweet Mary papaya popular among consumers. Papaya pulp is the edible part of the plant, and numerous studies have used it as a starting material due to the interest in its high content of bioactive compounds. However, papaya peel is considerably less studied than pulp and is usually discarded during the production process, generating waste and by-products. However, according to published studies, papaya peel is also a rich source of bioactive compounds, such as carotenoids [3]. Carotenoids are lipophilic compounds susceptible to oxidation and degradation, especially once extracted from their original tissues [4,5]. To exert their function in the human body, they must be released from the food matrix and become available for absorption. However, most of them present low availability (5–30%) compared to other food phytochemicals [6]. A previous study reported by our group [7] showed low bioaccessibility of the individual carotenoids from papaya pulp (Sweet Mary variety) exposed to the high hydrostatic pressure (HHP) process. In that work, the bioaccessibility values reported for β-cryptoxanthin, β-cryptoxanthin laurate, β-carotene, and lycopene ranged from 0 to 3% in the control and HHP-treated samples, respectively, indicating that there was no improvement in carotenoid bioaccessibility due to this process. Interest in incorporating carotenoids into functional food products has grown over the last few years. To design a functional ingredient with enhanced bioavailability of carotenoids, the formulation of O/W emulsions (micro- and nanoemulsions) or capsules for addition to foods could be an excellent strategy with promising results. A recent review on the application of advanced emulsion technology in the food industry [8] reported that advanced emulsions feature superior properties over conventional emulsions for delivering bioactive compounds, associated with higher retention efficiency, increased stability against environmental stresses, improved bioavailability, and the ability to control or trigger release. In the case of carotenoids, Xavier et al. [9] obtained a lutein bioaccessibility of up to 48% after incorporating a lutein emulsion into a yogurt, where the bioaccessibility value depended on the fat content of the yogurt. García-Cayuela et al. [10] reported that persimmon provitamin A carotenoids showed improved bioaccessibility by up to 38% when lyophilized fruit pulp was incorporated into whole cow's milk (3.6% fat). In general, emulsion-based delivery systems such as oil-in-water (O/W) emulsions are widely used to protect lipophilic bioactive compounds, i.e., carotenoids, from degradation and oxidation, enhancing their dispersion in aqueous media and increasing their bioaccessibility and stability under gastrointestinal conditions. During digestion, lipids are mandatory for the micellarization of carotenoids since they favor the dissolution of carotenoids into fat droplets in the gastric emulsion environment. Furthermore, the intake of fats may delay gastric emptying, with the fats forming more micelles. Moreover, the nature of triacylglycerol molecules (fatty acid chains and saturation degree) also influences the bioaccessibility of carotenoids [10,11]. Other studies showed that carotenes present higher bioaccessibility when they are consumed together with foods containing long-chain triacylglycerides, while xanthophylls need medium-chain triacylglycerides [12]. The presence of pectin also influences carotenoid bioaccessibility in different ways. In emulsions, pectin could help to reduce the interfacial tension between the lipid and the aqueous phases, increasing the viscosity of the aqueous phase. In food emulsions, pectin is used as an emulsifier and stabilizing agent [13], and some studies have reported the impact of food pectin on health, such as the satiety effect, prevention of some gastrointestinal diseases, and glycemia control [14]. Therefore, systems based on the formulation of carotenoid emulsions with pectin as an emulsifier could be studied under gastrointestinal conditions to evaluate their efficiency in improving their individual stability and bioaccessibility. Teixé-Roig et al. [15] reported improved bioaccessibility of β-carotene (standard) by up to ≈36% using nanoemulsions formulated with pectin. Other formulations have been used to encapsulate carotenoids. For example, in their study, Jain et al. [16] described the effect of modified rice-starch-based delivery systems on lycopene storage stability and bioaccessibility. In their work, lycopene showed a significantly higher bioaccessibility when it was encapsulated with the emulsions (20.2%) compared to the alginate beads (15.6%), whereas the chemical stability of lycopene was found to be significantly higher in the alginate beads (35.6%) than in the emulsions (29.5%). In the literature, some studies have reported the use of oil-in-water nanoemulsions with lactoferrin and different carrier oils to improve lycopene stability and bioaccessibility [17], but there are few published studies on the encapsulation of all carotenoids present in direct extracts (non-saponified) from papaya by-products with O/W emulsions. These studies would provide information about the stability and bioaccessibility of each individual carotenoid, including the xanthophyll esters, during in vitro gastrointestinal digestion.

In the present study, soybean oil and sunflower oil (rich in long-chain fatty acids) and coconut oil (TCM) (rich in medium-chain fatty acids) enriched with direct carotenoid extracts, obtained from a Sweet Mary papaya by-product (peels), were used to formulate oil-in-water (O/W) emulsions. The physical characteristics (particle size, zeta potential, and viscosity) and microstructure of the emulsions were studied. The digestive stability and bioaccessibility of the main individual O/W-encapsulated papaya carotenoids (including xanthophyll esters) were determined using the standardized in vitro INFOGEST gastrointestinal simulation method. In addition, the stability of the emulsion structures during in vitro digestion was studied using optical microscopy. This study could provide more information about the potential use of formulated papaya by-product extracts to obtain a healthy ingredient.

2. Materials and Methods

2.1. Chemicals, Standards, and Reagents

Ultrapure water was obtained from a Millipark® Express 40 system (Merk- Millipore, Darmstadt, Germany); methanol, diethyl ether, tetrahydrofuran (THF), methyl tert-butyl ether (MTBE), acetone, and petroleum ether were purchased from VWR International (Radnor, PA, USA). Butylated hydroxytoluene (BHT) and magnesium carbonate were obtained from Acros Organics (New Jersey, NY, USA). Anhydrous sodium sulfate, sodium chloride (NaCl), and potassium hydroxide (KOH) were purchased from Panreac Quimica (Barcelona, Spain). Standards for lycopene (L9879, ≥90%, from tomato), (*all*-E)-β-apo-8′-carotenal (10,810, ≥96%, (UV)), and lutein (X6250 from marigold) were purchased from Sigma-Aldrich (St. Louis, MO, USA). Standards for (*all*-E)-β-carotene (HPLC 96%, synth., cryst.), (*all*-E)-α-carotene (HPLC 97%, synth., cryst.) and (*all*-E)-β-cryptoxanthin (HPLC 97%, synth., cryst.), (*all*-E)-violaxanthin (HPLC 95%, isolated, cryst.), (*all*-E)-neoxanthin (HPLC 97%, isolated, cryst.), and (*all*-E)-zeaxanthin (HPLC 97%, synth., cryst.) were purchased from CaroteNature (Ostermundigen, Switzerland). Enzymes used in the simulated gastrointestinal in vitro digestion were acquired from Sigma-Aldrich (St. Louis, MO, USA). The following enzymes were used: α-amylase from porcine pancreas (10,080; 79 U mg^{-1}), pepsin from porcine gastric mucosa (P6887; 791 U mg^{-1}), pancreatin from porcine pancreas (P7545; 17 units TAME per mg), and bile salts from bovine and ovine origin (B8381). Other reagents used in the in vitro digestion assay were acquired from Sigma-Aldrich (St. Louis, MO, USA). Tween 20 was obtained from Sigma-Aldrich (St. Louis, MO, USA). Food-grade high-methoxyl pectin from citrus peel with 67% to 71% methylesterification was obtained from Acros Organics (Morris Plains, NJ, USA). Coconut oil (MCT, medium-chain triglyceride) was purchased from Shanghai YuanYe Biological Technology Co., Ltd., (Shanghai, China) and soybean (La Española Soy Plus®) and sunflower (Aliada®) refined oils were purchased from a local market in Madrid (Spain). The fatty acid composition of vegetable oils from Dubois et al. [17] is presented in Table S1 (Supplementary Materials).

2.2. Papaya By-Product

The Sweet Mary papaya (*Carica papaya* L.) fruit by-product used in the study was obtained from Tenerife (Canary Islands, Spain: 28°18′52″ north (N); 16°24′36″ west (W); 271 m above sea level). Papaya fruits that were not suitable for commercialization (not uniform in size) were washed and selected according to peel coloration and ripeness. After they were washed and drained, they were cut, the seeds were removed, and pulp and peel (<2 mm) tissues were separated. Uniform sections (1 cm^3) of papaya peel tissue were vacuum-packed in 200 mm × 300 mm plastic bags (Cryovac®, Sealed Air Corporation,

Madrid, Spain), frozen with liquid nitrogen, and freeze-dried for 5 days at −45 °C and 1.3×10^{-3} MPa (LyoBeta 15, Azbil Telstar SL, Terrasa, Spain). The freeze-dried material was ground by pulverizing (Grindomix GM200, Retsch, Haan, Germany) to a fine particle size (<2 mm), vacuum-packed in plastic bags, and stored at −80 °C until carotenoid analysis. For the present study, freeze-dried papaya peel was selected as a model papaya by-product to obtain direct carotenoid extracts to formulate O/W emulsions.

2.3. Carotenoid Extraction from Freeze-Dried Papaya Peel

Carotenoids and carotenoid esters were extracted according to the method of Cano et al. [18], with some modifications. The extraction was carried out under dim light, using amber flasks and avoiding long-term oxygen exposure. First, 1 g of freeze-dried sample was mixed with 0.5 g of magnesium carbonate. Then, 20 mL of tetrahydrofuran (THF) stabilized with 0.01% (*w*/*v*) butylated hydroxytoluene (BHT) was added for the extraction. The sample was homogenized in an Omnimixer (OMNI Macro S®, OMNI International, Kennesaw, GA, USA) for 3 min at 7000 rpm and placed in an ultrasonic water bath (3000514 model, 50/60 Hz, 360 W, J. P. Selecta S.A., Barcelona, Spain) for 30 min. The extract was centrifuged at 9000 rpm for 10 min at 4 °C, and the supernatant was collected. Next, 20 mL of acetone was added to the pellet and the sample was extracted again. This step was repeated until a colorless pellet was obtained. The supernatants were combined and placed in a separation funnel, adding 20 mL of diethyl ether. The funnel was shaken, and the organic phase was collected in a round amber flask. If an emulsion was formed, 10 mL of ultrapure water was added. This step was repeated twice. The organic phase was collected and dried with 2.5 g of anhydrous sodium sulfate for 10 min at room temperature. Then, the extract was filtered through Whatman No. 1 filter paper and vacuum-dried under a steam of nitrogen. Finally, the extract was kept at −80 °C until use in emulsion preparations and for HPLC analysis.

2.4. Preparation of Papaya By-Product Carotenoid O/W Microemulsions

The microemulsions were prepared using a reported method [15] with modifications. Elaboration of the O/W emulsions without carotenoid extracts was carried out in order to assess the best conditions to obtain stable emulsions (results not included) to prepare the O/W carotenoid-enriched emulsions.

Briefly, an oil-in-water emulsion containing papaya carotenoid extract was prepared in the dark to prevent the degradation of carotenoid by light. The oil phase was prepared by completely dissolving 7 mg of carotenoid extract in 1 g of vegetable oil (soybean, sunflower, or coconut oils). The aqueous phase consisted of pectin (1%, 2%, and 3%) dissolved in ultrapure water preheated at 70 °C 1% (*w/w*) and homogenized using an Ultraturrax homogenizer (T-25 Digital, IKA work Inc., Breisgau, Germany) at 9500 rpm for 5 min. Then, this aqueous phase was left for 1 h to cool at room temperature.

To obtain the coarse emulsion, in the first stage, the lipid phase (4% *w/w*), Tween 20 (4% *w*/*w*) as a low-molecular-weight emulsifier, and the aqueous phase containing pectin were homogenized at different speeds (9500, 12,000, 14,000, and 16,000 rpm) with 1 s ON/1 s OFF pulses and continuous mode for each speed, using an Ultraturrax homogenizer (T-25 Digital, IKA work Inc., Breisgau, Germany). In the second stage, a high-pressure homogenization process using an advance Panda PLUS 2000 GEA Niro Soavi (Parma, Italy) was employed. This device is composed of two valves. In valve no. 1, the pressure was set at 90 MPa to micronize the sample in stage 1. In valve no. 2, the operational pressure was set at 100 MPa to disperse the sample in stage 2, obtaining stable emulsions. After setting the parameters, the coarse emulsion was passed through the high-pressure homogenizer 5 times (5 cycles) under these conditions.

2.5. Characterization of O/W Microemulsions

2.5.1. Particle Size and Zeta Potential

The particle size of emulsions was measured using a Mastersizer 3000 (Malvern Instruments Ltd., Worcestershire, UK). Samples were diluted in ultrapure water and stirred in the dispersion unit at a constant speed of 2000 rpm. The particle size was expressed as average diameter volume-to-surface (d32) in micrometers (μm), using the refractive index 1.333 for water. The electrical charge (ζ-potential) was measured through phase-analysis light scattering (PALS) using a Zetasizer NanoZS (Malvern Instruments Ltd., Worcestershire, UK) to determine the surface charge at the interface of the droplets. Emulsions were diluted (1:10) in ultrapure water and placed in a capillary cell equipped with two electrodes to assess the electrophoretic mobility of the particles. The results are reported in millivolts (mV).

2.5.2. Microstructure of O/W Carotenoid Microemulsions and In Vitro Gastrointestinal Digestion Phases

Images of the coarse emulsion O/W, the O/W carotenoid microemulsions, and the corresponding O/W microemulsions during in vitro gastrointestinal digestion were obtained using an optical microscope Axioskop (Carl Zeiss, Germany) coupled to a Leica DMC 6200 pixel shift camera (Leica Microsystems, Germany), using a Zeiss Plan-Neofluar lens with a 100× objective and the addition of an immersion oil drop. Samples were observed using an open condenser with level 4 illumination, and no color filters were used. The color was manually adjusted to show real-time colors using Leica LAS v4.13 Application Suite Software.

Microstructural analysis was also performed using confocal laser scanning microscopy (CLSM). Neutral red solution (50 μg/mL in ethanol, Sigma, Deisenhofen, Germany) was used to dye the lipophilic phase of the emulsions. The excitation and emission peaks of the dye were at 515 nm and 585 nm, respectively. This analysis was achieved using a confocal multispectral TCS SP8 system (Leica Microsystems, Mannheim, Germany) at 40× with a Zeiss Plan-Neofluar lens.

2.6. Viscosity

The viscosity of O/W emulsions was measured using a Haake high-temperature viscometer (IKA®—Werke GmbH & Co., KG, Staufen, Germany) with an ME 1700 sensor system. The methods used are described in the ISO 7884-2 and ISO 7884-3 standards (www.iso.org, 2020).

2.7. Measurement of Encapsulation Rate

The encapsulation efficiency (EE) (%) was calculated as (the total carotenoids in the emulsion/total carotenoids in the extract used to formulate the emulsion) × 100. The total carotenoid content was determined both through the spectrophotometry method using β-carotene as the standard and though the sum of individual quantified carotenoids analyzed using HPLC. The extraction protocol to analyze the carotenoids encapsulated by the O/W emulsions is described in Section 2.8.

2.8. Carotenoid Extraction from O/W Microemulsions

The procedure reported by Teixé-Roig et al. [15] with some modifications was followed to extract the carotenoids from the O/W microemulsions. First, 5 mL of the O/W microemulsion was mixed with 5 mL of chloroform, and the mixture was centrifuged at 3000 rpm for 20 min at 4 °C. After centrifugation, the chloroform phase (orange-colored) was collected. This process was repeated 3 times. Then, the extract was dried in a rotatory evaporator at 20 °C, re-suspended in 2 mL of MeOH:MTBE:H_2O (45.5:52.5:2, *v/v/v*), filtered through a 0.45 μm filter, and analyzed through HPLC and by spectrophotometry for total carotenoid content.

2.9. Characterization and Quantification of Carotenoids from Papaya Peel (By-Product)

2.9.1. Analysis of Carotenoids using HPLC

The carotenoids (hydrocarbon carotenoids, free xanthophylls, and xanthophyll esters) were quantified simultaneously through high-performance liquid chromatography using a 1200 Series Agilent HPLC System (Agilent Technologies, Santa Clara, CA, USA) with a reverse-phase C30 column (YMC-Pack YMC C30, 250 × 4.6 mm inner diameter (i.d.), S-5 μm, YMC Co., Ltd., Kyoto, Japan) at 32 °C. The mobile phases used for the separation were MeOH:MTBE:H_2O (81:14:4, *v/v/v*, eluent A) and MeOH:MTBE (10:90, *v/v*, eluent B), both containing 0.1% ammonium acetate [19]. The UV–vis photodiode array detector was set at 450 nm. UV–vis spectra were recorded between 200 and 700 nm. The LC-MS/MS ($APCI^+$) analyses were performed using HPLC coupled to a mass spectrometry detector with an APCI source model G1947B (Agilent) compatible with LCMS SQ 6120 equipment [16]. Nitrogen was used as the drying gas at a flow rate of 60 L/min and as a nebulizing gas at a pressure of 50 psi. The nebulizer temperature was 350 °C, and a potential of +2779 kV was used on the capillary. Helium was the collision gas, and the fragmentation amplitude was 0.8–1.2 V. The vaporizer temperature was set at 400 °C, and the corona was 4000 nA as a positive ion mode. The positive ion mass spectra of the column eluted at 13,000 Th/s (peak width 0.6 Th, FWHM).

Each carotenoid was identified according to its elution time, chromatography with carotenoid standards, the UV–visible spectrum (λ_{max}, spectral fine structure (%III/II), peak cis intensity), and the mass spectrum as reported by Lara-Abia et al. [3]. The carotenoids were quantified using linear calibration curves prepared with concentrations of commercial standards in the range of 5–100 μg/mL of carotenoid stock solutions. Antheraxanthin and its isomers were quantified using the violaxanthin standard curve. Figure S1 (Supplementary Materials) shows the C30 HPLC reversed-phase chromatogram of the papaya peel carotenoids, and Table S2 displays all identified carotenoids present in the vegetable–papaya–carotenoid-enriched oils.

2.9.2. Analysis of Total Carotenoids through Spectrophotometry

Total carotenoids were analyzed using spectrophotometry at 420 nm. For this, the dried extract obtained from papaya peel was dissolved in hexane, β-carotene was used as the standard, and the β-carotene extinction coefficient was used in hexane 2560 [20]. This same papaya extract was used for the HPLC analysis of individual carotenoids, but in this case, the dried extract was dissolved in MeOH:MTBE:H_2O (45.5:52.5:2, *v/v/v*), as noted in Section 2.8.

2.10. Bioaccessibility Study of Papaya Carotenoid Extracts Encapsulated by O/W Microemulsions

2.10.1. In Vitro Gastrointestinal Digestion Assay

The in vitro gastrointestinal digestion assay was performed based on the standardized INFOGEST© protocol [21] with the adaptations for carotenoid studies reported by Cano et al. [19]. Microemulsions were digested immediately after their preparation.

Digestive solutions for the mouth (simulated saliva fluid, SSF), stomach (simulated gastric fluid, SGF), and small intestinal (simulated duodenal fluid, SDF) compartments were prepared following the methodology described by Eriksen et al. [22]. To avoid the loss of activity and denaturalization, enzyme solutions were prepared daily prior to the digestive assay. After each phase (oral, gastric, and intestinal) of the in vitro digestion, a digestive sample was obtained, named digesta. The digesta was ultra-centrifuged (L-70 Ultracentrifuge Beckman Coulter, Brea, CA, USA) at 20,000 rpm for 10 min at 4 °C. The supernatant was analyzed for carotenoid content, while the pellet was discarded. Carotenoid bioaccessibility values were calculated as the ratio between the carotenoid concentration in the supernatant (in the micellar fraction) of the intestinal fraction and its initial concentration in the emulsion prior to in vitro digestion (Equation (1)).

$$\text{Bioaccessibility (\%)} = (\text{Carotenoid content supernatant}/\text{Carotenoid content emulsion}) \times 100 \quad (1)$$

2.10.2. Carotenoid Extraction from In Vitro Digestion Phases

Carotenoids were extracted from the digesta according to the method described by Lara-Abia et al. [7]. The digesta was extracted with 15 mL of acetone and placed in an ultrasound bath for 15 min at 25 °C. Then, samples were centrifuged at 9000 rpm for 10 min at 4 °C, and the supernatant was recovered. The pellet was re-extracted twice with acetone and one last time with methanol. Afterward, in a separation funnel, 30 mL of petroleum ether/diethyl ether (1:1; *v/v*) was added to the combined supernatants. The organic phase was recovered and dried with anhydrous sodium sulfate (approx. 2.5 g), filtered, and evaporated until dry in a rotavapor at 30 °C. The sample was made up to 2 mL with MeOH:MTBE:H_2O (45.5:52.5:2, *v/v/v*), filtered through a 0.45 µm filter, and analyzed through HPLC.

Carotenoids were extracted from the supernatants (micellar fraction) of the digestive phases according to the method described by Petry and Mercadante [23], with modifications. Each supernatant obtained after ultra-centrifugation was extracted with 20 mL of diethyl ether by mixing with a vortex and centrifuging at 14,000 rpm for 10 min at 4 °C in an L-70 Ultracentrifuge (Beckman Coulter, Indianapolis, IN, USA). Then, the supernatant (micellar fraction) was placed in a separation funnel, 20 mL of diethyl ether was added, and the mixture was shaken vigorously. The organic phases were recovered, dried with anhydrous sodium sulfate, filtered, and vacuum-concentrated to be analyzed through HPLC.

2.11. Statistical Analysis

All experiments were carried out in duplicate, and three determinations of each analysis were performed on each parameter to obtain mean values. Data are expressed as the mean ± standard deviation. Significant differences were calculated through one-way analysis of variance (ANOVA) and a post hoc Tukey's test ($p < 0.05$). Student´s *t*-test was used to compare means of carotenoid bioaccessibility values and between the parameters (viscosity, particle diameter, and ζ- potential) of the soybean oil and sunflower oil emulsions. Statistical analyses were performed with IBM SPSS Statistics 23.0 (IBM Corp., Armonk, NY, USA).

3. Results and Discussion

3.1. O/W Papaya Carotenoid Microemulsion Optimization Process

Different ultra-homogenization conditions were evaluated using the Ultraturrax equipment to obtain the optimum coarse emulsion. Vegetable oils (sunflower and soybean oils) enriched with papaya peel direct extract (7 mg carotenoid/g oil, *w/w*) were used to optimize the O/W microemulsions. The carotenoid composition of this papaya peel extract is shown in Table S2 (Supplementary Materials), and the contents of different carotenoids in the papaya peel extract are displayed in Table S3. To achieve better conditions for procuring stable O/W emulsions, a two-step process was assayed, following the procedure of Teixé-Roig et al. [15] with some modifications. In the first stage of process validation, the papaya carotenoid extract dissolved in sunflower oil (7 mg carotenoid/g oil, *w/w*) was used to optimize the procurement of O/W coarse emulsion. The oil-in-water ratio was 1:24 (4% lipid phase and 96% aqueous phases). High-speed homogenization (rpm) was the initial parameter evaluated. No significant differences were found when using speeds of 9500, 12,000, 14,000, and 16,000 rpm in terms of the particle size and viscosity of the obtained O/W emulsions (Table 1). Therefore, 12,000 rpm was the speed selected for this step, since the distribution particle graph and the optical microscopy image showed good uniformity in the generated O/W emulsions (Figure 1). Then, the high-speed homogenization was evaluated in terms of the time taken (2, 3, 4, and 5 min). In this case, since no differences in the physicochemical characteristics (particle diameter, viscosity, ζ-potential, and in particle distribution and microscopy graphs) of the O/W emulsions were found, a time of 2 min was selected to continue with the second step of obtaining stable O/W emulsions. Finally, three different pectin concentrations (1%, 2%, and 3%) were assayed to formulate different

O/W coarse emulsions. The smallest particle diameter (2.7 ± 0.1 μm) was obtained using a 2% pectin concentration, in comparison with 1% (8.2 ± 0.2 μm) and 3% (4.6 ± 0.0 μm).

Table 1. Physicochemical characteristics (particle diameter (v, 0.5 μm), viscosity (mPa.s), and ζ-potential (mV)) of O/W sunflower oil emulsions obtained using different high-speed homogenization speeds (9500, 12,000, 14,000, and 16,000 rpm), time periods (2, 3, 4, and 5 min), and pectin concentrations (1%, 2%, and 3%) and of emulsion treated with high-pressure homogenization (HPH) at 100 MPa/5 cycles.

High Speed Homogenization Conditions		**Parameter**		
		Particle Diameter D (v, 0.5) (μm)	**Viscosity (mPa.s)**	**ζ-Potential (mV)**
Coarse emulsion	9500 rpm	10.6 ± 0.1 ^{b}	21.9 ± 0.2 ^{c}	−43.1 ± 1.1 ^{c}
2 min/2% pectin concentration	12,000 rpm	8.0 ± 0.1 ^{b}	22.0 ± 1.2 ^{c}	−41.9 ± 0.6 ^{c}
	14,000 rpm	7.2 ± 0.1 ^{b}	22.0 ± 1.5 ^{c}	−42.0 ± 0.2 ^{c}
	16,000 rpm	5.8 ± 0.1 ^{b}	21.4 ± 0.9 ^{c}	−32.6 ± 1.2 ^{b}
	2 min	8.0 ± 0.1 ^{b}	14.8 ± 0.3 ^{b}	−37.0 ± 0.4 ^{b}
12,000 rpm/2% pectin concentration	3 min	9.5 ± 0.8 ^{b}	14.9 ± 1.0 ^{b}	−35.7 ± 0.2 ^{b}
	4 min	9.0 ± 0.5 ^{b}	14.4 ± 0.2 ^{b}	−33.1 ± 1.8 ^{b}
	5 min	6.0 ± 0.1 ^{b}	13.2 ± 0.5 ^{b}	−31.8 ± 3.1 ^{b}
	1% pectin concentration	8.2 ± 0.2 ^{b}	5.7 ± 0.3 ^{a}	−40.4 ± 1.6 ^{c}
12,000 rpm/2 min	2% pectin concentration	8.0 ± 0.1 ^{b}	20.9 ± 1.3 ^{c}	−36.5 ± 1.4 ^{b}
Final Emulsion	3% pectin concentration	4.6 ± 0.0 ^{b}	48.8 ± 1.9 ^{d}	−27.8 ± 0.5 ^{a}
HPH treatment at 100 MPa/5 cycles [1]		0.5 ± 0.0 ^{a}	7.3 ± 0.2 ^{a}	−26.8 ± 0.2 ^{a}

[1] Final emulsion obtained using high-speed homogenization at 12,000 rpm for 2 min with 2% pectin concentration aqueous phase. The results are expressed as the mean ± standard deviation. Lowercase superscript letters indicate statistically significant differences ($p < 0.05$) between treatments in the same column.

No differences were found among the viscosity values (mPa.s) of the O/W coarse emulsions prepared with different concentrations of pectin, 1% (5.4 ± 0.0), 2% (5.4 ± 0.0), and 3% (5.5 ± 0.0), after the high-speed homogenization of the lipid phase at 12,000 rpm for 2 min (coarse emulsion) when coconut oil was used (Table 2).

After the selection of the optimum conditions (12,000 rpm for 2 min with 2% pectin concentration), Tween 20 was used as a second low-molecular-size emulsifier to formulate the coarse emulsion. Next, the final O/W emulsion was obtained using high-pressure homogenization (HPH) treatment at 100 MPa/5 cycles (see Materials and Methods section). The conditions selected for this last step were established according to the study reported by Teixé-Roig et al. [15] for the encapsulation of standard β-carotene, but in the present work, high-pressure homogenization (HPH) equipment was used instead of a microfluidizer. After HPH treatment, a particle size of 0.5 ± 0.0 μm, a viscosity of 7.3 ± 0.2 mPa.s, and a ζ-potential of −26.8 ± 0.2 were obtained in the corresponding final emulsion (Table 1). The particle distribution graph of this formulation showed great uniformity among the particles, and in the light microscopy image (Figure 1), a reduction in the emulsion size as well as pectin aggregation can be observed, possibly due to temperature increase during the HPH process (2.4 °C/100 bar).

Once the optimization of the O/W emulsion process was established for long-chain fatty acid vegetable oils (soybean and sunflower oils), coconut oil (medium-chain triglyceride) was assayed to formulate similar O/W emulsions (Table 2). The objective was to evaluate the possible differences in particle diameter, viscosity, ζ-potential, particle distribution, and stability of the final coconut oil O/W emulsions, compared to the long-chain triglyceride oil (soybean oil and sunflower oil) O/W emulsions. Table S2 shows the composition of the fatty acids in the assayed vegetable oils [18] to formulate the papaya carotenoid O/W emulsions. During the homogenization process, the high pressure applied pushes the emulsion through the small annular space, where the pressure is transformed into high velocity, generating extreme turbulence and cavitation, reducing the emulsion particle size [24].

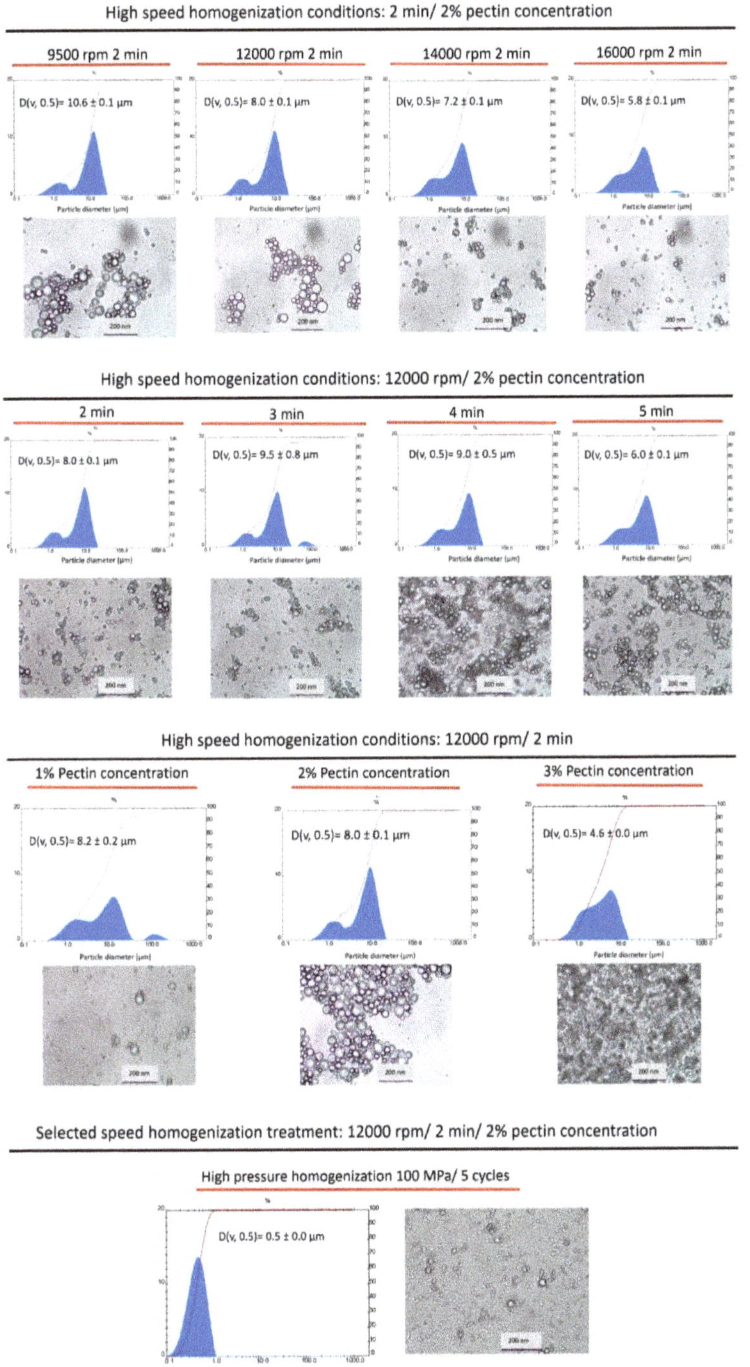

Figure 1. Particle size distribution and optical microscopy images of sunflower oil O/W emulsions obtained using different homogenization speeds (9500, 12,000, 14,000, and 16,000 rpm), time periods (2, 3, 4, and 5 min), and pectin concentrations (1%, 2%, and 3%). Scale bar = 200 µm.

Table 2. Physicochemical characteristics (particle diameter (v, 0.5 μm), viscosity (mPa.s), and ζ-potential (mV)) of MCT (coconut oil) emulsions obtained using different pectin concentrations (1%, 2%, and 3%).

	High-Speed Homogenization (12,000 rpm/2 min)			HPH (100 MPa/5 Cycles)		
Parameter	Pectin Concentration			Pectin Concentration		
	1%	2%	3%	1%	2%	3%
Viscosity (mPa.s)	5.4 ± 0.0^{b}	5.4 ± 0.0^{b}	5.5 ± 0.0^{b}	4.8 ± 0.0^{a}	5.7 ± 0.4^{b}	5.5 ± 0.0^{b}
ζ-potential (mV)	-35.0 ± 2.5^{d}	-24.8 ± 0.7^{c}	-15.6 ± 0.2^{a}	-19.8 ± 0.9^{b}	-18.1 ± 1.2^{b}	-19.9 ± 1.9^{b}
Particle diameter D (v, 0.5) (μm)	11.2 ± 0.2^{d}	8.0 ± 0.1^{c}	3.1 ± 0.1^{b}	0.5 ± 0.0^{a}	0.4 ± 0.0^{a}	0.6 ± 0.0^{a}

The results are expressed as the mean ± standard deviation. Lowercase superscript letters indicate statistically significant differences ($p < 0.05$) between 1%, 2%, and 3% pectin in the same row.

In the coconut O/W emulsions, as the pectin concentration increased, the ζ-potential increased as follows: −35.0 ± 2.5 mV (1% pectin), −24.8 ± 0.7 mV (2% pectin), and −15.6 ± 0.2 mV (3% pectin). The opposite effect occurred in terms of the particle diameter of O/W coarse emulsions, with the highest particle diameter (v, 0.5) at a pectin concentration of 1% (11.2 ± 0.2 μm), followed by 6.7 ± 0.0 μm at a 2% pectin concentration and 3.1 ± 0.1 μm (the smallest) at a 3% pectin concentration. These results show that stable and small-particle coarse emulsions were obtained when the concentration of pectin was 3%. However, after the HPH treatment at 100 MPa for five cycles, similar values of viscosity, particle diameter, and ζ-potential were observed in the final emulsions for all pectin concentrations (Table 2). These final O/W coconut oil emulsion data were also not significantly different to those observed for the O/W final emulsions formulated with long-chain triglycerides (soybean and vegetable oils) (Table 1).

In contrast, regarding the particle size distribution (Figure 2) and the optical microscopy images (Figure 3) of the coconut (MCT) O/W final emulsions, a notable coalescence effect between particles was observed, and instability became evident in high-pressure homogenization (HPH)-treated emulsions. The coalescence in the O/W emulsions could be due to different factors such as the required solid particles in the dispersed oil phase, the formation of irregular clusters, and the increased aggregation rate [24]. In the case of coconut (MCT) O/W emulsions, a greater electrostatic repulsion instead of a steric repulsion between the oil droplets and pectin molecules might be the factor leading to emulsion stability. Consequently, the particles tend to aggregate and produce the observed coalescence. For this reason, the formulation using coconut oil (MCT) was discarded before continuing the study for improved stability and bioaccessibility of papaya carotenoids using in vitro gastrointestinal assays.

3.2. *Physicochemical Characterization*

3.2.1. Particle Size

The coarse soybean oil and sunflower oil emulsions (before treatment with HPH) presented a higher mean particle size (6.6 ± 0.0 μm and 5.1 ± 0.0 μm, respectively) (Figure 1). After the high-pressure homogenization process, the particle sizes reduced to 0.6 ± 0.0 for the soybean oil emulsion and 0.4 ± 0.0 μm for the sunflower oil emulsion (Table 3). These differences observed in the particle sizes showed that in addition to reducing the particle size, the homogenization process created a microemulsion less polydisperse compared to the coarse emulsion (Figure 4). However, only particle sizes corresponding to microemulsions were obtained. Meanwhile, authors such as Teixé-Roig et al. [15] obtained lower emulsion particle sizes (439.3 ± 23.3 nm) using a corn oil O/W, 2% pectin, and a microfluidification process.

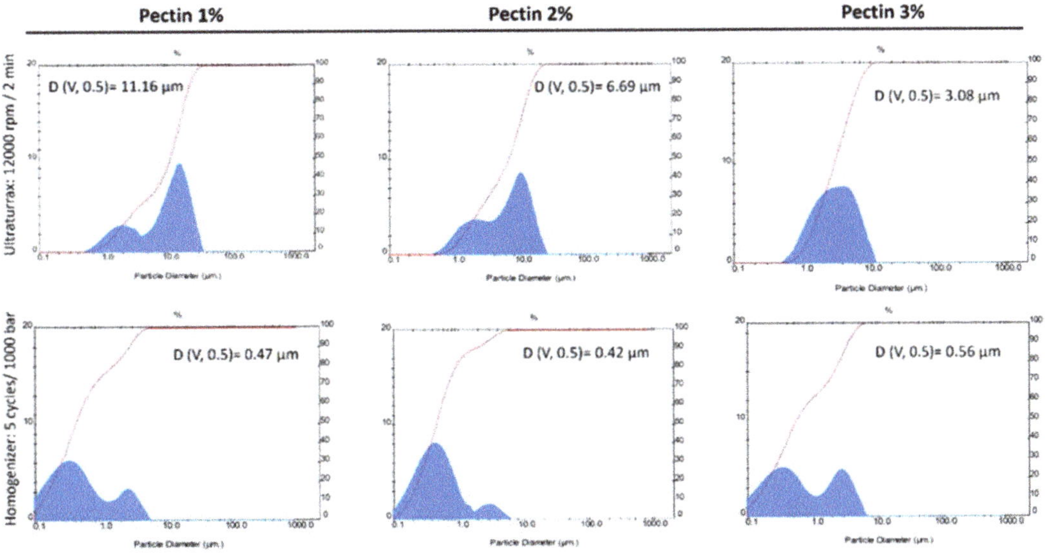

Figure 2. Particle distribution of coconut oil (MCT) emulsions formulated with high-speed homogenization at 12,000 rpm for 2 min (upper row) with 1%, 2%, and 3% pectin; the same emulsions after treatment with high-pressure homogenization (HPH) at 1000 bar (100 MPa) for 5 cycles (lower row).

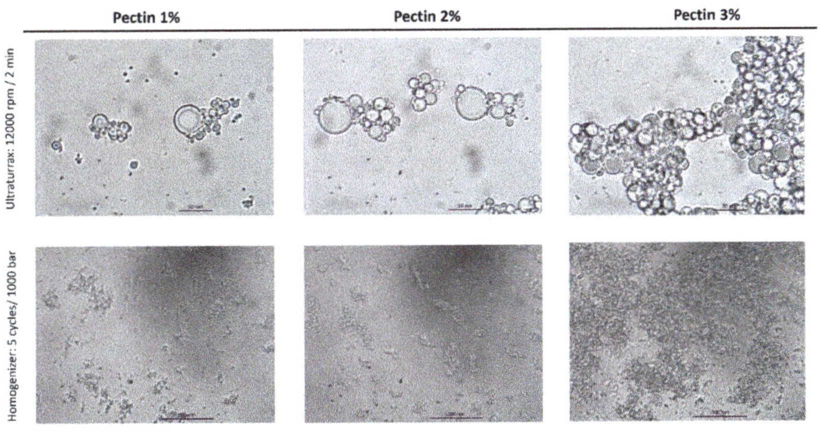

Figure 3. Optical microscopy of MCT (coconut oil) emulsions formulated using high-speed homogenization (Ultraturrax) at 12,000 rpm for 2 min (upper row) with 1%, 2%, and 3% pectin; the same emulsions after treatment with high-pressure homogenization (HPH) at 1000 bar (100 MPa) for 5 cycles (lower row). Scale bars = 50 µm for images in the upper row and 200 µm for images in the lower row.

During emulsification, only a small proportion of the pectin actually adsorbs on oil droplets to form a protective layer, which decreases the oil/water interfacial tension, reducing the emulsion droplet size, as observed in the present study. In the literature, it is reported that pectin, a plant cell wall polysaccharide, is a natural multifunctional ingredient that imparts textural and rheological properties to a wide range of food systems. The emulsion-stabilizing properties of pectin are controlled by the homogalacturonan (HG) domain and the neutral sugar side chains of the rhamnogalacturonan-I (RGI) structural element. However, the neutral sugar side chains might obstruct the accessibility of pectin hydrophobic species to the oil/water interface, hampering emulsification. In addition, the contribution of HG to emulsion stabilization might depend on the polymer HG:RGI ratio. Hence, the influence of the structural features of pectin on the polymer-emulsifying potential is yet to be fully unraveled, as identified in a published review [25]. The presence of Tween 20 together with pectin in the O/W emulsions may help reduce the particle size, since this surfactant has a lower molecular weight than pectin and could shift faster toward the interface in a competitive manner that may lead to small particle sizes, as reported previously [10]. The viscosity of the sunflower oil emulsion was significantly ($p < 0.05$) higher (6.2 ± 0.1 mPa.s) than that of the soybean oil emulsion (5.3 ± 0.0 mPa.s). This difference could enhance the reduction in the particle size by increasing the disruptive shear stress [26]. However, other authors [13] reported a higher viscosity, of 19.77 ± 0.26 mPa.s, when an O/W emulsion was formulated with corn oil and 2% of pectin to encapsulate standard β-carotene. This difference could be due to the different vegetable oil composition in fatty acids and to the technology employed to obtain the nanoemulsion, because they used microfluidizer equipment and in the present study, a high-pressure homogenizator was employed. These two pieces of equipment have different technical characteristics, and it is the interaction chamber that could influence the reduction in particle sizes in O/W emulsions.

Table 3. Particle, size, viscosity, ζ-potential, and bioaccessibility of soybean oil and sunflower oil emulsions with a selected pectin concentration of 2%.

Parameter	O/W Soybean Emulsion	O/W Sunflower Emulsion
Particle diameter D (v, 0.5)	0.6 ± 0.0 ^{a}	0.4 ± 0.0 ^{b}
Viscosity (mPa.s)	5.3 ± 0.0 ^{a}	6.2 ± 0.1 ^{b}
ζ-potential (mV)	−25.7 ± 1.0 ^{a}	−26.5 ± 1.3 ^{a}
Encapsulation efficiency (%)	91.0 ± 0.01 ^{a}	92.2 ± 0.0 ^{a}

The results are expressed as the mean ± standard deviation. Lowercase superscript letters indicate statistically significant differences ($p < 0.05$) between emulsions.

3.2.2. Electrical Charge

The electrical charge of the soybean emulsion was −25.7 ± 1.0 mV and that of the sunflower emulsion was −26.5 ± 1.3 mV, which are less negative values than the values before the homogenization process (−29.1 ± 1.6 mV for coarse soybean emulsion and −35.7 ± 2.9 mV for coarse sunflower emulsion) (Figure 5). These emulsions were formulated with Tween 20, which is a non-ionic surfactant. Therefore, it is expected that there will be no negative charge. Thus, the negative charges may be related to the preferential absorption of OH- species from water to the oil–water interface [26].

3.2.3. Encapsulation Efficiency

Table 3 presents the encapsulation efficiencies obtained in the present study.

These values were calculated using the total carotenoid content analyzed from the O/W microemulsions related to the content of each carotenoid in the papaya peel extract. The encapsulation efficiency was 91–92% for both O/W microemulsions. These data indicate that the formulations were effective in encapsulating the direct extract of papaya peel carotenoids. Relating to the main individual papaya carotenoid encapsulation values, a range between 85 and 97% was observed, with higher values for the hydrocarbon carotenoids, i.e., (*all-E*)-β-carotene (95%) and (*all-E*)-lycopene (97%), and lower values for the xanthophylls, i.e., (*all*-E)-β-cryptoxanthin (85%) and (*all-E*)-β-cryptoxanthin laurate (88%). In the literature, there are no data on the encapsulation efficiency of individual carotenoids from encapsulated direct papaya extracts, but in published studies about the encapsulation of individual carotenoid standards or of those isolated from fruit by-products, the encapsulation efficiencies reported range from 76.5% for β-lycopene extract encapsulated in gelatin and poly(γ-glutamic acid) as carriers [27] to 51.7% for β-carotene encapsulated in lipid microparticles stabilized with hydrolyzed soy protein isolate [28]. In the present work, the O/W microemulsions were efficient in encapsulating the main carotenoids from the papaya by-product (peel).

3.3. Behavior of the Papaya Carotenoid O/W Emulsions during In Vitro Gastrointestinal Digestion

3.3.1. Particle Size

Figures 4 and 5 show the mean particle size and particle size distribution, respectively, in the phases of in vitro gastrointestinal digestion. In the oral phase, no significant differences were observed in the mean particle sizes and particle size distribution between the O/W sunflower and soybean emulsions. In the gastric and intestinal phases, the particle size distribution became more polydisperse, indicating that many large particles were formed. This could be due to the use of Tween 20, which has been reported to provide greater stability in in vitro digestion compared to other surfactants, such as proteins or lecithin, which are less effective against flocculation [28]. Optical microscopy images of the in vitro gastrointestinal phases (Figure 5) showed flocculence and coalescence between particles, mainly in the gastric phases of both types of studied O/W soybean and sunflower emulsions, which could be the reason why the mean particle size increased significantly during these phases. Figure S4 (Supplementary Materials) shows the images taken with confocal microscopy of the (a) soybean and (b) sunflower final emulsions after they were processed through high-pressure homogenization (HPH) at 100 MPa for five cycles, where different distributions can be observed. In addition, the pectin present in the aqueous phase could promote this aggregation between molecules, generating gel-like pectin, namely Ca^{2+} cross-linkings with calcium ions present in the gastric fluids, which are clustering oil droplets [13]. In the intestinal phase, these oil droplets are digested by the intestinal enzymes. The noticeable increase in particle size at this stage could be related to the formation of mixed micelles. Figures 4 and 6 show the changes in the mean particle size of the O/W soybean and sunflower final emulsions in different phases of the in vitro gastrointestinal digestion. There could be some undigested oil droplets and some coalescence between molecules, since the interface stabilizers during the intestinal phase may be altered, losing their efficiency [29]. Images taken with a confocal microscope of the (a) soybean and (b) sunflower final emulsions after they were processed through high-pressure homogenization (HPH) at 100 MPa for five cycles are displayed in Figure S4 (Supplementary Materials).

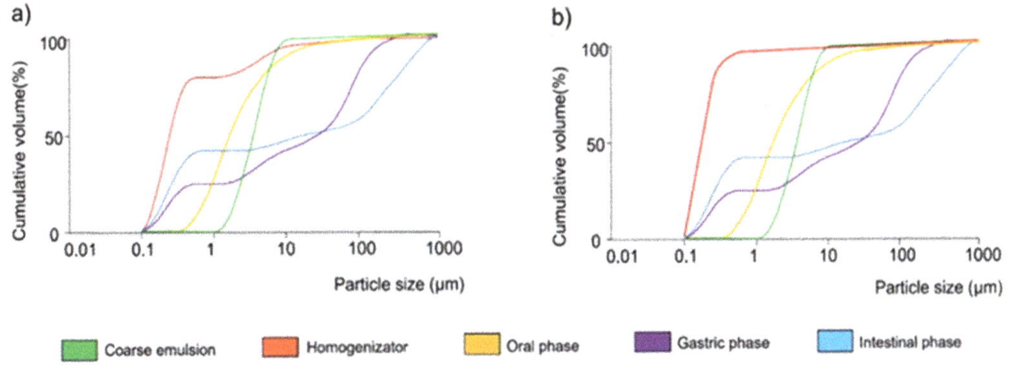

Figure 4. Particle size distribution of (**a**) soybean emulsions (coarse emulsion and HPH-treated emulsion) and (**b**) sunflower emulsions in different phases of the in vitro gastrointestinal digestion.

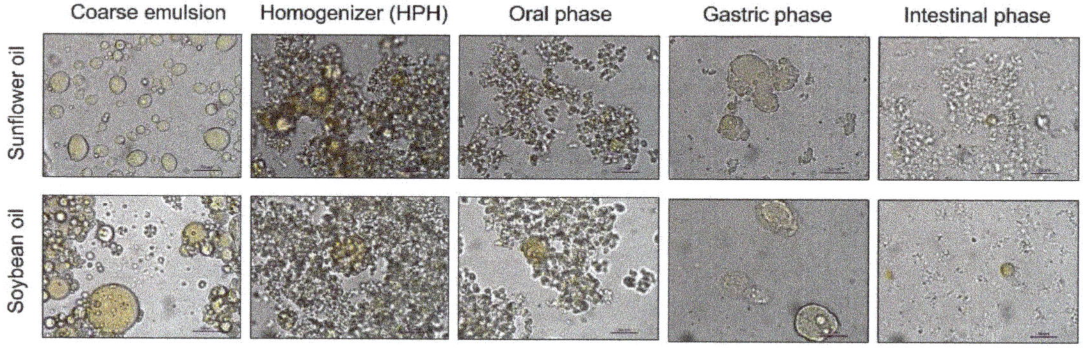

Figure 5. Images of the coarse emulsions and final emulsions in different phases of the in vitro digestion. Scale bar = 50 μm.

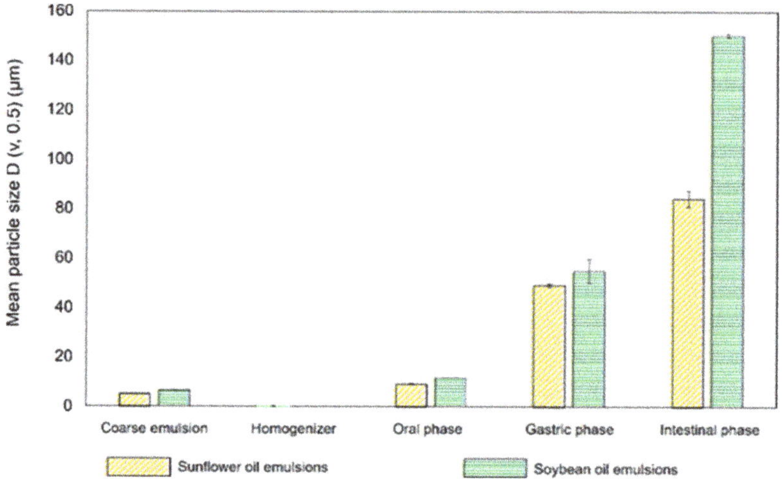

Figure 6. Mean particle size of O/W soybean and sunflower final emulsions (coarse emulsion and HPH-treated emulsion) in different phases of the in vitro gastrointestinal digestion.

3.3.2. Electrical Charge

During the oral and gastric phases, the ζ-potential remained constant for both the O/W emulsions (sunflower and soybean) but less negative than that for the coarse and the final emulsions before digestion (Figure 7). In the gastric phase, the presence of positive ions (Ca^{2+}, K^+, and H^+) could be responsible for the significant decrease in the negative electrical charge values. In addition, since most of the pectin carboxyl groups are protonated in the gastric phase, they cannot interact with the surrounding ions. In the intestinal phase, for both the O/W carotenoid soybean and sunflower emulsions, the negativity in the electrical charge increased almost to the same level as that for the coarse and final emulsions before digestion.

As Chang and McClements [30] reported, since the pH is neutral in the intestinal phase, the lipid droplets and the pectin present produced negative charges, creating repulsion between them, hindering the adsorption of the pectin on the interface. However, due to the high presence of negative molecules in this phase (e.g., bile salts), this effect was covered and was difficult to measure.

3.3.3. Carotenoid Stability and Bioaccessibility in O/W Microemulsions

The stability and bioaccessibility of the most abundant carotenoids ((*all-E*)-β-cryptoxanthin, (*all-E*)-β-cryptoxanthin laurate, (*all-E*)-β-carotene, and (*all-E*)- lycopene) found in the O/W papaya by-product carotenoid emulsions are displayed in Table 4.

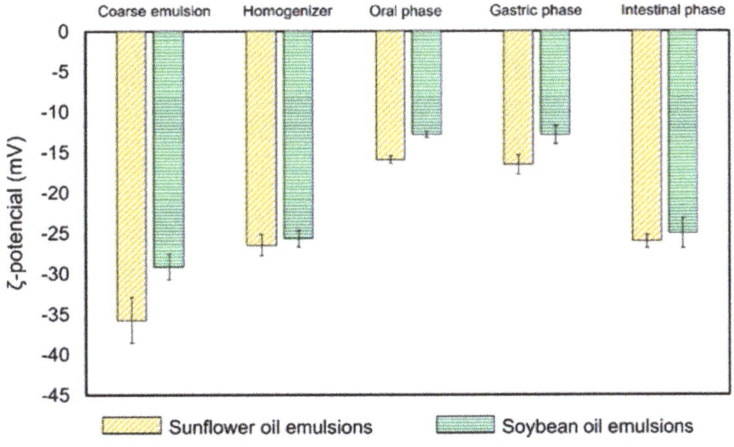

Figure 7. ζ-potential values of O/W carotenoid coarse and final emulsions in different phases of the in vitro gastrointestinal digestion.

Table 4. Major carotenoid stability (μg/g emulsion) and bioaccessibility (%) during oral phase (O), gastric phase (G), and intestinal phase (I) of in vitro gastrointestinal digestion of soybean and sunflower emulsions.

Carotenoid	O/W Soybean Emulsion					O/W Sunflower Emulsion				
	Stability (μg/g Emulsion) during In Vitro Digestion					Stability (μg/g Emulsion) during In Vitro Digestion				
	Oil Extract	Oral	Gastric	Intestinal	Bioaccessibility (%)	Oil Extract	Oral	Gastric	Intestinal	Bioaccessibility (%)
(*all-E*)-β-Cryptoxanthin	0.34 ± 0.0 ^{b}	0.47 ± 0.0 ^{c}	0.26 ± 0.0 ^{b}	0.12 ± 0.0 ^{a}	15.4 ± 0.4 ^{A}	0.23 ± 0.0 ^{a}	0.76 ± 0.0 ^{b}	0.56 ± 0.0 ^{c}	0.33 ± 0.0 ^{b}	15.0 ± 0.2 ^{A}
(*all-E*)-β-Cryptoxanthin laurate	2.60 ± 0.1 ^{b}	2.79 ± 0.1 ^{b}	2.28 ± 0.0 ^{b}	0.82 ± 0.0 ^{a}	3.5 ± 0.1 ^{A}	1.92 ± 0.0 ^{a}	2.36 ± 0.1 ^{b}	2.99 ± 0.3 ^{b}	0.98 ± 0.0 ^{a}	7.2 ± 0.9 ^{B}
(*all-E*)-β-Carotene	0.82 ± 0.0 ^{a}	1.19 ± 0.0 ^{b}	1.56 ± 0.0 ^{b}	1.57 ± 0.0 ^{b}	17.9 ± 0.7 ^{A}	0.77 ± 0.0 ^{a}	1.37 ± 0.0 ^{b}	0.77 ± 0.0 ^{a}	0.19 ± 0.1 ^{a}	18.4 ± 0.4 ^{A}
(*all-E*)-β-Lycopene	3.82 ± 0.3 ^{b}	7.14 ± 0.1 ^{c}	1.58 ± 0.0 ^{b}	1.58 ± 0.0 ^{b}	64.1 ± 3.7 ^{A}	3.80 ± 0.2 ^{b}	8.05 ± 0.1 ^{c}	3.75 ± 0.0 ^{b}	0.94 ± 0.0 ^{a}	71.4 ± 3.9 ^{B}

The results are expressed as the mean ± standard deviation. Lowercase superscript letters indicate statistically significant differences ($p < 0.05$) in the carotenoid content between emulsions, and uppercase superscript letters indicate statistically significant differences ($p < 0.05$) between the emulsion bioaccessibility values of the carotenoids.

Figure S1 (Supplementary Materials) shows the HPLC chromatograms of the carotenoids in the digesta fractions (carotenoids in each phase of the digestion) of all carotenoids (hydrocarbon carotenoids, xanthophylls, and xanthophyll esters) in the vegetable oil extracts, in the O/W formulated emulsions, and in the carotenoid O/W emulsions before the in vitro gastrointestinal digestion assay and during the in vitro digestion. The content of each main carotenoid in the digesta fraction provides information about the stability of the carotenoids during in vitro gastrointestinal digestion. Figure S2 (Supplementary Materials) shows the carotenoid compounds identified during the in vitro digestion of O/W soybean and sunflower emulsions in the micellar fractions (carotenoids that were in the micellar fraction of each digestion phase). The content of each carotenoid in the micellar fraction of the intestinal phase is the value used for calculating the bioaccessibility of carotenoids (see the Materials and Methods section). Table S2 shows the characterization of the carotenoids identified in the extracts of papaya by-product (peel) as reported by Lara-Abia et al. [7].

Regarding the obtained results, the bioaccessibility of (*all-E*)-β-cryptoxanthin in the soybean (15%) and sunflower (15%) O/W emulsions was the same (Table 4). However, significant differences ($p < 0.05$) were observed for its major xanthophyll ester, (*all*-E)-β-cryptoxanthin laurate, showing higher bioaccessibility in the O/W sunflower emulsion (7.2 ± 0.9%) than in the O/W soybean emulsion (3.5 ± 0.1%). With respect to (*all*-E)-β-carotene, higher bioaccessibility was observed in both the O/W soybean (17.9 ± 0.7%) and sunflower (18.4 ± 0.4%) emulsions compared to its bioaccessibility in a previous study of the in vitro digestion of un-encapsulated extracts [7]. (*all*-E)-Lycopene showed the highest bioaccessibility among all the carotenoids of the extracts of papaya encapsulated by O/W emulsions. Bioaccessibility values of 64 ± 4% for the O/W soybean and of 71 ± 4% for the sunflower emulsions were obtained for (*all*-E)-lycopene, possibly because this hydrocarbon carotenoid is highly soluble in the vegetable oils employed to formulate the O/W emulsions. The same reason may be applicable to the high (all-*E*)-β-carotene bioaccessibility, since hydrocarbon carotenoids are more hydrophobic than xanthophylls and xanthophyll esters. Zhao et al. [31] reported lower (all-*E*)-lycopene bioaccessibility (around 25%) in sesame oil and linseed oil nanoemulsions and bioaccessibility of 18% in walnut oil nanoemulsions when lactoferrin was used as an emulsifier, showing a similar trend with their stability. In addition, the results obtained in the present study showed higher bioaccessibility values for carotenoids in the O/W sunflower emulsion than for those in the O/W soybean emulsion (Table 4), possibly because the O/W sunflower emulsion presented smaller particle size (higher surface area), which could improve oil digestibility and the transfer of carotenoids to micelles during the gastrointestinal digestion (intestinal) phase.

The formation of pectin gels during the gastric phase may have a protective effect on carotenoids, enhancing the bioaccessibility of the carotenoids liberated from the O/W emulsions, facilitating their transport toward the intestinal phase. Other factors, such as the degree of oil unsaturation of the fatty components of vegetable oils and the carotenoid nature, affect their micellarization and bioaccessibility [11]. In this sense, monounsaturated fatty acids, such as the oleic acid present in sunflower oil, might be more hydrophobic in comparison to polyunsaturated fatty acids present in soybean oil (mainly composed of linoleic acid and linolenic acid), leading to better solubilization of hydrophobic carotenoids, such as β-cryptoxanthin and β-cryptoxanthin laurate, into the micelles. The stability (bioactive content) of (*all*-E)-β-cryptoxanthin, (*all*-E)-β-cryptoxanthin laurate, (*all*-E)-β-carotene, and (*all*-E)-lycopene during the in vitro gastrointestinal digestion of the soybean and sunflower emulsions is shown in Figure S3 (Supplementary Material), and the carotenoid content is shown in Table 4. A recently published study [3] reported low bioaccessibility of papaya carotenoid from the fruit pulp (edible portion) of the Sweet Mary papaya variety. In this previous study [3], the (*all*-E)-β-cryptoxanthin bioaccessibility value was ≈ 3.4 ± 0.1%, while in the present study, when the papaya extract was encapsulated by O/W emulsions, the bioaccessibility of this carotenoid showed an increase of more than 5 times (15.4%) for this xanthophyll for both formulated O/W emulsions (Table 4). Regarding the bioaccessibility of the most abundant xanthophyll ester of papaya peel extracts, (*all*-E)-β-cryptoxanthin laurate, in the encapsulated extract O/W emulsions, an increase of 3 times was observed with values of 3.5% for the O/W soybean emulsion and 7.2% for the O/W sunflower emulsion. Similar results were observed for the hydrocarbon carotenoids, (*all*-E)-β-carotene and (all-E)-β-lycopene. In general, the encapsulation of papaya by-product extract by O/W emulsions significantly improved the bioaccessibility of the carotenoids, with the greatest bioaccessibility for (*all*-E)-β-lycopene (with bioaccessibility values up to 63.8% in the O/W soybean emulsion and 71.1% in the O/W sunflower emulsion), while the bioaccessibility of this carotenoid in non-encapsulated papaya carotenoid extract was only 0.3%. Regarding (*all*-E)-β-carotene, the obtained bioaccessibility values for the carotenoid papaya extract encapsulated by O/W emulsions showed an improvement of up to 17.3% (soybean oil emulsion) and 17.8% (sunflower oil emulsion), while the bioaccessibility of (*all*-E)-β-carotene from the un-encapsulated papaya extract was only 0.6% [17]. However, the bioaccessibility values for (*all*-E)-β-carotene encapsulated by O/W emulsions in the present study using a 2% pectin concentration were lower than those obtained by Teixé-Roig et al. [14], who reported a bioaccessibility of ≈36% for encapsulated extract. This difference could be due to the use of different vegetable oils to produce the coarse emulsion: they used corn oil instead of the sunflower and soybean oils that were employed here to encapsulate direct carotenoid extract from the papaya by-product. However, in both studies, a significant improvement was observed in the bioaccessibility of (*all*-E)-β-carotene when O/W microemulsions were used.

4. Conclusions

The process for optimizing the formulation of O/W emulsions using soybean oil, sunflower oil, and coconut oil enriched with carotenoids from a papaya by-product (peel) was studied using high-speed and high-pressure homogenization processes and pectin. The obtained sunflower O/W microemulsion showed smaller mean particle size, higher negative ζ-potential, and higher viscosity than the soybean O/W microemulsion. The coconut O/W microemulsion was not stable and thus was not used for the digestion assays. Pectin could act as a protective barrier in the formulation of O/W microemulsions, increasing the bioaccessibility of encapsulated carotenoids. A reduction in the size of the particles in the sunflower O/W microemulsion favored the bioaccessibility of the studied carotenoids, with (*all*-E)-β-lycopene having higher bioaccessibility (64% and 71% in the soybean and sunflower O/W emulsions, respectively), followed by (all-E)-β-carotene, which showed similar bioaccessibility in the soybean (18%) and sunflower (18%) O/W emulsions. (*all*-E)-β-cryptoxanthin also had similar bioaccessibility for both the soybean (15%) and sunflower (15%) O/W microemulsions (15%), and (*all*-E)-β-cryptoxanthin laurate was the carotenoid that presented the lowest bioaccessibility values: 4% in the soybean emulsion and 7% in the sunflower emulsion. The bioaccessibility of (*all*-E)-β-carotene and (*all*-E)-β-lycopene also increased up to 17.3% and 63.8% in the soybean O/W microemulsions and 17.8% and 71.1% in the sunflower O/W microemulsions, respectively.

The high degree of unsaturation of sunflower oil and soybean oil fatty acids may promote the formation of small droplets and increase the interface flexibility in the emulsions due to unsaturated linkages. However, it is notable that there is scarce information on the influence of the degree of saturation of the vegetable oil employed to formulate the O/W emulsions on the bioaccessibility of carotenoids during gastrointestinal digestion. For this reason, further in vivo investigations are needed to understand the digestive behavior of O/W emulsions containing carotenoids to assess the real bioavailability of the encapsulated carotenoids from direct extracts.

Supplementary Materials: The following supporting information can be downloaded at https://www.mdpi.com/article/10.3390/foods12142654/s1: Figure S1: C30 reversed-phase chromatograms of carotenoids at 450 nm obtained from Sweet Mary papaya (*Carica papaya* L.) peel encapsulated by O/W emulsions in (a) carotenoid-enriched soybean oil (vegetable oil + papaya carotenoid extract (carotenoid μg/g vegetable oil), (b) carotenoid-enriched sunflower oil (vegetable oil + papaya carotenoid extract (μg carotenoid μg/g vegetable oil), (c) O/W soybean emulsion with encapsulated carotenoid extract, (d) O/W sunflower emulsion with encapsulated carotenoid extract, and in the digesta fractions during the in vitro digestion of (e) O/W soybean emulsion and (f) O/W sunflower oil emulsion in the oral phase, (g) O/W soybean emulsion and (h) O/W sunflower oil emulsion in the gastric phase, and (i) O/W soybean emulsion and (j) O/W sunflower oil emulsion in the intestinal phase. Peak identities in Table S1; Figure S2: C30 reversed-phase chromatograms of carotenoids at 450 nm obtained from Sweet Mary papaya (*Carica papaya* L.) peel encapsulated by O/W emulsions in the micellar fractions during the in vitro digestion of (a) soybean microemulsion and (b) sunflower oil microemulsion in the oral phase, (c) soybean microemulsion and (d) sunflower oil microemulsion in the gastric phase, and (e) soybean microemulsion and (f) sunflower oil microemulsion in the intestinal phase. Peak identities in Table S1; Figure S3: Stability of main papaya carotenoids (μg carotenoids/g emulsion) ((all-E)-β-cryptoxanthin, (all-E)-β-carotene, (all-E)-β-cryptoxanthin laurate, and (all-E)-lycopene) in (a) O/W soybean oil and (b) O/W sunflower oil emulsions before and after in vitro gastrointestinal digestion. Non-digested values refer to the content of each carotenoid in the carotenoid-enriched oil extract before encapsulation by O/W emulsions; Figure S4: Images taken with confocal microscope of (a) soybean and (b) sunflower final emulsions after being processed through high-pressure homogenization (HPH) at 100 MPa for 5 cycles. Scale bars are 5 μm long. Table S1: Percentage (%) fatty acid composition of edible vegetable oils (Dubois et al., 2007) [18]; Table S2: Chromatographic identification [a] of carotenoids and carotenoid esters from Sweet Mary papaya (*Carica papaya* L.) peel to formulate soybean oil and sunflower oil carotenoid-enriched emulsions. Table S3: Content of individual carotenoids and carotenoid esters in Sweet Mary papaya (*Carica papaya* L.) peel extracts.

Author Contributions: Conceptualization, M.P.C. and J.W.-C.; methodology, S.L.-A., M.P.C. and G.L.; validation, S.L.-A. and M.P.C.; formal analysis, S.L.-A. and N.P.-P.; investigation, S.L.-A., M.P.C. and J.W.-C.; resources, M.P.C. and J.W.-C.; data curation, S.L.-A. and M.P.C.; writing—original draft preparation, S.L.-A. and M.P.C.; writing—review and editing, M.P.C. and G.L.; supervision, M.P.C. and J.W.-C.; project administration, M.P.C. and J.W.-C.; funding acquisition, M.P.C. and J.W.-C. All authors have read and agreed to the published version of the manuscript.

Funding: This research was funded by the Grant PID2020-118300RB-C21 funded by MCIN/AEI/10.13039/501100011033. Spanish Ministry of Science and Innovation and Tecnologico of Monterrey (Mexico) (FunFoodEmertec) (Research Chair Funds CAT-200).

Data Availability Statement: Data is contained within the article or Supplementary Material.

Acknowledgments: S.L.-A. thanks CONACyT (Mexico) for her doctoral scholarship no. 895077.

Conflicts of Interest: The authors declare no conflict of interest.

References

1. Sancho, L.E.G.G.; Yahia, E.M.; González-Aguilar, G.A. Identification and quantification of phenols, carotenoids, and vitamin C from papaya (*Carica papaya* L., cv. Maradol) fruit determined by HPLC-DAD-MS/MS-ESI. *Food Res. Intern.* **2011**, *44*, 1284–1291. [CrossRef]
2. Schweiggert, R.M.; Steingass, C.B.; Heller, A.; Esquivel, P.; Carle, R. Characterization of chromoplasts and carotenoids of red-and yellow-fleshed papaya (*Carica papaya* L.). *Planta* **2011**, *234*, 1031–1044. [CrossRef]
3. Lara-Abia, S.; Lobo-Rodrigo, G.; Welti-Chanes, J.; Cano, M.P. Carotenoid and carotenoid ester profile and their deposition in plastids in fruits of new papaya (*Carica papaya* L.) varieties from the Canary Islands. *Foods* **2021**, *10*, 434. [CrossRef]
4. Boon, C.S.; McClements, D.J.; Weiss, J.; Decker, E.A. Factors influencing the chemical stability of carotenoids in foods. *Crit. Rev. Food Sci. Nutr.* **2010**, *50*, 515–532. [CrossRef] [PubMed]
5. Atencio, S.; Verkempinck, S.H.; Reineke, K.; Hendrickx, M.; Van Loey, A. Heat and Light Stability of Pumpkin-Based Carotenoids in a Photosensitive Food: A Carotenoid-Coloured Beverage. *Foods* **2022**, *11*, 485. [CrossRef] [PubMed]
6. Bohn, T. Metabolic fate of bioaccessible and non-bioaccessible carotenoids. Non-Extractable Polyphenols and Carotenoids: Importance in Human Nutrition and Health. In *Book Non-Extractable Polyphenols and Carotenoids: Importance in Human Nutrition and Health*; Saura-Calixto, F., Pérez-Jiménez, J., Eds.; Royal Society of Chemistry: London, UK, 2018; pp. 165–200.
7. Lara-Abia, S.; Welti-Chanes, J.; Cano, M.P. Effect of High Hydrostatic Pressure on the Extractability and Bioaccessibility of Carotenoids and Their Esters from Papaya (*Carica papaya* L.) and Its Impact on Tissue Microstructure. *Foods* **2021**, *10*, 2435. [CrossRef]
8. Tan, C.; McClements, D.J. Application of Advanced Emulsion Technology in the Food Industry: A Review and Critical Evaluation. *Foods* **2021**, *10*, 812. [CrossRef]
9. Xavier, A.A.O.; Mercadante, A.Z.; Garrido-Fernandez, J.; Perez-Galvez, A. Fat content affects bioaccessibility and efficiency of enzymatic hydrolysis of lutein esters added to milk and yogurt. *Food Res. Intern.* **2014**, *65*, 171–176. [CrossRef]
10. García-Cayuela, T.; Nuño-Escobar, B.; Welti-Chanes, J.; Cano, M.P. In Vitro bioaccessibility of individual carotenoids from persimmon (*Diospyros kaki*, cv. Rojo Brillante) used as an ingredient in a model diary food. *J. Sci. Food Agric.* **2018**, *98*, 3246–3254. [CrossRef]
11. Verkempinck, S.H.E.; Salvia-Trujillo, L.; Moens, L.G.; Carrillo, C.; Van Loey, A.M.; Hendrickx, M.E.; Grauwet, T. Kinetic approach to study the relation between In Vitro lipid digestion and carotenoid bioaccessibility in emulsions with different oil unsaturation degree. *J. Func. Foods* **2018**, *41*, 135–147. [CrossRef]
12. Yuan, X.; Liu, X.; McClements, D.J.; Cao, Y.; Xiao, H. Enhancement of phytochemical bioaccessibility from plant-based foods using excipient emulsions: Impact of lipid type on carotenoid solubilization from spinach. *Food Funct.* **2018**, *9*, 4352–4365. [CrossRef] [PubMed]
13. Cervantes-Paz, B.; de Jesús Ornelas-Paz, J.; Ruiz-Cruz, S.; Rios-Velasco, C.; Ibarra-Junquera, V.; Yahia, E.M.; Gardea-Béjar, A.A. Effects of pectin on lipid digestion and possible implications for carotenoid bioavailability during pre-absorptive stages: A review. *Food Res. Intern.* **2017**, *99*, 917–927. [CrossRef] [PubMed]

14. Anderson, J.W.; Baird, P.; Davis, R.H.; Ferreri, S.; Knudtson, M.; Koraym, A.; Williams, C.L. Health benefits of dietary fiber. *Nutr. Rev.* **2009**, *67*, 188–205. [CrossRef] [PubMed]
15. Teixé-Roig, J.; Oms-Oliu, G.; Ballesté-Muñoz, S.; Odriozola-Serrano, I.; Martín-Belloso, O. Improving the In Vitro Bioaccessibility of β-Carotene Using Pectin Added Nanoemulsions. *Foods* **2020**, *9*, 447. [CrossRef]
16. Jain, S.; Winuprasith, T.; Suphantharika, M. Encapsulation of lycopene in emulsions and hydrogel beads using dual modified rice starch: Characterization, stability analysis and release behaviour during in-vitro digestion. *Food Hydroc.* **2020**, *104*, 105730. [CrossRef]
17. Zhao, C.; Wei, L.; Yin, B.; Liu, F.; Li, J.; Liu, X.; Wang, J.; Wang, Y. Encapsulation of lycopene within oil-in-water nanoemulsions using lactoferrin: Impact of carrier oils on physicochemical stability and bioaccessibility. *Inter. J. Biol. Macrom.* **2020**, *153*, 912–920. [CrossRef]
18. Dubois, V.; Breton, S.; Linder, M.; Fanni, J.; Parmentier, M. Fatty acid profiles of 80 vegetable oils with regard to their nutritional potential. *Eur. J. Lipid Sci. Technol.* **2007**, *109*, 710–732. [CrossRef]
19. Cano, M.P.; Gómez-Maqueo, A.; Fernández-López, R.; Welti-Chanes, J.; García-Cayuela, T. Impact of high hydrostatic pressure and thermal treatment on the stability and bioaccessibility of carotenoid and carotenoid esters in astringent persimmon (*Diospyros kaki Thunb*, var. Rojo Brillante). *Food Res. Intern.* **2019**, *123*, 538–549. [CrossRef]
20. Jaeger de Carvalho, L.M.; Barros Gomes, P.; Luiz de Oliveira Godoy, R.; Pacheco, S.; Fernandes do Monte, P.H.; Viana de Carvalho, J.L.; Regini Nutti, M.; Lima Neves, A.C.; Rodrigues Alves Vieira, A.C.; Ramalho Ramos, S.R. Total carotenoid content, α-carotene and β-carotene, of landrace pumpkins (*Cucurbita moschata* Duch): A preliminary study. *Food Res. Intern.* **2011**, *47*, 337–340. [CrossRef]
21. Minekus, M.; Alminger, M.; Alvito, P.; Ballance, S.; Bohn, T.; Bourlieu, C.; Brodkorb, A. A standardised static in vitro digestion method suitable for food–an international consensus. *Food Funct.* **2014**, *5*, 1113–1124. [CrossRef]
22. Eriksen, J.N.; Luu, A.Y.; Dragsted, L.O.; Arrigoni, E. Adaption of an in vitro digestion method to screen carotenoid liberation and In Vitro accessibility from differently processed spinach preparations. *Food Chem.* **2017**, *224*, 407–413. [CrossRef] [PubMed]
23. Petry, F.C.; Mercadante, A.Z. Impact of in vitro digestion phases on the stability and bioaccessibility of carotenoids and their esters in mandarin pulps. *Food Funct.* **2017**, *8*, 3951–3963. [CrossRef] [PubMed]
24. Liu, C.; Zheng, Z.; Cao, C.; Liu, Y. The partial coalescence behavior of oil-in-water emulsions: Comparison between refrigerated and room temperature storage. *Food Chem.* **2019**, *300*, 125219. [CrossRef] [PubMed]
25. Ngouémazong, E.D.; Christiaens, S.; Shpigelman, A.; Van Loey, A.; Hendrickx, M. The emulsifying and emulsion-stabilizing properties of pectin: A review. *Compreh. Rev. Food Sci. Food Saf.* **2015**, *14*, 705–718. [CrossRef]
26. Maa, Y.F.; Hsu, C.C. Performance of sonication and microfluidization for liquid–liquid emulsification. *Pharmac. Develop. Techn.* **1999**, *4*, 233–240. [CrossRef]
27. Qian, C.; McClements, D.J. Formation of nanoemulsions stabilized by model food-grade emulsifiers using high-pressure homogenization: Factors affecting particle size. *Food Hydroc.* **2011**, *25*, 1000–1008. [CrossRef]
28. Mun, S.; Decker, E.A.; McClements, D.J. Influence of emulsifier type on In Vitro digestibility of lipid droplets by pancreatic lipase. *Food Res. Intern.* **2007**, *40*, 770–781. [CrossRef]
29. Chiu, Y.T.; Chiu, C.P.; Chien, J.T.; Ho, G.H.; Yang, J.; Chen, B.H. Encapsulation of Lycopene Extract from Tomato Pulp Waste with Gelatin and Poly(γ-glutamic acid) as Carrier. *J. Agric. Food Chem.* **2007**, *55*, 5123–5130. [CrossRef]
30. Brito-Oliveira, T.C.; Molina, C.V.; Netto, F.M.; Pinho, S.C. Encapsulation of Beta-carotene in Lipid Microparticles Stabilized with Hydrolyzed Soy Protein Isolate: Production Parameters, Alpha-tocopherol Coencapsulation and Stability Under Stress Conditions. *J. Food Sci.* **2017**, *82*, 659–669. [CrossRef]
31. Chang, Y.; McClements, D.J. Influence of emulsifier type on the in vitro digestion of fish oil-in-water emulsions in the presence of an anionic marine polysaccharide (fucoidan): Caseinate, whey protein, lecithin, or Tween 80. *Food Hydroc.* **2016**, *61*, 92–101. [CrossRef]

foods

MDPI

Article

Experimental Investigation and Modeling for the Influence of Adding Date Press Cake on Drinkable Yogurt Quality

Nashi K. Alqahtani [1,2], Tareq M. Alnemr [2,3], Abdullah K. Alsalem [2], Mutlaq M. Alotaibi [2] and Maged Mohammed [1,4,*]

1 Date Palm Research Center of Excellence, King Faisal University, Al-Ahsa 36362, Saudi Arabia; nalqahtani@kfu.edu.sa
2 Department of Food and Nutrition Sciences, College of Agricultural and Food Sciences, King Faisal University, P.O. Box 400, Al-Ahsa 31982, Saudi Arabia; talnemr@kfu.edu.sa
3 Department of Dairy Science and Technology, Faculty of Agriculture (El-Shatby), University of Alexandria, Alexandria 21545, Egypt
4 Department of Agricultural and Biosystems Engineering, Faculty of Agriculture, Menoufia University, Shebin El Koum 32514, Egypt
* Correspondence: memohammed@kfu.edu.sa

Abstract: The extraction of date syrup produces a large quantity of by-product known as date press cake (DPC). This study aimed to utilize valuable ingredients of the DPC by adding 0 (Control), 2, 4, and 6% (g/100 g) of its powder to drinkable yogurt before fermentation. The physicochemical properties, texture profile, and sensory evaluation of the treated DPC-based drinkable yogurt (DPC drinkable yogurt) were measured after fermentation and 5, 10, and 15 days of storage at 4 °C. The modeling of the most critical quality attributes, i.e., pH, acidity, syneresis, water holding capacity (WHC), viscosity, and color difference (ΔE), was conducted to predict their values based on the DPC percentage and storage period. The DPC drinkable yogurt's total solids, protein, and fat ranged between 11.19–11.83, 3.10–3.42, and 2.26–2.34%, respectively. Adding 2–6% DPC slightly increased the pH of DPC drinkable yogurt and decreased its acidity ($p > 0.05$) during storage. Increasing the DPC percent in DPC drinkable yogurt decreased the syneresis value, and WHC increased during storage. The color parameters and viscosity of DPC drinkable yogurt recorded the highest value at the end of the storage period for all treatments and increased steadily with the increase in DPC. The evaluation of the prediction models indicated that the predicted values were close to the actual experimental values for pH ($R^2 = 0.779$), acidity ($R^2 = 0.973$), syneresis ($R^2 = 0.961$), WHC ($R^2 = 0.989$), viscosity ($R^2 = 0.99$), L* ($R^2 = 0.919$), a* ($R^2 = 0.995$), b* ($R^2 = 0.922$), and ΔE ($R^2 = 0.921$). The textural analysis indicated that increasing the concentration of DPC in the DPC drinkable yogurt increased hardness (g), springiness, cohesiveness, and gumminess and decreased adhesiveness and resilience during cold storage. The evaluation of sensory acceptance during the cold storage of the DPC drinkable yogurt was conducted by 30 expert panelists. Each panelist received four cups of 10 mL drinkable yogurt treatments at 5–10 °C. The evaluation results indicated that adding 2% of DPC was closest in overall sensory acceptability to the control sample ($p < 0.05$). This study revealed the potential use of DPC in drinkable yogurt as a natural, functional, and low-cost ingredient to improve the fiber content, physicochemical properties, and overall acceptability. Therefore, the fermented DPC-based yogurt drink has the potency to be a practical, value-added, and novel alternative to dairy-based yogurt.

Keywords: dairy alternatives; design expert; food quality; texture profile analysis; sensory evaluation; prediction model; food waste upgrading

Citation: Alqahtani, N.K.; Alnemr, T.M.; Alsalem, A.K.; Alotaibi, M.M.; Mohammed, M. Experimental Investigation and Modeling for the Influence of Adding Date Press Cake on Drinkable Yogurt Quality. *Foods* **2023**, *12*, 1219. https://doi.org/10.3390/foods12061219

Academic Editors: Stanisław Kowalski and Dorota Gumul

Received: 8 February 2023
Revised: 7 March 2023
Accepted: 9 March 2023
Published: 13 March 2023

1. Introduction

Food functionalization is one of the ever-growing markets, which require new bioactive ingredients. In addition, the bioactive components can be used to develop innovative functional products [1]. However, with the limited natural sources of functional compounds to achieve this effect, most productive research studies have turned to natural alternatives to maintain the increasing demand for the ingredients added to those foods. Therefore, the world is now turning to the exploitation of biological agricultural wastes, which constitute a great environmental burden [2]. However, agricultural waste ingredients contain significant active functional additives, such as dietary fiber, sugars, proteins, phenols, antioxidants, anti-carcinogens, pressure reducers, and cholesterol. In addition, these products are produced without additional production costs [2,3].

Furthermore, recent research studies have shown that agricultural by-products or agricultural wastes and their extracts can be successfully incorporated as functional ingredients for developing new food products [4–6]. For example, the extraction of date syrup (Dips) produces large quantities of by-products known as date press cake (DPC), which contain high-fiber waste retained after filtering the Dips. DPC is a major waste and an environmental burden due to its high content of sugars and moisture. DPC is regarded as one of the functional raw materials, which can be developed and used as an added value to produce healthy foods [7]. The dips industry generates about 17–28% of DPC, often used in animal feeds as a filler [8] or disposed of in burial or sewage, which constitutes environmental pollution [9]. The lack of optimal use of that by-product may be due to the lack of chemical composition, health benefits, and knowledge of its nutritional applications when used in foods. It was only previously used as a raw material in the production of active carbon or feed ingredients [8]. Research studies indicated that DPC of the Khalas variety of date fruit contains the essential constituents (7.16% moisture, 2.82 ash, 2.65% protein, 0.81% fat, 86.6% carbohydrates, 8.5% fructose, 5.6% glucose), and soluble fiber content was 6.5%. The insoluble fiber was 49.99%, while DPC per 100 g contained 194.39 mg calcium, 10.73 mg iron, 443.38 mg potassium, and 97.41 mg magnesium. Its water absorption was 1.98 g/g, and fat constituted 0.67 g/g. The emulsifying capacity was 56.17. The emulsifying stability was 71.46% [9,10].

The consumer orientation emphasizes healthy foods, which contain functional products. Dairy products are a fertile field in terms of the possibility of developing them by integrating many available added-value ingredients with milk products, such as yogurt. Yogurt is already considered a healthy food because it contains viable bacteria but does not contain dietary fiber compounds [11,12]. Yogurt is an exceedingly consumed dairy food globally, renowned for its health benefits, nutritional value, and digestibility [13]. Yogurts are vastly consumed across the globe and are produced by fermenting different types of milk with bacteria, such as *Streptococcus thermophilus, Lactobacillus bulgaricus,* resulting in a product with creamy characteristics and a slightly acidic taste [14]. Yogurt is becoming more vogue due to probiotics, higher digested nutrients, gel-like structure, taste, and mouthfeel [15].

The most important features for the consumer acceptance of yogurts are texture and firmness, related to viscosity and stability. In addition, adding other materials, such as pectin, gelatin, inulin, or dietary fiber, may enhance some sensory acceptance of yogurt and decrease syneresis [16]. Therefore, the possibility of improving and modifying the characteristics of those products to avoid excessive use of additives is achieved for the same purpose. Furthermore, yogurt is one of the most famous milk products fermented with lactic acid bacteria, with a wide production and consumer acceptance of it and its health benefits. Therefore, it was one of the essential milk industries to develop, especially for enhancing it with upgraded agricultural waste ingredients [17]. In addition, enriching yogurt with dietary fiber sources is of growing interest in the creation of functional foods with health benefits, perfecting their functionality, and enhancing their nutritional value [18].

The quality attributes of yogurt that can be controlled with stabilizers or fibers include acidification, sensory aspects, gel structure, and syneresis [19–21]. In addition, such control allows for modifying the viscosity and lowering syneresis [21,22]. The "clean label" trend popular among consumer producers encourages the use of natural plant materials for stabilization purposes, the presence of which in a product does not raise any health controversies or concerns [23]. The results of some studies indicated the possibility of combining yogurt with vegetable fiber residues, such as date fruit, mango peel powder, citrus peel, persimmon and its powder, apple peel powder, tomato pomace, apricot press cake, sweet lupine husks, potato, cranberry fiber, and others [6,24–29]. This study aimed to investigate the effect of DPC powder addition on drinkable yogurt quality attributes before fermentation, model the influence of DPC addition and storage time on the most important physicochemical properties, and evaluate the textural and sensory properties of the fermented DPC-based yogurt drink.

2. Materials and Methods

2.1. Materials

DPC of date fruit (Khalas cv.) was obtained from Aldahaby Dates Factory, Al-Ahsa, Saudi Arabia. The DPC samples were dried under vacuum at 48 °C in an electric vacuum drying oven (LVO-2041P, Daihan Labtech Co., Ltd., Namyangju-si, Gyeonggi-do, Korea) for 72 h. They were then ground and sifted to obtain granules of 250 µm [30].

The drinkable yogurt was manufactured at the Agricultural Research Station pilot plant of King Faisal University. The freeze-dried starter culture used to ferment the yogurt was YC-X11 (Chr. Hansen company, Hørsholm, Denmark), containing *Lactobacillus delbrueckii subsp. Bulgaricus* and *Streptococcus thermophiles*. Cow's milk was prepared in equal quantities for each treatment. The pH and fat of the cow's milk used were 6.76 ± 0.04 and 2.26 ± 0.05 g/100 g, respectively. The DPC powder was added to T_0, T_1, T_2, and T_3 with concentrations (g/100 g) of 0, 2, 4, and 6%, respectively. Then, the pasteurization process was performed at 90 °C for 10 min, followed by the sudden cooling process to 40 °C. According to the starter data sheet, a starter culture (50 units) was added to the previous treatments, then incubated at 42 °C for 3 h (fermentation complete). Finally, the fermentation was stopped by cooling at 4 °C. The measurements were taken immediately after fermentation and after 5, 10, and 15 days of storage at 4 °C [31].

2.2. DPC, Milk, and DPC Drinkable Yogurt Analysis

All measurements were carried out in three replicates for the milk used, DPC, and DPC drinkable yogurt treatments. Total solids (TS), protein, fat, ash, and pH were determined for the milk used for yogurt preparation by Lactoscan Funke-Gerber D-12105, Berlin, Germany. DPC drinkable yogurt treatments were analyzed in terms of total solids, protein, and fat, while DPC was analyzed in terms of moisture, protein, fat, ash, crude fiber, and water binding capacity. The solubility and color parameters were determined according to the standard methods of AOAC [32]. Meanwhile, the moisture, fat, ash, and dietary fibers were estimated by the gravimetric method No. AOAC 934.01. Kieldahl measured protein No. AOAC 976.05, and the obtained values were expressed as total nitrogen multiplied by 6.38 and 6.25 to obtain the total protein content in DPC drinkable yogurt treatments and the DPC. Fully automated grude and detergent fibre analysis (Fibertec^{TM} 8000, FOSS, Hilleroed, Denmark) was used to determine dietary fiber. The total carbohydrates were calculated mathematically (100 − (moisture% + fat% + protein% + ash%). The changes in chemical characteristics were tracked by analyzing the DPC drinkable yogurt treatments immediately after fermentation and cooling and then after 5, 10, and 15 days of storage at 4 °C in terms of each of the following.

2.2.1. pH and Acidity

pH was measured with a Thermo Orion 3 Star pH Benchtop Meter (Fisher Scientific, Instruments, Pittsburgh, PA, USA), which was calibrated with a pH 4.00 and 7.00 buffer solution (Thermo Fisher Scientific, Waltham, MA, USA). Titratable acidity as a percentage expressed as lactic acid was also estimated according to the method described in AOAC [32].

2.2.2. Syneresis, Water Holding Capacity, and Viscosity

Syneresis or whey separation was determined according to the method described by Cichońska et al. [33] using a centrifuge (Hermel-Z233 M-2, Hermle Labortechnik GmbH, Wehingen, Germany). The 40 g samples were mixed and centrifuged at 4 °C and 16,125× *g* (× g is times gravity) for 20 min. The relationship between the centrifuge speed in revolutions per minute (RPM) and relative centrifugal force (RCF) was calculated as follows:

$$g = \left(1.118 \times 10^{-5}\right) R\, S^2 \quad (1)$$

where g is the relative centrifugal force (RCF), R is the rotor radius in centimeters, and S is the centrifuge speed in RPM.

After complete expulsion, the separated serum was poured and weighed. The syneresis values of three replicates for each treatment were calculated as a percentage according to the following equation:

$$S = \frac{m_1}{m_2} \times 100 \quad (2)$$

where S is the syneresis percentage, m_1 is the mass of the separated serum after centrifugation in grams, and m_2 is the initial mass of the yogurt before centrifugation in grams.

Water holding capacity (WHC) values of the yogurt treatments were estimated according to the method described by Feng et al. [34]. First, a yogurt sample (10 g) is centrifuged for the three replicates at 1500 × *g* for 10 min; then, the filtrate is poured, and the precipitate is weighed. Then, *WHC* is calculated according to the following equation:

$$WHC = \frac{m_{ys} - m_s}{m_{ys}} \times 100 \quad (3)$$

where *WHC* is the water binding capacity (%), m_{ys} is the mass of the yogurt sample, and m_s is the mass of the sediment.

The viscosity values were estimated according to the method described by Cichońska et al. [33] using rotary viscometer LV DV-II+Pro (Brookfield Engineering, Middleboro, MA, USA) at 4 °C and a spindle (S64) with a rotation of 100 rpm by applying a constant shear speed (0.05 s^{-1}). The readings were taken in the 15th second of measurement. Three replicates of yogurt treatments were taken in centipoise units (cP).

2.2.3. Color Parameters

The color characteristics of the drinkable yogurt treatments were measured according to the method described by Hunter and Harold [35] using the Hunter Lab color meter (Hunter Associates Laboratory Inc., Reston, VA, USA). The Hunter Lab color meter was calibrated before the measurements with black and white plates, where the reading was taken in (L*, a*, b*). The value of L* indicates the extent of lightness or luminance/darkness, ranging between the value of black and 100 for white. a* value expresses redness/greenness; the positive value of redness falls into negative greenness values. b* expresses yellowness/blueness and the positive values for yellow and negative for blueness. L* (brightness, 100 = white, 0 = black), a* (+, red; −, green), and b* (+, yellow; −, blue). The color difference (ΔE) was calculated based on the International Commission on Illumination (CIE) lab using the following equation:

$$\Delta E = \sqrt{\left(L_2^* - L_1^*\right)^2 + \left(a_2^* - a_1^*\right)^2 + \left(b_2^* - b_1^*\right)^2} \quad (4)$$

where ΔE is the color difference of the fruit, L* is fruit lightness, a* is greenness–redness, b* is blueness–yellowness.

2.2.4. Texture Profile Analysis (TPA)

Texture profile analysis (TPA) features, which include hardness, adhesiveness, springiness, cohesiveness, gumminess, and resilience [36], were assessed by a double stress test using a texture analyzer (model: TA.XTplusC, Stable Micro Systems Ltd., Godalming, UK). Before the TPA analysis, the test samples were left at 10°. A 25 mm diameter perplex cylindrical probe was used to measure the textural profile of the yogurt samples at 10 ± 0.5 °C. The TPA analysis was performed by compressing twice using the probe for 10 mm penetration. In the first stage, the samples were compressed, and the probe's speed was fixed at 5 mm/s during the samples' pretest, compression, and relaxation. The load cell was 5 kg, and the trigger force was 0.1 N. The typical textural profile (force–time) curve was obtained with one complete run. The hardness, adhesiveness, springiness, cohesiveness, gumminess, and resilience of yogurt samples were calculated by the software included with the texture meter used.

2.3. Sensory Evaluation

Sensory attributes were assessed by 30 panelists (10 women and 20 men, aged 20 to 60 years) according to the method described by Wang et al. [37]. The acceptance test was conducted under the illumination of the sensory evaluation room maintained at 25 °C in the Department of Food and Nutritional Sciences, College of Agricultural and Food Sciences, King Faisal University. Sensory evaluations were carried out by 30 professional panelists, including teaching staff of the food science and nutrition department, university dairy pilot plant staff, and selected staff of dairy companies in Al-Ahsa 31982, Saudi Arabia. Each panelist received four cups of 10 mL drinkable yogurt treatments at a temperature of 5–10 °C, a sensory assessment sheet, and a water bottle for mouthwash provided between each sample assessment. A 5-point hedonic scale (1 = I don't like it, and 5 = I really like it) was used. Samples were evaluated based on color, texture, flavor, and overall acceptability [38,39].

2.4. Statistical Analysis

The results were statistically analyzed through the SAS program in four analyses during the storage period (0, 5, 10, and 15 days) of the prepared drinkable yogurt treatments (0, 2, 4, and 6% DPC) following a randomized complete block design. Duncan's multiple range test (MRT) was used to determine the variance between treatments within the 0.05 level of significance, where the averages in the same columns with capital letters denote a significant difference within the level of significance ($p < 0.05$). Design Expert software (DX Version 13, Stat-Ease, Inc., Minneapolis, MN, USA) was used to graphically analyze the experimental data and model the influence of DPC addition and storage time on the physicochemical properties of the DPC drinkable yogurt.

3. Results and Discussion

3.1. Physicochemical Analysis of the Milk Used and DPC

As the results reveal in Table 1, the total solid, protein, fat, and ash contents (g/100 g) of the cow's milk used in preparing yogurt drink treatments were added to the DPC. Additionally, DPC contents (gm/100 g) of moisture, total protein, fat, ash, crude fiber, total sugars, water holding capacity (WHC), solubility, pH, and color determinants were listed. The values of color are consistent with the shape appearance of the DPC used during drying and after grinding (Figure 1).

Table 1. Physicochemical analysis of milk used and DPC.

Materials	Properties	Values
Milk	TS (g/100 g)	11.19 ± 0.11
	Protein (g/100 g)	2.97 ± 0.02
	Fat (g/100 g)	2.26 ± 0.05
	Ash (g/100 g)	0.68 ± 0.01
	pH	6.76 ± 0.04
DPC	Moisture (g/100 g)	7.40 ± 0.12
	Protein (g/100 g)	4.52 ± 0.21
	Fat (g/100 g)	2.69 ± 0.03
	Ash (g/100 g)	2.26 ± 0.15
	Crude fiber (g/100 g)	8.36 ± 0.11
	* Total saccharides (g/100 g)	79.13 ± 2.35
	WHC (g/100 g)	3.49 ± 0.11
	Solubility (g/100 g)	15.46 ± 0.42
	pH	5.32 ± 0.05
	L*	64.52 ± 8.79
	a*	11.19 ± 1.20
	b*	27.28 ± 3.14

The indicative values of the parameters are the means (±SD) of three replicates. * Calculated according to the equation = 100 – (moisture + protein + fat + ash).

DPC after drying and grinding   DPC during the drying process   Raw DPC

Figure 1. The appearance of DPC as raw material during drying and after drying and grinding.

The results of Khalas DPC analysis agreed with Al-Farsi et al. [40], d the DPC protein of Omani varieties ranged from 3.62 g/100 gm in Al-Shahal DPC to 5.23 g/100 gm in Mabseeli DPC. In contrast, the fat content ranged from 5.02 g/100 g in the Mabassili DPC to 5.90 g/100 g in the Um-sellah DPC. The dietary fiber ranged between 77.75 and 80.15 g/100 g, respectively. On the other hand, Hashim and Khalil [41] recorded the chemical compositions and physicochemical properties of Lulu, Khalas, and Barhi DPC, where the moisture content range was 8.73, 7.16, 6.14; ash—2.15, 2.82, 2.98; protein—2.18, 2.56, 3.09; fat—1.04, 0.81; fat—0.95; and carbohydrates—85.9, 86.56, 86.84, respectively. The study also recorded the values of colorimetric parameters for the same varieties in terms of the extent of whiteness or luminance/darkness L*, redness/greenness a*, and yellowing/blueness b*, where the values of L* were 48.64, 55.51, 54.76; the values of a*—7.63, 8.02, 7.34; and b*—17.78, 18.33, 18.22, respectively. The study also recorded the water holding capacity (WHC), which reached (g/g) 1.96, 1.98, and 2.0, respectively. Majzoobi et al. [42] mentioned that Iranian Shahani DPC had 13.37% moisture, 4.92% fat, 6.35% protein, 11.74% crude fiber, and 79.06% carbohydrates. The Khalas DPC contained 20.5% moisture, 72.7% carbohydrates, 3.12% protein, 0.31% fat, and 1.72% ash [9,10]. A study by Sheir [7] recorded Egyptian Saidi DPC's contents, with moisture content of 6.11%, fat of 5.12%, protein of 7.4, crude fiber of 12.38%, ash of 2.78%, and carbohydrates of 66.21%.

3.2. Physicochemical Analysis of the Treated DPC Drinkable Yogurt

3.2.1. Total Solids, Protein, and Fat of DPC Drinkable Yogurt

DPC drinkable yogurt treatments include the T_0 treatment (the control sample free of DPC), while T_1, T_2, and T_3 treatments contain 2%, 4%, and 6% (g/100 g) of DPC, respectively. The results in Table 2 show that the percentages (g/100 g) of total solids, protein, and fat for T_0 were 11.19, 2.97, and 2.26; T_1—11.83, 3.10, and 2.34; T_2—12.92, 3.3, and 2.3; and T_3—13.45, 3.42, and 2.35, respectively. Significant increases ($p > 0.05$) in total solids for DPC drinkable yogurt treatments due to the solid contents of DPC and the slight increases in fat and protein are due to the few ranges of DPC. These results agree with the studies, which supplemented yogurt with different sources of dietary fiber [11,43–45].

Table 2. Physicochemical analysis of DPC drinkable yogurt treatments during storage period at 4 °C.

Treatments	Total Solids (g/100 g)	Protein (g/100 g)	Fat (g/100 g)
T_0	11.19 ± 0.11^{d}	2.97 ± 0.02^{d}	2.26 ± 0.05^{b}
T_1	11.83 ± 0.03^{c}	3.10 ± 0.10^{c}	2.34 ± 0.01^{a}
T_2	12.92 ± 0.05^{b}	3.30 ± 0.04^{b}	2.34 ± 0.02^{a}
T_3	13.45 ± 0.13^{a}	3.42 ± 0.01^{a}	2.35 ± 0.02^{a}

The indicative values of the parameters are the means (±SD) of three replicates. Lowercase letters a, b, c, and d for the horizontal comparison between the treatments and the significance of the difference within ($p < 0.05$) limits. T_0, T_1, T_2, and T_3 treatments refer to the percentage of DPC added to drinkable yogurt at 0, 2, 4, and 6%, respectively.

3.2.2. pH and Acidity of DPC Drinkable Yogurt

Increasing the percentage of DPC between 2 and 6% caused slight increases in pH (Figure 2a) and a slight decrease ($p > 0.05$) in acidity (Figure 2b). Generally, storage over 14 days affected the pH and acidity values, as the pH values decreased, and the acidity increased. The pH values for T_0, T_1, T_2, and T_3 were 4.51, 4.53, 4.55, and 4.63 at the beginning of the storage period, while they were 4.37, 4.39, 4.41, and 4.43, respectively, at the end of the storage period. The acidity of T_1 had the greatest value and that of T_3 the lowest value at the beginning and end of storage compared to the control sample, which recorded 0.61 and 0.76% at the beginning of storage and 0.77 and 0.81% at the end of storage, respectively. Compared to the control sample, with increasing the DPC addition, the pH values increased, and the acidity ratios decreased with the increase in DPC concentrations. However, with the progression of the storage period, the pH values decreased, and the acidity percentages increased. This increase in pH with DPC addition can be attributed to its ability to retain water and thus dilute the concentrations of lactic acid and other organic acids produced by the starter culture. This may increase the pH, especially in yogurt fortified with dietary fiber [29,46]. The acidity estimation results for all treatments were kept at 0.87%, as recommended for yogurt [47]. The pH values of all yogurt treatments increased with increased DPC and water absorption from its fibers. Some research studies have reported that adding different dietary fiber sources affected the pH and acidity of fortified yogurt [44,48]. On the other hand, not all the fruit peels or their residues affected yogurt fermentation. No differences were recorded in the acidity degree of the yogurt to which papaya peel flour was added during storage [49]. At the same time, the acidity of yogurt containing passion fruit peel powder was much higher than in the respective controls, which is the behavior expected by the metabolism of lactic acid bacteria [50]. Factors such as total soluble solids, storage temperature, and additives can also reduce the pH of yogurt [51] due to post-acidification and increased activity of lactic acid bacteria. Adding fruits or pulp to yogurt stimulates starter culture bacteria to increase acidity and lower the pH compared to some peels or fibers of these fruits [48,52].

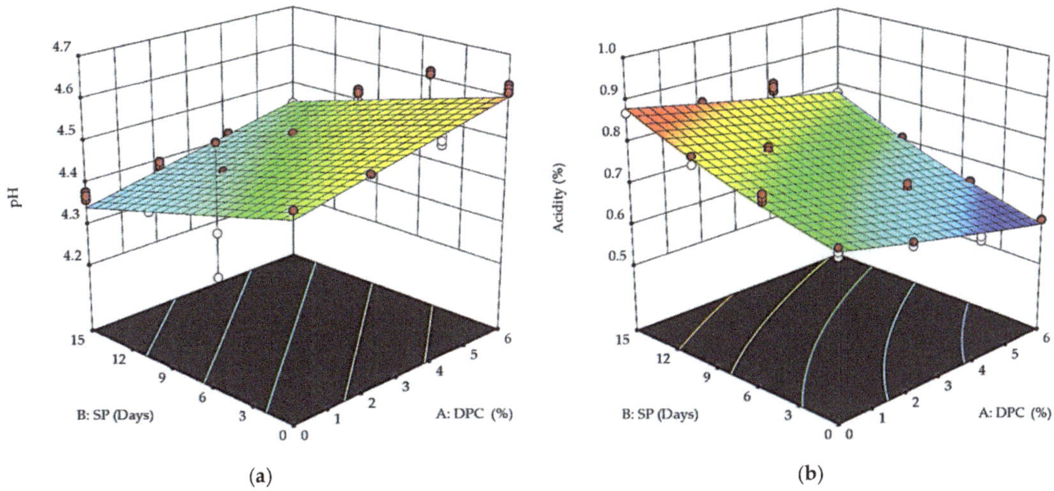

Figure 2. Effect of DPC addition percent (A: DPC) and storage period (B: SP) on pH (**a**) and acidity (**b**) of DPC drinkable yogurt treatments during storage period at 4 °C.

3.2.3. Syneresis, WHC, and Viscosity of DPC Drinkable Yogurt

The results in Figure 3a indicate that the storage time and DPC addition slightly affected the syneresis values, which decreased with the increase in DPC addition and during the storage period. The highest ability to retain whey (the lowest degree) was noted immediately after fermentation. T_0 was significantly higher for syneresis ($p < 0.05$) than T_1, T_2, and T_3 during storage. Syneresis values increased during storage time, but the rate of increase was lower for DPC drinkable yogurt treatments. The rate ranged during the beginning and end of the storage period between 50.26 and 57.06 for the control sample, while for the T_1, T_2, and T_3 samples, the ranges were 48.03–53.26, 45.03–52.23, and 43.1–51.33, respectively. Thus, the increased amounts of DPC added reduced the syneresis. T_3 was the lowest syneresis value ($p < 0.05$); this decrease may be attributed to an increase in total solids, as mentioned by Mahdian and Tehrani [53]. Recent studies [6,11,24,54,55] reported that fortification of yogurt with different dietary fibers supports the viscosity and thickening properties of the yogurt gel. The decrease in syneresis in the DPC drinkable yogurt treatments may be attributed to gummy sugars in the fibers, which can trap water and be released to the DPC during the milling process [25]. These results are in agreement with the results of Arabshahi-Delouee et al. [56] for yogurt with flaxseed press cake, Karaca et al. [24] for yogurt with apricot press cake, Pérez-Chabela et al. [26] for yogurt with mango and potato peels powder, Rojas-Torres et al. [57] for yogurt with butternut squash, and Diep et al. [27] for yogurt with tamarillo.

Figure 3b display the effect of DPC treatments and storage period on the WHC. With the increase in DPC percent, the WHC values decreased during storage. For example, the WHC values for treatments T_0, T_1, T_2, and T_3 were 51.8, 56.76, 58.66, and 58.9 at the beginning of the storage period, while they were 47, 48.96, 52.1, and 54.13 at the end of the storage period ($p < 0.05$). These results are in agreement with the results of Güler-Akın et al. [58] for yogurt with oat and inulin fibers, Karaca et al. [24] for yogurt with apricot press cake, and Diep et al. [27] for yogurt with tamarillo.

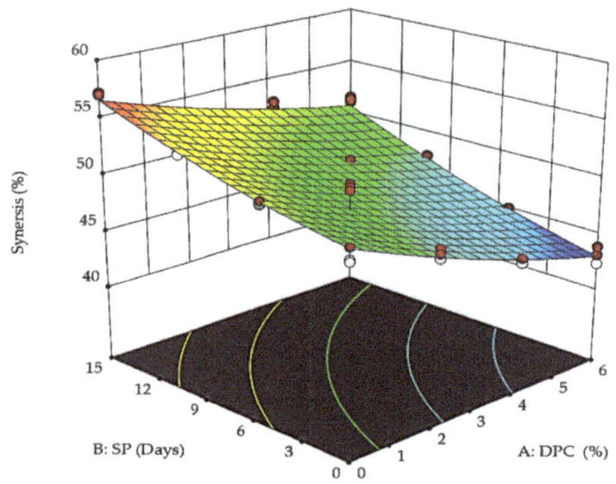

(a)

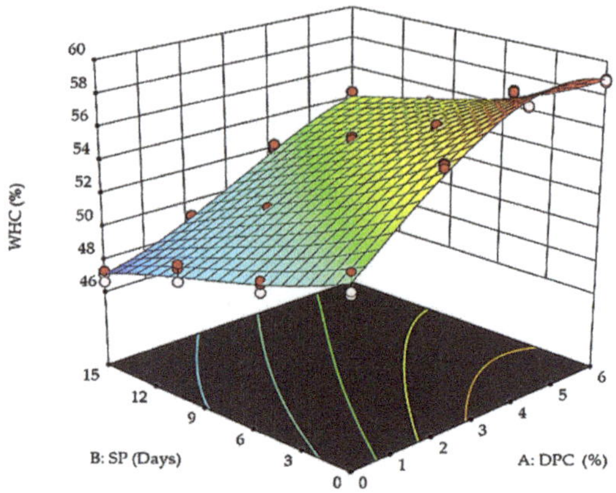

(b)

Figure 3. Effect of DPC addition percent (A: DPC) and storage period (B: SP) on syneresis (**a**) and WHC (**b**) of DPC drinkable yogurt treatments during the storage period at 4 °C.

The viscosity results (Figure 4) showed the highest values at the end of the storage period for all DPC yogurt treatments—T_1, T_2, and T_3—compared to the control sample T_0. In addition, the viscosity values increased directly with the DPC increase. For example, T_3 recorded 2452.33 and 2962.66 cp at the beginning and end of the storage period, respectively. The findings are in agreement with the results of Karaca et al. [24] for yogurt with apricot press cake, Varnait et al. [6] for yogurt with blackberry press cake, and Diep et al. [27] for yogurt with tamarillo.

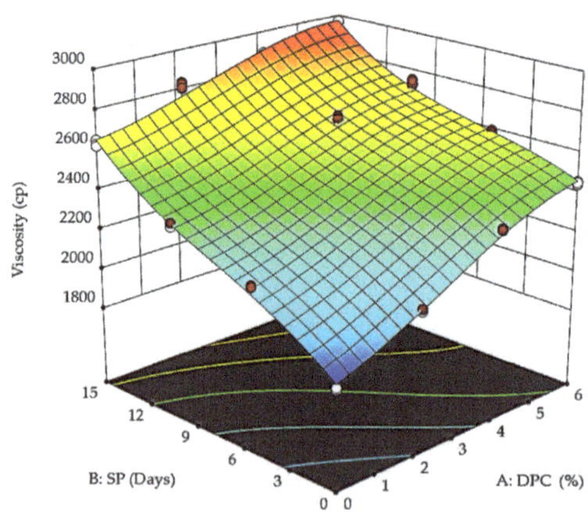

Figure 4. Effect of DPC percent (A: DPC) and storage period (B: SP) on the viscosity of DPC drinkable yogurt during the storage period at 4 °C.

On the other hand, the results of the study conducted by Tseng et al. [59] found that the use of 3% of red grape peels after fermentation weakened the viscosity of yogurt. As the addition to yogurt takes place after or before the fermentation process, Cichonska et al. [33] stated that the statistical analysis of their results showed the differences in the viscosity of yogurt when milled flaxseed was added after fermentation and before fermentation compared to the control sample. The viscosity of yogurt significantly increased when ground flaxseeds were added after fermentation, while the viscosity decreased when they were added before fermentation.

3.2.4. Color Parameters of DPC Drinkable Yogurt

The color measurement value determinants given by Hunter Lab Device L*, a*, b* for DPC drinkable yogurt treatments during the storage period are shown in Figures 5 and 6. The L* values ranged between 69.83 and 78.45 for DPC yogurt treatments vs. the control treatment of 96.75 in the first storage period (Figure 5). In addition, they ranged between 45.92 and 57.54 compared to the control treatment and 72.6 at the end of the storage period. T_3 containing 6% of DPC was the darkest treatment according to the L* values, which amounted to 69.83 at the beginning of the storage period. The degrees of whiteness decreased significantly during the storage period for all treatments ($p < 0.05$).

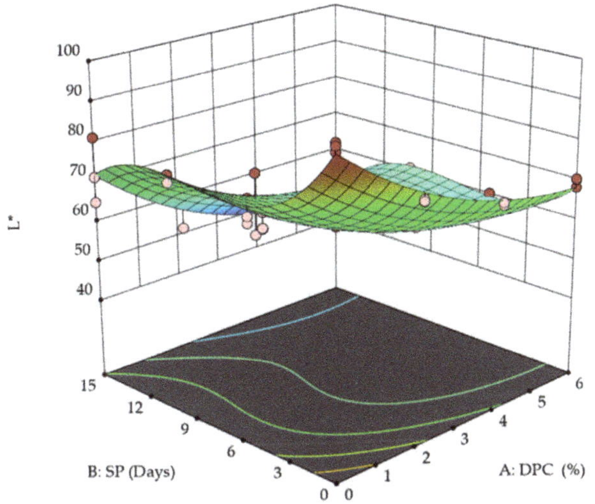

Figure 5. Effect of DPC addition percent (A: DPC) and storage period (B: SP) on L* of DPC drinkable yogurt treatments during the storage period at 4 °C.

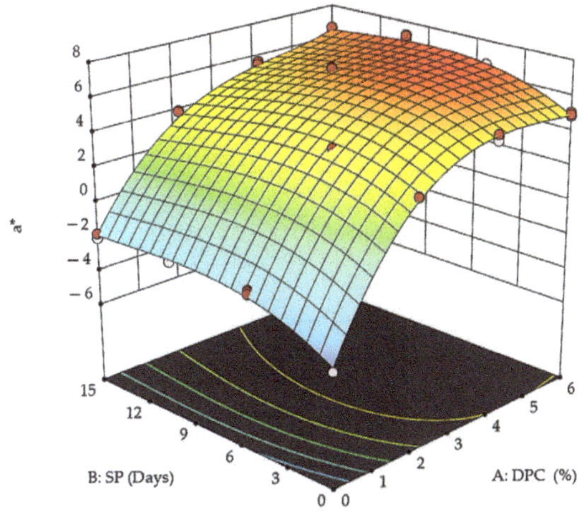

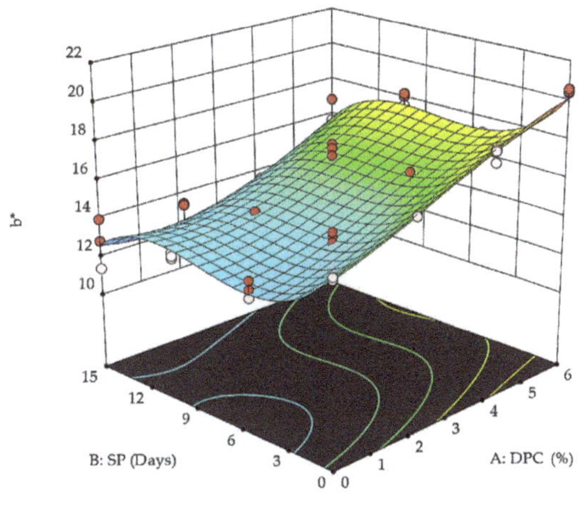

Figure 6. Effect of DPC addition percent (A: DPC) and storage period (B: SP) on a* and b* of DPC drinkable yogurt treatments during the storage period at 4 °C.

Regarding a*, the results indicated that the control sample during the first and last storage period was negative, ranging from −4.24 to −1.91, which indicated its tendency toward greenness, and the degree of greenness decreased at the end of the storage period compared to its beginning (Figure 6a). On the other hand, all DPC yogurt treatments—T_1, T_2 and T_3—recorded significant ($p < 0.05$) positive values—3.3, 5.04, and 4.98, for T_1, T_2 and T_3, respectively—at the beginning of the storage period, and the values were 3.69, 5.46, and 6.35 at the end of the storage period. This indicates a tendency toward redness, which may be attributed to the reddish-brown color of the DPC. Thus, the storage period significantly ($p < 0.05$) affected the a* values, as the values increased, as did their tendency to redden more at the end of the storage period compared to its beginning.

The b* values of DPC drinkable yogurt treatments revealed yellowing at the beginning and end of the storage period (Figure 6b). The control sample recorded significantly lower values ($p < 0.05$) than all treatments at the beginning and end of the storage period, whereby it scored 15.35 at the beginning of storage and 12.72 at the end of it. T_3 was the most yellow, with 20.47, followed by T_2 with 18.49 and T_1 with 16.9, compared to the control T_0, which reached 15.35 at the beginning of the storage period.

Regarding the color difference (ΔE) value, the results indicated that the samples in the first storage period were zero because this is the baseline (control), as shown in Figure 7, accompanied by the visual color at the beginning and end of the storage period (Figure 8). The highest values were found at the end of the storage period for all DPC drinkable yogurt treatments. In addition, ΔE values increased directly with increasing DPC percentage and storage period. For example, after five days, T_0, T_1, T_2, and T_3, ΔE recorded 21.13, 30.7, 37.52, and 40.39. This did not change significantly subsequently for another 10 days, and then, values of 24.42, 40.1, 51.79, and 48.18, respectively, were recorded after 15 days. Generally, all DPC yogurt treatment values—T_1, T_2, and T_3—were significantly ($p < 0.05$) decreased with increasing DPC and storage period at 4 °C. Figure 8 shows the visual color of DPC drinkable yogurt treatments during the storage period.

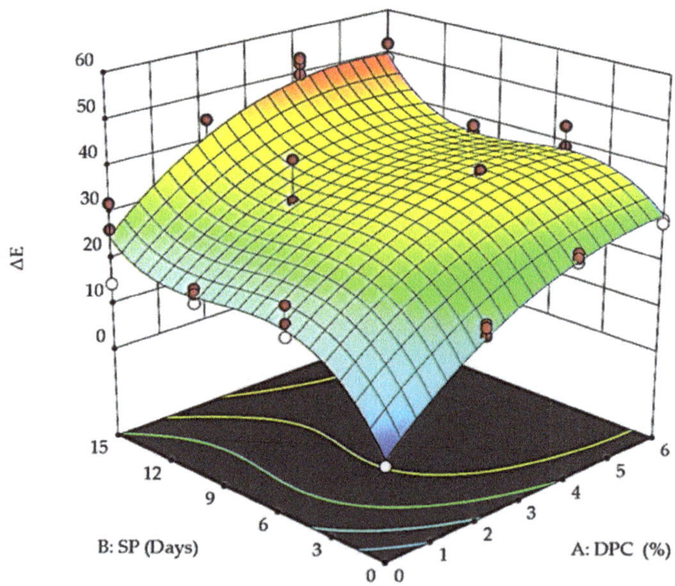

Figure 7. Effect of DPC addition percent (A: DPC) and storage period (B: SP) on the color difference (ΔE) value of DPC drinkable yogurt treatments during the storage period at 4 °C.

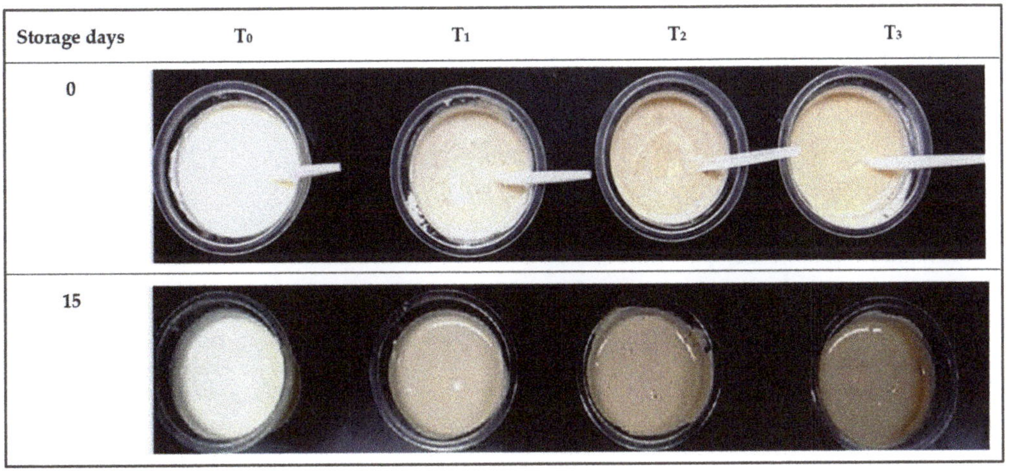

Figure 8. Visual color of DPC drinkable yogurt treatments during the storage period. T_0, T_1, T_2, and T_3 treatments refer to the percentage of DPC added to drinkable yogurt at 0, 2, 4, and 6%, respectively.

Rojas-Torres et al. [57] indicated that yogurt samples supplemented with thickeners from butternut squash seeds were affected in terms of glossiness, which they attributed to the size of the fat granules and protein. Łopusiewicz et al. [11] found that adding Camelina press cake to yogurt affected the a* (redness) value of the yogurt treatments, which the study attributed to the presence of a carotenoid pigment; however, the values decreased during the storage period due to the decrease in the stability of the dye.

Additionally, a decrease in a* (redness) values was observed in yogurt supplemented with flaxseed press cake as a result of fermentation [60]. In a study where ginseng root extract was added to yogurt, the results showed that the color values were affected by the addition, as the L* values decreased from 93.96 to 90.99, while a* and b* values increased from 2.92 and 2.30 to 5.91 and 11.11, respectively. Darkness and decreased luminosity L* compared to the control samples combined with increased b* values resulted in a yellowish color [55]. Alqahtani et al. [61] mentioned the effect of yogurt with tomato pomace on color measurements during storage. The results indicated a decrease in the whiteness values of L* and increases in the values of a* and b*. The results of Mkadem et al. [28] in enhancing the fermented milk drink with dried dates showed its effects on the color characteristics, especially a* and b*, where the a* value increased from -1.8 ± 0.8 to 4.5 ± 0.2, while the b* value increased from 3.1 ± 0.1 to 12.1 ± 0.1. In contrast, the value of the gloss L* decreased, which was attributed to the dyes of the date dryer. Adding dried tamarillo plant fiber (5–15%) to yogurt decreased the L* value, while the value of a* increased with the increase in the addition percentage [27]. The study attributed this to the natural pigments of the anthocyanins in tamarillo. In a study mixing microencapsulated guava pomace fibers with yogurt, the L* value was unaffected, and the values of a* and b* within the treatments were close; this was due to the contents of beta-carotene inside the guava pomace fibers used [62].

3.3. Correlation between the Factors and Physicochemical Properties of DPC Drinkable Yogurt

Figure 9 displays the correlation between the DPC treatments, storage period, and physicochemical properties of DPC drinkable yogurt. Significant positive correlations exist between the DPC treatments and pH, WHC, viscosity, a*, b*, and the DPC drinkable yogurt's color difference (ΔE).

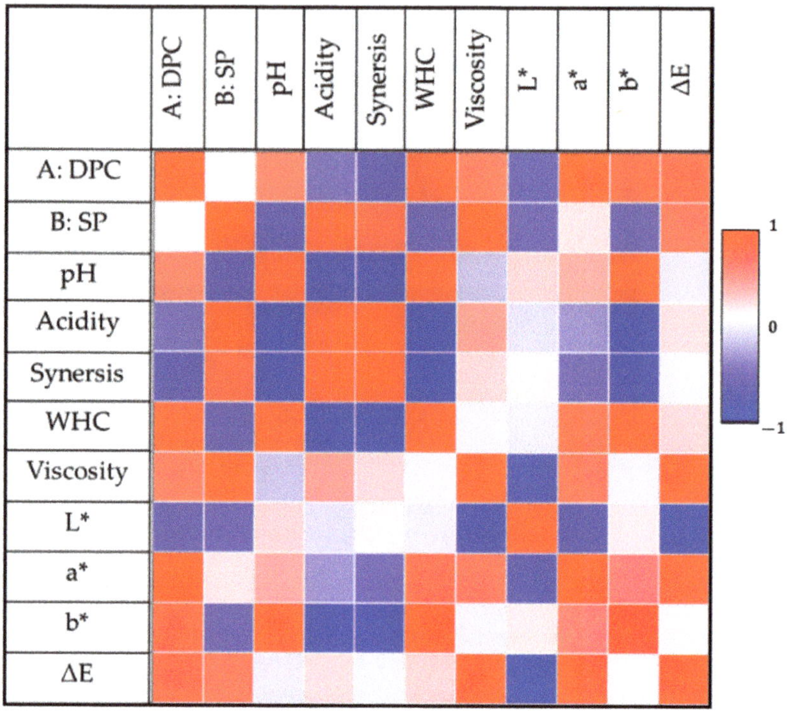

Figure 9. The correlation between the DPC treatments (A: DPC), storage period (B: SP), and physicochemical properties of DPC drinkable yogurt. WHC, L*, a*, b*, and ΔE, refers to water holding capacity, lightness, redness/greenness, yellowness/blueness, and color difference, respectively.

On the other hand, there is a significant negative correlation between the DPC treatments and acidity, syneresis, and L* value. Regarding the storage period, there is a significant positive correlation between the DPC treatments and acidity, syneresis, viscosity, and ΔE of the DPC drinkable yogurt and a significant negative correlation between the DPC treatments and pH, WHC, L*, and b* value of the DPC drinkable yogurt. In addition, a slight positive correlation existed between the a* value and storage period.

3.4. Modeling of Physicochemical Properties of DPC Drinkable Yogurt

In order to model the physicochemical properties of DPC drinkable yogurt, the experiment outcomes were input into the Design Expert software for additional data analysis. The data of the two factors (DPC percent and storage period) were fitted with different models, i.e., linear, quadratic, and cubic, in order to model the physicochemical properties of DPC drinkable yogurt. The quadratic polynomial models most appropriately described the physicochemical properties of DPC drinkable yogurt, i.e., pH, acidity, syneresis, WHC, viscosity, and ΔE. The final predictive models for the two essential factors for pH, acidity, syneresis, WHC, viscosity, and ΔE are presented as Equations (5)–(13), respectively. These equations can be used to predict the mentioned physicochemical properties based on the input values for the actual percentage of the DPC and storage time (Day).

$$\text{pH} = 4.49 + 0.006\,\text{DPC} - 0.011\,\text{SP} - 0.001\,\text{DPC} \times \text{SP} + 0.001\,\text{DPC}^2 + 0.001\,\text{SP}^2 \quad (5)$$

$$\text{Acidity} = 0.722 - 0.022\,\text{DPC} + 0.007\,\text{SP} + 0.001\,\text{DPC} \times \text{SP} + 0.001\,\text{DPC}^2 + 0.001\,\text{SP}^2 \quad (6)$$

$$\text{Syneresis} = 50.892 - 1.698\,\text{DPC} + 0.217\,\text{SP} + 0.023\,\text{DPC} \times \text{SP} + 0.064\,\text{DPC}^2 + 0.011\,\text{SP}^2 \quad (7)$$

$$\text{WHC} = 51.96517 + 2.237\,\text{DPC} - 0.481\,\text{SP} - 0.137400\,\text{DPC} \times \text{SP} + 0.001\,\text{DPC}^2 + 0.025\,\text{SP}^2 \quad (8)$$

$$\text{Viscosity} = 1915.18 + 70.869\,\text{DPC} + 68.915\,\text{SP} - 5.233\,\text{DPC} \times \text{SP} + 16.398\,\text{DPC}^2 - 4.497\,\text{SP}^2 \quad (9)$$

$$L^* = 95.321 - 9.31\ DPC - 7.189\ SP + 0.374\ DPC \times SP + 0.97\ DPC^2 + 0.891\ SP^2 \quad (10)$$

$$a^* = -4.019 + 5.267\ DPC + 0.61\ SP - 0.081\ DPC \times SP - 0.977\ DPC^2 - 0.051\ SP^2 \quad (11)$$

$$b^* = 15.413 + 0.546\ DPC - 1.108\ SP - 0.068\ DPC \times SP + 0.144\ DPC^2\ 0.159\ SP^2 \quad (12)$$

$$\Delta E = 1.514 + 10.344\ DPC + 7.107\ SP - 0.406\ DPC \times SP - 1.165\ DPC^2 - 0.869\ SP^2 \quad (13)$$

where DPC is the date press cake percentage, and SP is the daily storage period.

The standard deviation (Std. Dev.), mean value, coefficient of variation percentage (C.V.%), coefficient of determination (R^2), adjusted R^2, and predicted R^2 Adeq precision criteria were used to evaluate the selected predicted models (Equations (5)–(13)) in terms of the pH, acidity, syneresis, WHC, viscosity, L*, a*, b*, and ΔE of DPC drinkable yogurt. The quadratic model emerged as the best because it exhibited a low standard deviation, high R-squared values close to 1, and low PRESS. The evaluation criteria of the prediction models are shown in Table 3. The evaluation criteria indicated that the selected prediction models could efficiently describe the experiments. Therefore, these models can be used to navigate the design space for the target physicochemical properties of DPC drinkable yogurt responses. The results of the quadratic models agree with Mohammed et al. [63].

Table 3. The evaluation criteria, i.e., standard deviation (Std. Dev.), mean value, coefficient of variation percentage (C.V.%), coefficient of determination (R^2), adjusted R^2, predicted R^2, and Adeq precision criteria, for the selected quadratic models for the target physicochemical properties of DPC drinkable yogurt.

Criteria	Physicochemical Properties of DPC Drinkable Yogurt								
	pH	Acidity	Syneresis	WHC	Viscosity	L*	a*	b*	ΔE
Std. Dev.	0.040	0.014	0.783	0.405	30.85	3.916	0.264	0.706	3.99
Mean	4.47	0.738	49.77	53.35	2492.85	66.86	3.364	15.48	31.02
C.V.%	0.894	1.86	1.57	0.759	1.24	5.858	7.875	4.565	12.85
R^2	0.779	0.973	0.961	0.989	0.990	0.919	0.995	0.922	0.921
Adjusted R^2	0.752	0.969	0.956	0.987	0.988	0.868	0.994	0.905	0.902
Predicted R^2	0.717	0.964	0.949	0.983	0.985	23.78	0.993	0.874	0.870
Adeq Precision	18.316	58.098	48.999	64.421	74.943	3.916	89.907	23.717	26.563

WHC, L*, a*, b*, and ΔE, refers to water holding capacity, lightness, redness/greenness, yellow-ness/blueness, and color difference, respectively.

Figure 10a–e show the scatter plots of the predicted values generated by the developed models of the pH, acidity, syneresis, WHC, viscosity, and ΔE of DPC drinkable yogurt vs. the actual experimental values. Figure 11a–d show the scatter plots of the predicted values generated by the developed models of the L*, a*, b*, and ΔE of DPC drinkable yogurt vs. the actual experimental values. The results indicated that the predicted values were close to the observed values. Therefore, the prediction model is suitable for responses to the target physicochemical properties of DPC drinkable yogurt.

3.5. *Texture Profile Analysis of DPC Drinkable Yogurt*

The texture profile analysis results (Table 4) indicated that hardness (g) increased with the increase in the addition of DPC, and sample T_3 recorded a value of 491.83 compared to 316.23 for the control sample at the beginning of the storage period, while the value was 581.33 compared to 506.6 for the control sample at the end of the storage period. On the other hand, the adhesiveness of treatments decreased with an increase in the addition of DPC (negative increasing numbers), with treatments T_3 and T_2 obtaining the lowest scores of −368 and −371.33 compared with sample T_1, which scored −349, and the control sample, which scored −286 at the beginning of storage. The values at the end of storage for the same samples were −455.33, −453.33, −445, and 411.33, respectively.

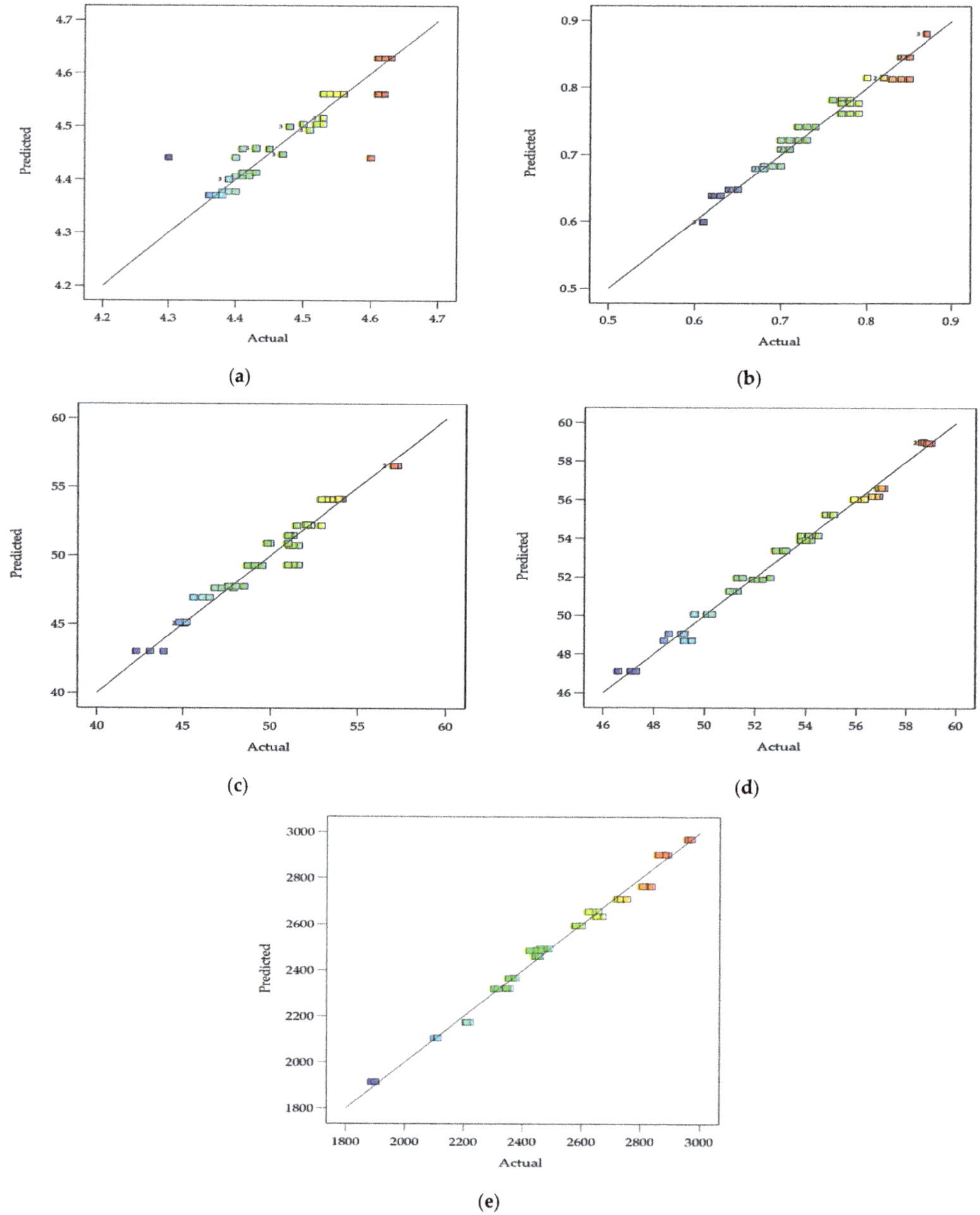

Figure 10. Scatter plots of the actual vs. predicted values generated by the prediction models of the pH (**a**), acidity (**b**), syneresis (**c**), WHC (**d**) and viscosity (**e**) of DPC drinkable yogurt.

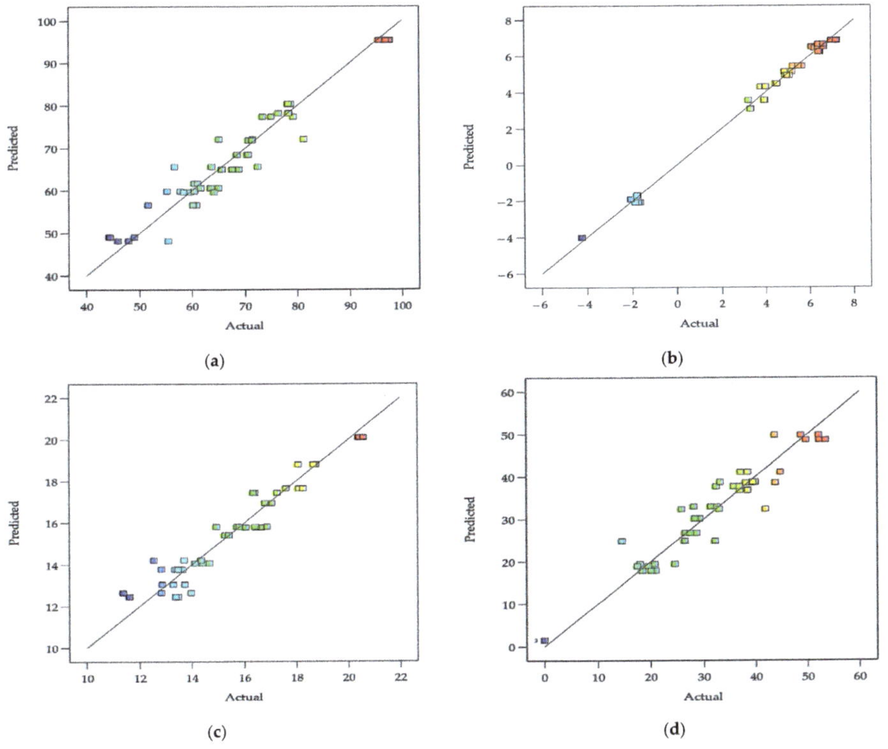

Figure 11. Scatter plots of the actual vs. predicted values generated by the prediction models of the L (**a**), a* (**b**), b* (**c**), ΔE (**d**) of DPC drinkable yogurt.

Table 4. Texture profile analysis of DPC drinkable yogurt treatments during storage.

Texture Parameters	Storage Days	T_0	T_1	T_2	T_3
Hardness, g	0	316.23 ± 8.28 ^{Ad}	432.73 ± 2.51 ^{Bc}	451.83 ± 1.60 ^{Bb}	491.83 ± 3.68 ^{Ba}
	15	506.60 ± 7.93 ^{Bd}	524.50 ± 9.35 ^{Ac}	547.76 ± 4.28 ^{Ab}	581.33 ± 2.06 ^{Aa}
Adhesiveness, g s	0	−286.00 ± 4.35 ^{Aa}	−349.00 ± 8.54 ^{Ab}	−371.33 ± 1.52 ^{Ac}	−368.00 ± 5.29 ^{Ac}
	15	−411.33 ± 1.52 ^{Ba}	−445.00 ± 6.08 ^{Bb}	−453.33 ± 2.51 ^{Bc}	−455.33 ± 5.03 ^{Bc}
Springiness	0	0.83 ± 0.00 ^{Ac}	0.86 ± 0.02 ^{Ab}	0.92 ± 0.01 ^{Aa}	0.93 ± 0.005 ^{Aa}
	15	0.71 ± 0.005 ^{Bb}	0.73 ± 0.01 ^{Bb}	0.75 ± 0.04 ^{Bab}	0.79 ± 0.01 ^{Ba}
Cohesiveness	0	0.80 ± 0.005 ^{Ad}	0.84 ± 0.005 ^{Ac}	0.86 ± 0.005 ^{Ab}	0.97 ± 0.01 ^{Aa}
	15	0.54 ± 0.04 ^{Bb}	0.57 ± 0.01 ^{Bb}	0.70 ± 0.005 ^{Ba}	0.73 ± 0.02 ^{Ba}
Gumminess	0	399.10 ± 0.85 ^{Ab}	142.23 ± 0.32 ^{Bd}	367.66 ± 0.49 ^{Ac}	578.56 ± 0.51 ^{Aa}
	15	282.43–0.30 ^{Bc}	244.80 ± 0.26 ^{Ad}	366.5 ± 0.50 ^{Ba}	291.16 ± 1.04 ^{Bb}
Resilience	0	0.64 ± 0.02 ^{Aa}	0.52 ± 0.02 ^{Ab}	0.55 ± 0.04 ^{Ab}	0.31 ± 0.01 ^{Ac}
	15	0.53 ± 0.02 ^{Ba}	0.34 ± 0.04 ^{Bb}	0.15 ± 0.04 ^{Bc}	0.10 ± 0.01 ^{Bc}

The indicative values of the parameters are the means (±SD) of three replicates. The uppercase letters A and B, represent the vertical comparison between storage periods. The lowercase letters a, b, c, and d represent the horizontal comparison between the treatments. Treatment T_0, T_1, T_2, and T_3 refers to the percentage of DPC added to drinkable yogurt at 0, 2, 4, and 6%, respectively.

Although the springiness values increased with the increase in DPC, the increase was significantly inversely proportional ($p < 0.05$) to the progression of the storage period. Treatment T_3 recorded the highest value of 0.93 vs. 0.83 for the control sample at the beginning of the storage period, and the value was 0.79 vs. 0.71 for the control sample at the end of the storage period. This approach to characterization of springiness values was followed for each of the values of cohesiveness and gumminess. At the same time, the resilience values decreased with the increase in the addition of DPC and the progression of the storage period, with treatment T_3 recording a value of 0.31 vs. 0.64 for the control sample at the beginning of the storage period and 0.1 vs. 0.51 for the control sample at the end of the storage period. The results of the texture profile analysis in this study agree with those in previous studies [24,36,44,61,62,64].

3.6. Sensory Evaluation of DPC Drinkable Yogurt

The results of sensory acceptance regarding the color, texture, flavor, and general acceptance of DPC drinkable yogurt during the storage period at 4 °C showed that samples T_1 followed by T_2 and T_3 were the closest in overall sensory traits to the control sample ($p < 0.05$) over the progression of the storage period (Table 5 and Figure 12). T_1 recorded values of sensory preference at the beginning of the storage period in terms of color, texture, flavor, and general acceptance of 4, 3.76, 3.73, and 3.86 vs. the control sample with 4.3, 4, 4.1, and 4.26; at the end of the storage period, it reached values of 3.66, 3.73, 3.53, and 3.46 vs. the control sample with 4.30, 3.90, 3.80, and 3.83, respectively. These results are consistent with Łopusiewicz et al. [11]—who used Camelina press cake with yogurt—regarding the color effect. Karaca et al. [24] stated that the increase in the addition of apricot press cake fibers affected the arbitrators' preference for yogurt samples due to the increase in viscosity. At the same time, the results differed in terms of the effect of the storage period. Brodziak et al. [65] mentioned that adding sea buckthorn fruit to yogurt affected the organoleptic characteristics of addition and storage. The sensory characteristics were improved in yogurt samples with thickeners added from butternut squash seeds [57]. In another study [55] attempting to enhance yogurt with hydroponic ginseng extracts, the results showed that the treatment in terms of color was superior to the control sample at 1% concentration. At the same time, the preference decreased with increasing concentrations of it. At the same time, the texture and flavor were not affected by the addition of ginseng. With the addition of moringa to yogurt, the results of the study mentioned by Mendoza-Taco et al. [66] showed that the yogurt samples were not affected in relation to the sensory traits assessed. Meanwhile, in a study [67] attempting to enhance the vitality-boosting fermented camel milk drink with sugary date fibers and demonstrate its effects on sensory traits, the results indicated that adding 12.5% of sugary dates fiber was preferred by sensory traits assessors.

Table 5. Sensory evaluation of DPC drinkable yogurt treatments during storage period at 4 °C.

Sensory Parameters	Storage Days	T_0	T_1	T_2	T_3
Color	0	4.30 ± 1.34 ^{Aa}	4.00 ± 1.08 ^{ABa}	3.20 ± 0.88 ^{Bb}	2.90 ± 1.26 ^{Bb}
	5	4.70 ± 0.74 ^{Aa}	4.26 ± 0.94 ^{Aab}	4.40 ± 0.85 ^{Aab}	4.16 ± 1.01 ^{Ab}
	10	3.80 ± 0.84 ^{Ba}	3.53 ± 0.77 ^{Cab}	3.43 ± 0.72 ^{Bab}	3.16 ± 0.98 ^{Bb}
	15	4.30–0.59 ^{Aa}	3.66 ± 0.66 ^{BCbc}	3.26 ± 1.04 ^{Bc}	3.80 ± 0.80 ^{Ab}
Texture	0	4.00 ± 1.28 ^{Ba}	3.76 ± 1.16 ^{Aab}	3.46 ± 1.10 ^{ABab}	3.26 ± 1.11 ^{Bb}
	5	4.63 ± 0.55 ^{Aa}	3.86 ± 1.04 ^{Ab}	3.70 ± 1.11 ^{Ab}	3.70 ± 1.41 ^{ABb}
	10	3.50 ± 0.86 ^{Ca}	3.50 ± 0.97 ^{Aa}	3.03 ± 0.88 ^{Ba}	3.23 ± 1.00 ^{Ba}
	15	3.90 ± 0.66 ^{BCa}	3.73 ± 0.90 ^{Aa}	3.10 ± 0.92 ^{Bb}	3.86 ± 1.10 ^{Aa}
Flavor	0	4.10 ± 1.29 ^{Ba}	3.73 ± 1.31 ^{Aab}	3.10 ± 1.29 ^{Bbc}	3.00 ± 1.48 ^{Bc}
	5	4.56 ± 0.62 ^{Aa}	3.80 ± 1.39 ^{Ab}	4.10 ± 0.88 ^{Aab}	3.90 ± 1.34 ^{Ab}
	10	2.96 ± 0.92 ^{Ca}	3.10 ± 0.88 ^{Ba}	2.90 ± 0.99 ^{Ba}	2.76 ± 1.04 ^{Ba}
	15	3.80 ± 0.61 ^{Ba}	3.53 ± 0.93 ^{ABab}	3.03 ± 1.12 ^{Bb}	3.40 ± 1.19 ^{ABab}

The indicative values of the parameters are the means (±SD) of three replicates. The uppercase letters A, B, and C represent the vertical comparison between storage periods, while the lowercase letters a, b and c represent the horizontal comparison between the treatments, and the difference is significant within ($p < 0.05$). Treatment T_0, T_1, T_2, and T_3 refers to the percentage of DPC added to drinkable yogurt at 0, 2, 4, and 6%, respectively.

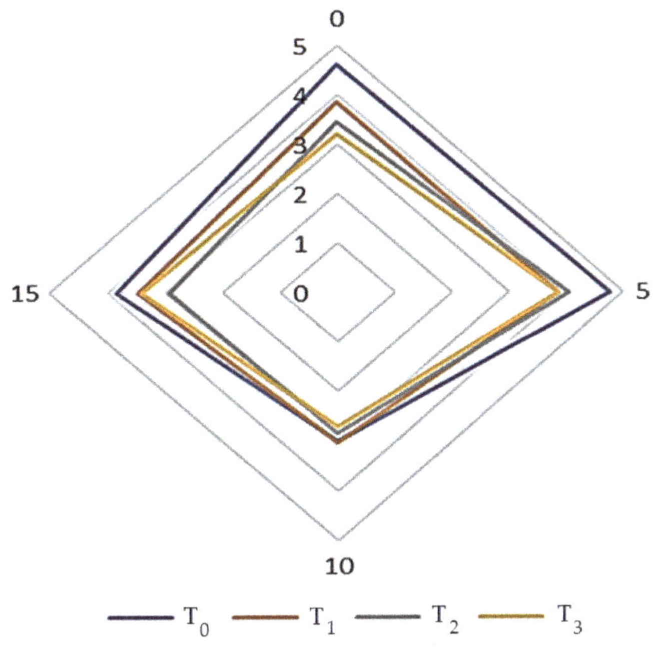

Figure 12. Overall acceptability of DPC drinkable yogurt treatments during storage period at 4 °C. Treatments T_0, T_1, T_2, and T_3 refer to the percentage of DPC added to drinkable yogurt at 0, 2, 4, and 6%, respectively.

The addition of 0.5% of dried orange peel fibers did not have any significant effects ($p < 0.05$) on the sensory acceptance of the viability-boosting yogurt drink treatments compared to the control sample, while increasing the concentrations of dried orange peel fibers up to 2% had a negative effect on the sensory properties [68].

4. Conclusions

Adding DPC powder before the fermentation of DPC yogurt slightly increased the pH and decreased acidity during the storage period. DPC addition slightly reduced syneresis values and increased the WHC during storage. Viscosity recorded the highest value at the end of the storage period for all treatments. Yogurt containing 6% DPC was the highest in viscosity. The L* values decreased significantly ($p < 0.05$) with increasing DPC, and the degrees of whiteness or glossiness decreased significantly during the storage period for all treatments. The a* values were positive for all DPC yogurt samples, and the b* values showed yellowness in all DPC treatments at the beginning and end of the storage. Texture profile data indicated that increased DPC and the storage period's progression increased the hardness (g), springiness, cohesiveness, and gumminess. Adding DPC reduced adhesiveness and resilience during storage. Yogurt containing 2% was the closest in overall sensory acceptability to the control sample ($p < 0.05$) and during the storage. The predicted values of the quadratic models were close to the actual observed values for the pH, acidity, syneresis, WHC, viscosity, and ΔE of the treated DPC drinkable yogurt. Generally, the results indicated the possibility of enriching yogurt drinks by 2–4% with DPC as an innovative functional additive with general acceptance. The results also support the idea of expanding the uses of DPC in yogurt, enhancing the circular economy for waste upgrading, and DPC can be proposed as a new functional fiber-based yogurt. Overall, these results gave a new idea regarding several aspects related to the upcycling vision, ranging from technological strategies to reusing agrifood by-products and obtaining functional ingredients for high-value-added food production. The sensory evaluation may require more participants in a further study before the proposed product can be applied on a large scale.

Author Contributions: Conceptualization, N.K.A., T.M.A., M.M., M.M.A. and A.K.A.; methodology, T.M.A., M.M. and A.K.A.; software, M.M.; validation, M.M.; formal analysis, T.M.A., N.K.A., A.K.A. and M.M.; data curation, T.M.A., N.K.A., M.M.A., M.M. and A.K.A.; writing—original draft preparation, T.M.A., N.K.A., M.M., M.M.A. and A.K.A.; writing—review and editing, T.M.A., M.M., N.K.A., M.M.A. and A.K.A.; visualization, M.M. and T.M.A.; project administration, T.M.A., N.K.A., M.M.A. and A.K.A.; supervision, T.M.A., N.K.A. and M.M.A.; funding acquisition, T.M.A., N.K.A., M.M.A. and A.K.A. All authors have read and agreed to the published version of the manuscript.

Funding: This research was supported by a project (DPRC-01-2022) financed by Date Palm Research Center of Excellence, King Faisal University, Saudi Arabia.

Institutional Review Board Statement: Not applicable.

Informed Consent Statement: Not applicable.

Data Availability Statement: Data are contained within the article.

Acknowledgments: The authors gratefully acknowledge the Date Palm Research Center of Excellence and the Department of Food and Nutrition Sciences, College of Agricultural and Food Sciences, King Faisal University, Saudi Arabia, for their sincere help and support.

Conflicts of Interest: The authors declare no conflict of interest.

References

1. Grimaldi, M.; Pitirollo, O.; Ornaghi, P.; Corradini, C.; Cavazza, A. Valorization of Agro-Industrial Byproducts: Extraction and Analytical Characterization of Valuable Compounds for Potential Edible Active Packaging Formulation. *Food Packag. Shelf Life* **2022**, *33*, 100900. [CrossRef]
2. Tapia-Quirós, P.; Montenegro-Landívar, M.F.; Reig, M.; Vecino, X.; Cortina, J.L.; Saurina, J.; Granados, M. Recovery of Polyphenols from Agri-Food By-Products: The Olive Oil and Winery Industries Cases. *Foods* **2022**, *11*, 362. [CrossRef] [PubMed]
3. Bhardwaj, K.; Najda, A.; Sharma, R.; Nurzyńska-Wierdak, R.; Dhanjal, D.S.; Sharma, R.; Manickam, S.; Kabra, A.; Kuča, K.; Bhardwaj, P. Fruit and Vegetable Peel-Enriched Functional Foods: Potential Avenues and Health Perspectives. *Evidence-Based Complement. Altern. Med.* **2022**, *2022*, 8543881. [CrossRef]
4. Zhu, Z.; Gavahian, M.; Barba, F.J.; Roselló-Soto, E.; Bursać Kovačević, D.; Putnik, P.; Denoya, G.I. Valorization of Waste and By-Products from Food Industries through the Use of Innovative Technologies. In *Agri-Food Industry Strategies for Healthy Diets and Sustainability*; Barba, F.J., Putnik, P., KovaÄ, D.B., Eds.; Elsevier: Amsterdam, The Netherlands, 2020; pp. 249–266.
5. Galanakis, C.M. Food Use for Social Innovation by Optimizing Food Waste Recovery Strategies. In *Food Waste Recovery*; Galanakis, C.M., Ed.; Academic Press: Cambridge, MA, USA; Elsevier: Amsterdam, The Netherlands, 2022; pp. 209–227.
6. Varnaitė, L.; Keršienė, M.; Šipailienė, A.; Kazernavičiūtė, R.; Venskutonis, P.R.; Leskauskaitė, D. Fiber-Rich Cranberry Pomace as Food Ingredient with Functional Activity for Yogurt Production. *Foods* **2022**, *11*, 758. [CrossRef]
7. Sheir, M.A. Innovative Use of Date (*Phoenix dactylifera* L.) Press Cake in the Food Industry. *Foods Raw Mater.* **2022**, *10*, 2–9. [CrossRef]
8. Attia, A.I.; Reda, F.M.; Patra, A.K.; Elnesr, S.S.; Attia, Y.A.; Alagawany, M. Date (*Phoenix dactylifera* L.) by-Products: Chemical Composition, Nutritive Value and Applications in Poultry Nutrition, an Updating Review. *Animals* **2021**, *11*, 1133. [CrossRef]
9. Oladzad, S.; Fallah, N.; Mahboubi, A.; Afsham, N.; Taherzadeh, M.J. Date Fruit Processing Waste and Approaches to Its Valorization: A Review. *Bioresour. Technol.* **2021**, *340*, 125625. [CrossRef]
10. Rambabu, K.; Bharath, G.; Banat, F.; Hai, A.; Show, P.L.; Phong Nguyen, T.H. Ferric Oxide/Date Seed Activated Carbon Nanocomposites Mediated Dark Fermentation of Date Fruit Wastes for Enriched Biohydrogen Production. *Int. J. Hydrogen Energy* **2021**, *46*, 16631–16643. [CrossRef]
11. Łopusiewicz, Ł.; Waszkowiak, K.; Polanowska, K.; Mikołajczak, B.; Śmietana, N.; Hrebień-Filisińska, A.; Sadowska, J.; Mazurkiewicz-Zapałowicz, K.; Drozłowska, E. The Effect of Yogurt and Kefir Starter Cultures on Bioactivity of Fermented Industrial By-Product from Cannabis Sativa Production—Hemp Press Cake. *Fermentation* **2022**, *8*, 490. [CrossRef]
12. Zadeh, N.S.; Tedesco, M.; Basso, S.; Ghirardello, D.; Giovando, S.; Battaglia, M.; Zeppa, G. Effect of Yogurt Enrichment with Wood Tannin during Shelf Life: Focus on Physicochemical, Microbiological, and Sensory Characteristics. *Foods* **2023**, *12*, 405. [CrossRef]
13. Desobry-Banon, S.; Vetier, N.; Hardy, J. Health Benefits of Yogurt Consumption. A Review. *Int. J. Food Prop.* **1999**, *2*, 1–12. [CrossRef]
14. Andrade, R.D.; Arteaga, M.R.; Simanca, M.M. Efecto Del Salvado de Trigo En El Comportamiento Reológico Del Yogurt de Leche de Búfala. *Inf. Tecnol.* **2010**, *21*, 117–124. [CrossRef]

15. Mittal, M.; Thakur, A.; Kaushik, R.; Chawla, P. Physicochemical Properties of Ocimum Sanctum Enriched Herbal Fruit Yoghurt. *J. Food Process. Preserv.* **2020**, *44*, e14976. [CrossRef]
16. Sigdel, A.; Ojha, P.; Karki, T.B. Phytochemicals and Syneresis of Osmo-Dried Mulberry Incorporated Yoghurt. *Food Sci. Nutr.* **2018**, *6*, 1045–1052. [CrossRef]
17. Hadjimbei, E.; Botsaris, G.; Chrysostomou, S. Beneficial Effects of Yoghurts and Probiotic Fermented Milks and Their Functional Food Potential. *Foods* **2022**, *11*, 2691. [CrossRef] [PubMed]
18. Khalaf, A.T.; Wei, Y.; Alneamah, S.J.A.; Al-Shawi, S.G.; Kadir, S.Y.A.; Zainol, J.; Liu, X. What Is New in the Preventive and Therapeutic Role of Dairy Products as Nutraceuticals and Functional Foods? *Biomed Res. Int.* **2021**, *2021*, 8823222. [CrossRef]
19. Sanyal, M.K.; Pal, S.C.; Gangopadhyay, S.K.; Dutta, S.K.; Ganguli, D.; Das, S.; Maiti, P. Influence of Stabilizers on Quality of Sandesh from Buffalo Milk. *J. Food Sci. Technol.* **2011**, *48*, 740–744. [CrossRef]
20. Bruzantin, F.P.; Daniel, J.L.P.; da Silva, P.P.M.; Spoto, M.H.F. Physicochemical and Sensory Characteristics of Fat-Free Goat Milk Yogurt with Added Stabilizers and Skim Milk Powder Fortification. *J. Dairy Sci.* **2016**, *99*, 3316–3324. [CrossRef]
21. Waszkiewicz-Robak, B. Food Additives Applied in Milk Fermented Beverages. *Ferment. Fruits Veg. Ind.* **2012**, *56*, 4–6.
22. Soomro, A.H. Effect of Milk Source and Stabilizers on the Compositional and Sensorial Quality of Yoghurt. *Pure Appl. Biol.* **2016**, *5*, 1316–1322. [CrossRef]
23. Asioli, D.; Aschemann-Witzel, J.; Caputo, V.; Vecchio, R.; Annunziata, A.; Næs, T.; Varela, P. Making Sense of the "Clean Label" Trends: A Review of Consumer Food Choice Behavior and Discussion of Industry Implications. *Food Res. Int.* **2017**, *99*, 58–71. [CrossRef]
24. Karaca, O.; Güzeler, N.; Tangüler, H.; Yaşar, K.; Akın, M. Effects of Apricot Fibre on the Physicochemical Characteristics, the Sensory Properties and Bacterial Viability of Nonfat Probiotic Yoghurts. *Foods* **2019**, *8*, 33. [CrossRef]
25. Tosif, M.M.; Najda, A.; Bains, A.; Kaushik, R.; Dhull, S.B.; Chawla, P.; Walasek-Janusz, M. A Comprehensive Review on Plant-Derived Mucilage: Characterization, Functional Properties, Applications, and Its Utilization for Nanocarrier Fabrication. *Polymers* **2021**, *13*, 1066. [CrossRef]
26. de Pérez-Chabela, M.L.; Cebollón-Juárez, A.; Bosquez-Molina, E.; Totosaus, A. Mango Peel Flour and Potato Peel Flour as Bioactive Ingredients in the Formulation of Functional Yogurt. *Food Sci. Technol.* **2021**, *42*. [CrossRef]
27. Diep, T.T.; Yoo, M.J.Y.; Rush, E. Effect of Tamarillo Fortification and Fermentation Process on Physicochemical Properties and Nutrient and Volatiles Content of Yoghurt. *Foods* **2021**, *11*, 79. [CrossRef]
28. Mkadem, W.; Belguith, K.; Ben Zid, M.; Boudhrioua, N. Fortification of Traditional Fermented Milk "Lben" with Date Powder: Physicochemical and Sensory Attributes. In Proceedings of the 1st International Electronic Conference on Processes: Processes System Innovation, Online, 17 May 2022; MDPI: Basel, Switzerland; p. 43.
29. Dello Staffolo, M.; Bertola, N.; Martino, M.; Bevilacqua, A. Influence of Dietary Fiber Addition on Sensory and Rheological Properties of Yogurt. *Int. Dairy J.* **2004**, *14*, 263–268. [CrossRef]
30. Alqahtani, N.K.; Alnemr, T.M.; Ahmed, A.R.; Ali, S. Effect of Inclusion of Date Press Cake on Texture, Color, Sensory, Microstructure, and Functional Properties of Date Jam. *Processes* **2022**, *10*, 2442. [CrossRef]
31. Mkadem, W.; Belguith, K.; Semmar, N.; Ben Zid, M.; ElHatmi, H.; Boudhrioua, N. Effect of Process Parameters on Quality Attributes of Lben: Correlation between Physicochemical and Sensory Properties. *LWT* **2022**, *155*, 112987. [CrossRef]
32. AOAC Association of Official Analytical Chemists. *Association of Official Analytical Chemists Gaithersburg (Maryland)*, 19th ed.; AOAC International: Washington, DC, USA, 2012.
33. Cichońska, P.; Pudło, E.; Wojtczak, A.; Ziarno, M. Effect of the Addition of Whole and Milled Flaxseed on the Quality Characteristics of Yogurt. *Foods* **2021**, *10*, 2140. [CrossRef]
34. Li, L.; Zhao, W.; Feng, X.; Chen, L.; Zhang, L.; Zhao, L. Changes in Fruit Firmness, Cell Wall Composition, and Transcriptional Profile in the Yellow Fruit Tomato 1 (Yft1) Mutant. *J. Agric. Food Chem.* **2019**, *67*, 463–472. [CrossRef]
35. Hunter, R.S.; Harold, R.w. *The Measurement of Appearance*; John Wiley and Sons: Hoboken, NJ, USA, 1987; Volume 29.
36. Mousavi, M.; Heshmati, A.; Daraei Garmakhany, A.; Vahidinia, A.; Taheri, M. Texture and Sensory Characterization of Functional Yogurt Supplemented with Flaxseed during Cold Storage. *Food Sci. Nutr.* **2019**, *7*, 907–917. [CrossRef]
37. Wang, C.; Gao, F.; Zhang, T.; Wang, Y.; Guo, M. Physiochemical, Textural, Sensory Properties and Probiotic Survivability of Chinese Laosuan Nai (Protein-Fortified Set Yoghurt) Using Polymerised Whey Protein as a Co-Thickening Agent. *Int. J. Dairy Technol.* **2015**, *68*, 261–269. [CrossRef]
38. Chambers IV, E. Analysis of Sensory Properties in Foods: A Special Issue. *Foods* **2019**, *8*, 291. [CrossRef] [PubMed]
39. Lawless, H.T.; Heymann, H. *Sensory Evaluation of Food: Principles and Practices*; Springer: New York, NY, USA, 2010; Volume 2.
40. Al-Farsi, M.; Alasalvar, C.; Al-Abid, M.; Al-Shoaily, K.; Al-Amry, M.; Al-Rawahy, F. Compositional and Functional Characteristics of Dates, Syrups, and Their by-Products. *Food Chem.* **2007**, *104*, 943–947. [CrossRef]
41. Hashim, I.; Khalil, A. Composition and Functional Properties of the Date Fruit Residue a By-Product of Date Syrup/Debis Production. *Nutr. Food Technol.* **2015**, *1*, 2. Available online: https://www.sciforschenonline.org/journals/nutrition-food/article-data/NFTOA-1-108/NFTOA-1-108.pdf (accessed on 1 January 2023). [CrossRef]
42. Majzoobi, M.; Karambakhsh, G.; Golmakani, M.T.; Mesbahi, G.R.; Farahnaky, A. Chemical Composition and Functional Properties of Date Press Cake, an Agro-Industrial Waste. *J. Agric. Sci. Technol.* **2019**, *21*, 1807–1817.
43. Alqahtani, N.K.; Barbabosa-Pliego, A.; Alnemr, T.M. Influence of Sweet Lupine Husk's Addition on Fat-Free Stirred Yogurt Properties. *Int. J. Dairy Sci.* **2021**, *16*, 41–47. [CrossRef]

44. Alqahtani, N.K.; Darwish, A.A.; El-Menawy, R.K.; Alnemr, T.M.; Aly, E. Textural and Organoleptic Attributes and Antioxidant Activity of Goat Milk Yoghurt with Added Oat Flour. *Int. J. Food Prop.* **2021**, *24*, 433–445. [CrossRef]
45. Kong, X.; Xiao, Z.; Du, M.; Wang, K.; Yu, W.; Chen, Y.; Liu, Z.; Cheng, Y.; Gan, J. Physicochemical, Textural, and Sensorial Properties of Soy Yogurt as Affected by Addition of Low Acyl Gellan Gum. *Gels* **2022**, *8*, 453. [CrossRef]
46. Bhat, S.V.; Deva, A.M.; Amin, T. Physicochemical and Textural Properties of Yogurt Fortified with Psyllium (*Plantago ovate*) Husk. *J. Food Process. Preserv.* **2017**, *42*, e13425. [CrossRef]
47. Eke, M.O.; Olaitan, N.I.; Sule, H.I. Nutritional Evaluation of Yoghurt-like Product from Baobab (*Adansonia digitata*) Fruit Pulp Emulsion and the Micronutrient Content of Baobab Leaves. *Adv. J. Food Sci. Technol.* **2013**, *5*, 1266–1270. [CrossRef]
48. Meenakshi, V.; Suganya, S.; Umamaheswari, T. Formulation of Value Enriched Probiotic Fruit Yoghurt. *Int. J. Curr. Microbiol. Appl. Sci.* **2018**, *7*, 1440–1450. [CrossRef]
49. Manzoor, S.; Yusof, Y.A.; Chin, N.L.; Mohamed Amin Tawakkal, I.S.; Fikry, M.; Chang, L.S. Quality Characteristics and Sensory Profile of Stirred Yogurt Enriched with Papaya Peel Powder. *Pertanika J. Trop. Agric. Sci.* **2019**, *42*, 519–533.
50. do Espírito Santo, A.P.; Perego, P.; Converti, A.; Oliveira, M.N. Influence of Milk Type and Addition of Passion Fruit Peel Powder on Fermentation Kinetics, Texture Profile and Bacterial Viability in Probiotic Yoghurts. *LWT* **2012**, *47*, 393–399. [CrossRef]
51. Dabija, A.; Oroian, M.A.; Mironeasa, S.; Sidor, A.M. Physictochemical and Sensory Properties of Yogurt with Seabuckthorn Powder, Rosehip Powder and Grape Seed Extract during Storage. In Proceedings of the 18th International Multidisciplinary Scientific GeoConference SGEM2018, Albena, Bulgaria, 30 June–9 July 2018; Volume 18, pp. 429–436. [CrossRef]
52. Matter, A.; Mahmoud, A.M.; Zidan, N.S. Fruit Flavored Yoghurt: Chemical, Functional and Rheological Properties. *Int. J. Environ. Agric. Res.* **2016**, *2*, 1–11.
53. Mahdian, E.; Tehrani, M.M. Evaluation the Effect of Milk Total Solids on the Relationship Between Growth and Activity of Starter Cultures and Quality of Concentrated Yoghurt. *J. Agric. Environ. Sci.* **2007**, *2*, 587–592.
54. Garcia-Amezquita, L.E.; Tejada-Ortigoza, V.; Serna-Saldivar, S.O.; Welti-Chanes, J. Dietary Fiber Concentrates from Fruit and Vegetable By-Products: Processing, Modification, and Application as Functional Ingredients. *Food Bioprocess Technol.* **2018**, *11*, 1439–1463. [CrossRef]
55. Lee, H.S.; Song, M.W.; Kim, K.-T.; Hong, W.-S.; Paik, H.-D. Antioxidant Effect and Sensory Evaluation of Yogurt Supplemented with Hydroponic Ginseng Root Extract. *Foods* **2021**, *10*, 639. [CrossRef]
56. Arabshahi-Delouee, S.; Ghochani, S.R.; Mohammadi, A. Effect of Flaxseed (*Linum usitatissimum*) Mucilage Mucilage on Physicochemical and Sensorial Properties of Semi-Fat Set Yoghurt. *J. Food Biosci. Technol.* **2020**, *10*, 91–100.
57. Rojas-torres, S.A.; Quintana, S.E.; García-zapateiro, L.A. Natural Yogurt Stabilized with Hydrocolloids from Butternut Squash (*Cucurbita moschata*) Seeds: Effect on Physicochemical, Rheological Properties and Sensory Perception. *Fluids* **2021**, *6*, 251. [CrossRef]
58. Güler-Akın, M.B.; Ferliarslan, I.; Serdar Akın, M. Apricot Probiotic Drinking Yoghurt Supplied with Inulin and Oat Fiber. *Adv. Microbiol.* **2016**, *06*, 999–1009. [CrossRef]
59. Tseng, A.; Zhao, Y. Wine Grape Pomace as Antioxidant Dietary Fibre for Enhancing Nutritional Value and Improving Storability of Yogurt and Salad Dressing. *Food Chem.* **2013**, *138*, 356–365. [CrossRef] [PubMed]
60. Łopusiewicz, Ł.; Drozłowska, E.; Siedlecka, P.; Mężyńska, M.; Bartkowiak, A.; Sienkiewicz, M.; Zielińska-Bliźniewska, H.; Kwiatkowski, P. Development, Characterization, and Bioactivity of Non-Dairy Kefir-Like Fermented Beverage Based on Flaxseed Oil Cake. *Foods* **2019**, *8*, 544. [CrossRef]
61. Alqahtani, N.K.; Helal, A.; Alnemr, T.M.; Marquez, O. Influence of Tomato Pomace Inclusion on the Chemical, Physical and Microbiological Properties of Stirred Yoghurt. *Int. J. Dairy Sci.* **2020**, *15*, 152–160. [CrossRef]
62. Otálora, M.C.; Wilches-Torres, A.; Gómez Castaño, J.A. Evaluation of Guava Pulp Microencapsulated in Mucilage of Aloe Vera and Opuntia Ficus-Indica as a Natural Dye for Yogurt: Functional Characterization and Color Stability. *Foods* **2022**, *11*, 2380. [CrossRef] [PubMed]
63. Mohammed, M.; Alqahtani, N.K. Design and Validation of Automated Sensor-Based Artificial Ripening System Combined with Ultrasound Pretreatment for Date Fruits. *Agronomy* **2022**, *12*, 2805. [CrossRef]
64. Hussain, S.; Mohamed, A.A.; Alamri, M.S.; Saleh, A.; Ibraheem, M.A.; Qasem, A.A.A.; Shamlan, G.; Ababtain, I.A. Rheological, Textural, and Sensory Properties of Non-Fat Yogurt Containing Cress (*Lepidium sativum*) Seed Gum and Various Starches. *Food Sci. Technol.* **2022**, *42*. [CrossRef]
65. Brodziak, A.; Król, J.; Matwijczuk, A.; Czernecki, T.; Glibowski, P.; Wlazło, Ł.; Litwińczuk, A. Effect of Sea Buckthorn (*Hippophae rhamnoides* L.) Mousse on Properties of Probiotic Yoghurt. *Appl. Sci.* **2021**, *11*, 545. [CrossRef]
66. Mendoza-Taco, M.M.; Cruz-Hernández, A.; Ochoa-Flores, A.A.; Hernández-Becerra, J.A.; Gómez-Vázquez, A.; Moo-Huchin, V.M.; Piñeiro-Vázquez, Á.; Chay-Canul, A.J.; Vargas-Bello-Pérez, E. Physicochemical Characteristics of Yogurt from Sheep Fed with Moringa Oleifera Leaf Extracts. *Animals* **2022**, *12*, 110. [CrossRef]

67. Aljutaily, T.; Barakat, H.; Moustafa, M.M.A.; Rehan, M. Incorporation of Sukkari Date in Probiotic-Enriched Fermented Camel Milk Improves the Nutritional, Physicochemical, and Organoleptical Characteristics. *Fermentation* **2022**, *8*, 5. [CrossRef]
68. Fathy, H.M.; Abd El-Maksoud, A.A.; Cheng, W.; Elshaghabee, F.M.F. Value-Added Utilization of Citrus Peels in Improving Functional Properties and Probiotic Viability of Acidophilus-Bifidus-Thermophilus (ABT)-Type Synbiotic Yoghurt during Cold Storage. *Foods* **2022**, *11*, 2677. [CrossRef] [PubMed]

foods

MDPI

Article

Apple Pomace as an Ingredient Enriching Wheat Pasta with Health-Promoting Compounds

Dorota Gumul [1], Marek Kruczek [1], Eva Ivanišová [2], Jacek Słupski [3] and Stanisław Kowalski [1,*]

1 Department of Carbohydrates Technology and Cereal Processing, Faculty of Food Technology, University of Agriculture in Krakow, Balicka Street 122, 31-149 Krakow, Poland
2 Institute of Food Science, Faculty of Biotechnology and Food Sciences, Slovak University of Agriculture, Trieda A. Hlinku 2, 949 76 Nitra, Slovakia
3 Department of Plant Products Technology and Hygiene Nutrition, Faculty of Food Technology, University of Agriculture in Krakow, Balicka Street 122, 31-149 Krakow, Poland
* Correspondence: rrkowals@cyf-kr.edu.pl

Abstract: The global overproduction of apples is associated with large amounts of post-production waste, for which new forms of utilization should be sought. Therefore, we aimed to enrich wheat pasta with apple pomace in various percentages (10, 20, 30 and 50%). The content of total polyphenols, individual polyphenols (using UPLC-PDA-MS/MS methods) and dietary fibre, chemical composition and physical properties of the resulting pasta were determined. The addition of apple pomace to pasta resulted in increased levels of pro-health compounds: total polyphenols, phenolic acids, quercetin derivatives, flavon-3-ols and dihydrochalcones as well as dietary fibre. Decreases in hardness and maximum cutting energy were also observed in pasta supplemented with apple pomace compared to control pasta. Water absorption capacity was not influenced by the addition of apple pomace, with the exception of pasta made with 50% apple pomace.

Keywords: apple pomace; bioactive compounds; cooking properties; dietary fibre; pasta; polyphenol compounds; texture; dietary fibre

Citation: Gumul, D.; Kruczek, M.; Ivanišová, E.; Słupski, J.; Kowalski, S. Apple Pomace as an Ingredient Enriching Wheat Pasta with Health-Promoting Compounds. *Foods* **2023**, *12*, 804. https://doi.org/10.3390/foods12040804

Academic Editors: Claudia Monika Haros and Costas Stathopoulos

Received: 9 January 2023
Revised: 6 February 2023
Accepted: 10 February 2023
Published: 13 February 2023

1. Introduction

The global annual production of apples reached more than 93 million tons in 2021. The main producers of apples worldwide are China, India, the USA, Turkey and Poland. Poland is the largest producer of apples in the European Union. About 30% of apples are processed into products such as juice, cider and dried products. These technological processes generate up to 30% waste, i.e., apple pomace [1]. It is estimated that plant raw materials such as vegetables and fruits are responsible for more than 20% of losses in the supply chain, becoming bio-waste [2]. According to numerous authors [3,4], about 12% of apple pomace in Poland goes to landfills, which results in its contamination, while the rest is used as a raw material, mainly for the production of pectin, compost for soil fertilization or as a compound feed for animals. Apple pomace can also be applied as a raw material for the production of biogas, ethyl alcohol and organic acids, including citric acid produced by *Aspergillus niger* or as fibre preparations [5–8].

Apple pomace is a heterogeneous mass consisting of skin and flesh (95%), seeds (2–4%) and stems (1%) [9]. It contains numerous compounds, the quantity and quality of which depend on the variety of apples, climatic and soil conditions, the type of apple processing technology and the method of obtaining pomace [10]. However, apple pomace is a valuable source of health-promoting compounds, mainly dietary fibre (DF) (35–65 g/100 g dry matter DM) and bioactive substances from the group of polyphenols (262–856 mg/100 g DM) [11,12].

Such compounds present in apple pomace have demonstrated, among other properties, hypoglycaemic, hypocholesterolaemic and anti-cancerogenic activity. They reduce postprandial glucose levels and hypertension, have anti-inflammatory, antiviral, antibacterial, antiallergic and anticoagulant effects and also reduce the risk of such diseases as atherosclerosis and other cardiovascular diseases, cataract, diabetes, genetic damage, bone degeneration and neurodegenerative diseases including Alzheimer's disease [13–19].

It was only in the last decade that a number of publications suggested the possibility of enriching traditional products by adding by-products. Fruit and vegetable pomaces in particular have been used to enrich products. Among fruit pomaces, apple pomace deserves attention due to the global overproduction of apples [1,2]. Research on apple pomace concerned their use as a fortifying agent in the production technology of biscuits and muffins [20], where 10% and 15% of wheat flour was replaced with apple pomace. Apple pomace in the amount of 3 to 9% was used to produce gluten-free brown rice crackers [21]. Kirbas et al. [22] replaced 5 to 15% of rice flour with apple pomace flour to make batter and cakes. Meanwhile, Drozdz et al. [23] and Reis et al. [24] obtained extruded snacks with 10 to 30% apple pomace.

Pasta is a traditional cereal product which is readily accepted by consumers, because of its ease of preparation and sensory values [25,26]. An additional asset of pasta is the versatility of its use. A valuable feature of pasta is its low cost of production, making it easily accessible to every household, regardless of their budget. It should be emphasized that the nutritional value of pasta will vary depending on the recipe. Egg pasta will have a higher protein content than pasta based on semolina and water alone, and the amount of dietary fibre depends mainly on its content in semolina [25,26]. Studies on increasing the nutritional and, above all, health-promoting value of pasta have been reported, but it should be noted that the addition of gluten-free ingredients to pasta dough contributes to the weakening of the gluten network and thus the structure of the product [25,26]. This can adversely affect the quality, and above all, the physical characteristics of the pasta; alternatively, it may improve the quality of the product when the ingredients are properly selected, especially those with which the product will be fortified [25–27]. Recently, publications on the possibility of producing pasta using fruit or vegetable pomace appeared. However, they mainly focused on the physical characteristics of products, while paying less attention to the health-promoting qualities of this popular product [27–33]. Therefore, suggesting that apple pomace is a source of health-promoting compounds, we decided to comprehensively investigate pasta with apple pomace, both in terms of the content of nutrients and health-promoting ingredients and the physical characteristics. In this study, we produced, via the process of low-temperature extrusion, egg-wheat pasta fortified with apple pomace. It enabled the utilization of a by-product, and the resulting pasta was enriched with polyphenol compounds (such as quercetin derivatives, chlorogenic acid and phloridzin) present in the apple pomace. The aim of the study was to determine the influence of the addition of different amounts of dried apple pomace (replacing 10, 20, 30 and 50% of wheat flour in recipe) on the content of health-promoting compounds from the polyphenol group, as well as on the texture, quality and functional features of wheat pasta.

2. Materials and Methods

2.1. Materials

Materials in this work were dried and milled apple pomace (ZPOW Hortino Leżajsk Sp.z.o.o Poland) and egg-wheat pasta prepared with dried apple pomace replacing flour in various percentages: 10, 20, 30 and 50%; see Table 1. Samples were coded as follows:

Control—control wheat pasta (no dried apple pomace added).
P 10%—wheat pasta with 10% apple pomace.
P 20%—wheat pasta with 20% apple pomace.
P 30%—wheat pasta with 30% apple pomace.
P 50%—wheat pasta with 50% apple pomace.

Table 1. Composition of the mixture used to prepare wheat pasta.

	Wheat Flour (g)	Apple Pomace (g)	Distilled Water (mL)	Egg Mass (g)	Salt (g)
Control	500	0	150	56	5
P 10%	450	50	160	56	5
P 20%	400	100	170	56	5
P 30%	350	150	210	56	5
P 50%	250	250	340	56	5

2.2. Pasta Preparation

Wheat pasta with dried apple pomace (10, 20, 30 and 50%) and a control sample were obtained by mixing the ingredients according to the recipe (Table 1) in a rotary-roller mixer (Laboratory Spiral Mixer SP 12, Diosna, Germany) for about 20 min at low speed. Pasta production was carried out using a Gina low pressure extruder (Ostoni, Italy). The length of the screw (diameter of 5.5 cm) was 30 cm, and it ended with a forming nozzle (diameter: 1.7 mm). The conditions applied during extrusion were as follows: pressure was about 3.4×10^5 Pa, and temperature was 50 °C. The extruded pasta was dried in a drying chamber (8 h at 40 °C) with up to 12.5% moisture.

About 110 g of prepared pasta was boiled in 1000 mL of distilled water for 8 min. The cooking time has been previously determined in order to obtain *al dente* pasta according to Hirawan et al. [34]. After cooling, the pasta was frozen (−20 °C) and then freeze-dried for 24 h in a freeze dryer (Labconco FreeZone 6, USA) at a temperature of −47 °C and pressure of 37 Pa. The freeze-dried pasta was stored at room temperature for further analysis. Before analysis, the pasta was ground into a powder using a Laboratory Mill 3100 (Perten Instruments, Springfield, IL, USA) equipped with a 0.88 mm mesh.

2.3. Chemical Composition of Apple Pomace and Pasta

Protein (Nx5.7) was determined using the Kjeldahl method (AOAC method No. 920.87) using the Kjeltec 2200 extraction unit (Foss, Hillerød, Denmark), fat according to the Soxhlet method (AOAC method No. 953.38) using Soxtec Avanti 2055 (Foss, Denmark), and contents of ash and reducing sugars were determined according to AOAC (2006) methods (AOAC method: 920.183 and AOAC method No. 930.05) [35]. The contents of non-starch polysaccharides, i.e., total (TDF), soluble (SDF) and insoluble (IDF) dietary fibre, were determined using method 32-07 AACCI. TDF was calculated as the sum of soluble and insoluble fractions. Ground samples were dispersed in water and treated with alpha-amylase, protease and glucosidase to remove starch and protein. The residue was precipitated with ethanol, filtered and washed with ethanol and acetone, and dried. TDF was calculated as the mass of the residue minus the protein and ash content [36]. Pectin concentration was determined using the carbazole method. Briefly, 2 g of pomace was weighed, 40 mL of ethanol (80%) was added, and then they were heated under reflux (30 min) and filtered. The filter with the precipitate was transferred to a flask, 50 mL of distilled water was added, brought to the boil, filtered hot and the filtrate was made up to 100 mL. The resulting extract contained pectins [37]. Each of the above-mentioned determinations was performed in at least 2 replicates.

2.4. Content of Antioxidant in Apple Pomace and Pasta

Determination of bioactive compounds was performed using spectrophotometric methods. Antioxidant constituents were determined in the ethanol extracts; 0.6 g of the sample was dissolved in 30 mL of 80 g/100 g ethanol, shaken in the dark for 120 min (electric shaker: type WB22, Memmert, Schwabach, Germany), and centrifuged (15 min, 4500 rpm. 1050× *g*) in a centrifuge (type MPW-350, MPW MED. Instruments, Warsaw, Poland). The supernatant was decanted and stored at −20 °C for further analyses.

Determination of total polyphenol content (TPC) was performed with the spectrophotometric method using Folin–Ciocalteu reagent (F–C reagent), in accordance with Singleton et al. [38]; 5 mL of the extract was diluted to a volume of 50 mL with distilled water, and 5 mL of the diluted extract was combined with 0.25 mL of F–C reagent (previously diluted with distilled water in the proportion 1:1 *v/v*) and 0.5 mL of 7% Na_2CO_3. The contents were vortexed (WF2, Janke & Kunkel, Staufen, Germany) and stored for 30 min in a dark place. The absorbance was measured using a Helios Gamma 100–240 (Thermo Fisher Scientific, Runcorn, UK), at the wavelength λ = 760 nm. The results were converted to mg catechin/100 g DM.

Determination of flavonoids was performed using the spectrophotometric method, in accordance with El Hariri et al. [39]; 0.5 mL of the extract was combined with 1.8 mL of distilled water and 0.2 mL of 2-aminoethyldiphenylborate reagent in a test tube. The contents were vortexed, and the absorbance was measured at the wavelength λ = 404 nm. Flavonoid content was expressed as mg of rutin/100 g DM.

Determination of polyphenols individual was performed using ultra-performance liquid chromatography/photodiode array detection/tandem mass spectrometry (UPLC-PDA-MS/MS). Samples of the raw material (about 1 g) were extracted with 10 mL of a mixture containing methanol of HPLC purity level (30 mL/100 mL), ascorbic acid (2.0 g/100 mL) and acetic acid in the amount of 1.0 mL/100 mL of the reagent. Extraction was carried out twice by incubation for 20 min under sonication (Sonic 6D, Polsonic, Warsaw, Poland) and occasionally mixing. The suspension was then centrifuged at 19,000× *g* for 10 min and the supernatant was filtered through a 0.20 µm Hydrophilic PTFE membrane (marble filter Simplicity Millex, Merck, Darmstadt, Germany) and used for analysis. Phenolic compounds were analysed using an Acquity Ultra-Performance Liquid Chromatograph equipped with a Binary Solvent Manager (BSM), Sample Manager (SM) combined with a PDA detector and quadrilateral time of flight (Q-TOF) (Waters, Manchester, United Kingdom). The analysis was carried out on a 2.1 × 100 mm UPLC BEH C18 column containing 1.7 µm particles (Waters, Manchester, UK). Data were collected and analysed using MassLynx v 4.1 (Waters) software. Anthocyanins were analysed in the positive ion mode and the remaining polyphenols in the negative ion mode. Quantification of phenolic compounds was performed using external standard curves, using reference compounds selected on the basis of the target analyte/structure standard (chemical structure or functional group). The standards were prepared in concentrations ranging from 0.05 to 5 mg/mL. The resulting correlation coefficient was $R^2 \leq 0.9998$. The results were expressed in mg per 100 g DM.

2.5. Texture Profile Analysis: Determination of Maximum Cutting Force and Energy

The maximum cutting force and energy for freshly cooked pasta were determined on a TAXT2 plus texture analyser (Stable Micro Systems, Godalming, UK) using a Warner-Bratzle adapter with a flat knife for cutting the sample at a speed of 3 mm/s. Exponent v. 4.0.13.0 software was used to collect data. The measurements were made in seven replications; two extreme results were discarded, and the others were used to calculate the arithmetic mean.

2.6. Water Absorption of Pasta

The water absorption of pasta was determined in accordance with the methodology of Tudoric et al. [40]. Briefly, dry pasta (10 g) was weighed, cooked in 500 mL of water, drained and then reweighed. The measurement was repeated three times.

Calculation of the water absorption (WA) of the pasta (%) was made on the basis of the formula:

$$WA = (a - b)/a, \tag{1}$$

where:

a—is the weight of cooked pasta (g);

b—is the weight of pasta before cooking (g) [40].

2.7. Statistical Analysis

The experimental data were subjected to analysis of variance (Duncan's test), at the confidence level of 0.05, using Statistica v. 8.0 (StatSoft, Inc., Tulsa, OK, USA). All measurements were made at least in duplicate. Pearson's correlation coefficient was calculated at alpha = 0.01.

3. Results and Discussion

3.1. Apple Pomace Characteristics

The chemical composition of apple pomace is presented in Figure 1. According to Wang et al. [41], the amount of protein was 3.8 g/100 g DM, fat was 3.8 g/100 g DM, TDF was about 26.5 g/100 g DM, and ash was 1.8 g/100 g DM in apple pomace. Jin et al. [42] determined the amount of protein, fat and ash at the levels of 4.7, 4.2 and 1.5 g/100 g DM. Jannati et al. [43] determined the amounts of protein, fat, TDF and ash as 1.2, 0.6, 14.5 and 2.5 g/100 g DM, respectively. Ktenioudaki et al. [44] reported that the amounts of protein, fat, TDF and ash were 2.4, 2.7, 42.5 and 1.7 g/100 g DM. According to Pieszka et al. [45] and Leyva-Corral et al. [46], the amounts of protein, fat, TDF and ash were, respectively, 3.73–3.8 g/100 g DM, 1.8–2 g/100 g DM, 36–45 g/100 g DM and 1.88–2 g/100 g DM.

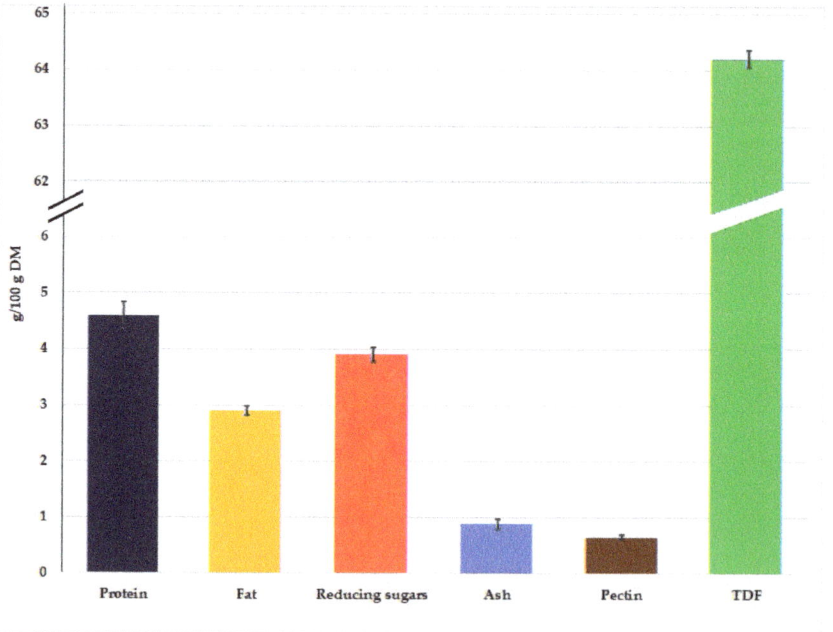

Figure 1. Basic chemical composition of apple pomace.

This study showed that the constituent with the largest share in the dry matter of apple pomace is fibre, which was also confirmed by the above-mentioned authors [41,43,45,46]. It should also be emphasized that the results of our research were consistent with the results of the above authors, and the few discrepancies may result from climatic, soil, agrotechnical conditions or apple variety and the method of obtaining pomace [45]. In our study, it is important that we had a mixture without varieties of apple pomace from various varieties of apples grown in eastern Poland and the method of obtaining the pomace (methods of press and number of press cycles). According to Constenla et al. [10] and Kieliszek et al. [47], the above-mentioned factors may affect the composition of pomace, because the amount of protein may be in the range of 2.7–5.7 g/100 g DM, ash 1.1–2.2 g/100 g DM, carbohydrate 20–57.4 g/110 g DM and dietary fibre 43–61.6 g/100 g DM.

It should be noted that the dietary fibre (DF) of apple pomace constitutes two-thirds of the insoluble fraction, which consists mainly of cellulose, hemicellulose and lignin, and pectins are the dominant component of the soluble fibre fraction [48]. The amount of pectins recorded in this work in apple pomace is 0.65 g/100 g DM. Pectins are very important compounds because of their various physiological properties. They exert prebiotic effects and are also fermented in the large intestine by the local microflora, resulting in the formation of short-chain fatty acids (SCFA) which are absorbed and converted into colonic mucosa, liver or peripheral tissues. A relationship has been identified between the consumption of pectins and maintenance of regular blood cholesterol concentrations, and the reduction of post-prandial glycemic responses [48].

Apple pomace is a valuable source of polyphenols and flavonoids, the amounts of which determined in this research were 89.39 mg gallic acid/100 g DM and 94.27 mg rutin/100 g DM. According to the study by Ćetković et al. [11], the total amount of polyphenols ranged in apple pomace from 420 to 867 mg chlorogenic acid/100 g DM, whereas flavonoids in apple pomace ranged from 45 to 119 mg rutin/100 g DMDM [11]. Persic et al. [49] noted the total polyphenol content in the range of 19–50 mg gallic acid/100 g DM [49]. The content of total polyphenols in plant material is influenced not only by the extraction conditions, but also by the various ways in which the results were expressed (e.g., another type of phenolic compound used to calculate polyphenols) [50].

UPLC-PDA-MS/MS analysis of the profile of individual phenolic compounds present in apple pomace was performed (Figure 2). It was found that among phenolic acids, chlorogenic acid had the largest share (20.65 mg/100 g DM Figure 2), which is similar to the other authors' results of 92–104 mg/100 g DM [51]. Other phenolic acids were also identified; cryptochlorogenic acid was determined at 1.03 mg/100 g DM and p-coumaroylquinic acid at 0.16 mg/100 g DM (Figure 2). According to Kammerer et al. [12], the amount of p-coumaroylquinic acid was 0.18 mg/100 g DM [12]. Among the analysed flavonols, quercetin derivatives had a large proportion, while quercetin-3-O-galactoside, quercetin-3-O-rhamnoside and quercetin-3-O-xyloside were dominant (Figure 2). Other authors have also noted similar results for quercetin-3-O-glucoside: 28.6–61.0 mg/100 g DM [11] and 52.1–68.1 mg/100 g DM [52]. Flavan-3-ols and dihydrochalcones are also a very important group of phenolic compounds in apple pomace. Among flavan-3-ols, catechin at 1.40 mg/100 g DM, procyanidin B2 at 2.53 mg/100 g DM and epicatechin at 0.70 mg/100 g DM were noted (Figure 2). The catechin content was previously reported to be 1.7–12.7 mg/100 g DM [11], 0.24 mg/100 g DM [12] and 0.94–1.4 mg/100 g DM [52]. The content of epicatechin in the apple pomace assessed by other authors was: 2.4–17.3 mg/100 g DM [11]; 0.93 mg/100 g DM [12]; 14–19 mg/100 g DM [52]; 12.23 mg/100 g DM [46]. They also determined the amount of procyanidin B2: 2.3–10 mg/100 g DM [51], 0.93 mg/100 g DM [12] and 9.3–16 mg/100 g DM [52]. Among dihydrochalcones in apple pomace, phloridzin was dominant, which was determined at 15.47 mg/100 g DM (Figure 2). Similar values (17.97 mg/100 g DM) were reported by Leyva-Corral et al. [46]. In the study by Lyu et al. [9], the content of dihydrochalcones in apple pomace ranged from 68.8 to 253.5 mg/100 g DM.

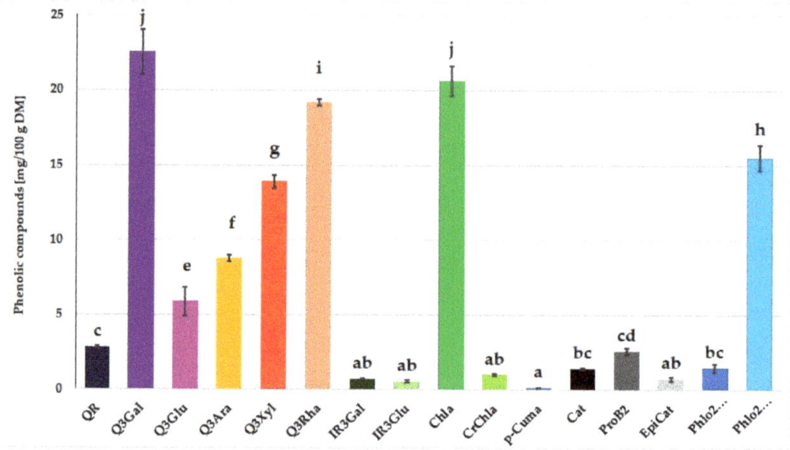

Figure 2. Quality and quantity of phenolic compounds in apple pomace: QR (quercetin-O-rutinoside), Q3Gal (quercetin-3-O-galactoside), Q3Glu (quercetin-3-O-glucoside), Q3Ara (quercetin-3-O-arabinoside), Q3Xyl (quercetin-3-O-xyloside), Q3Rha (quercetin-3-O-rhamnoside), IR3Gal (isorhamnetin-3-O-galactoside), IR3Glu (isorhamnetin-3-O-glucoside), Chla (chlorogenic acid), CrChla (cryptochlorogenic acid), p-Cuma (p-coumaroylquinic acid), Cat ((+) catechin), ProB2 (procyanidin B2), EpiCat ((-) epicatechin), Phlo2Xyl (phloretin-2-O-xylosylglucoside), Phlo2Gluc (phloretin 2-O-glucoside (phloridzin)). Different letters over the bars represent the statistically significant difference of average values (α = 0.05).

3.2. Characteristics of Pasta with Different Percentage Content of Apple Pomace

Table 2 contains the results of the determination of total polyphenols, flavonoids, proteins, fat, reducing sugars and ash in cooked pasta with or without apple pomace. The control sample contained 11.44 g/100 g DM protein, and the addition of apple pomace reduced the amount of this ingredient from 6% to 19% compared to the control (Table 2). The partial replacement of wheat flour (which is a source of protein) with apple pomace and also the cooking process probably resulted in a dilution effect of this important nutrient [53].

Table 2. Total polyphenols, flavonoid content and chemical composition of gluten pasta with apple pomace.

Sample	Total Polyphenols	Flavonoids	Protein	Fat	Reducing Sugars	Ash
	(mg catechin/100 g DM)	(mg rutin/100 g DM)	(g/100 g DM)			
Control	21.87 ± 0.92 a *	10.01 ± 1.76 a	11.44 ± 0.09 e *	2.25 ± 0.05 e	0.84 ± 0.03 a	0.55 ± 0.01 b
P 10%	47.79 ± 0.00 b	19.03 ± 1.46 b	10.78 ± 0.06 d	2.05 ± 0.05 d	1.07 ± 0.01 b	0.50 ± 0.01 a
P 20%	68.03 ± 1.11 c	38.18 ± 1.95 c	10.51 ± 0.12 c	1.92 ± 0.06 c	1.07 ± 0.01 b	0.63 ± 0.01 d
P 30%	89.06 ± 1.59 d	53.11 ± 1.29 d	10.14 ± 0.03 b	1.73 ± 0.01 b	1.07 ± 0.01 b	0.59 ± 0.01 c
P 50%	111.28 ± 0.00 e	87.48 ± 1.46 e	9.23 ± 0.04 a	1.39 ± 0.05 a	1.05 ± 0.01 b	0.81 ± 0.01 e

* Different letters in the column represent the statistically significant difference of average values (α = 0.05).

In the case of fat, the highest content was observed in the control sample (2.25 g/100 g DM), and with increasing apple pomace content, the amount of fat decreased in the range from 9% to 38% as compared to the control. Pasta with the highest percentage of apple pomace contained the lowest amount of fat (1.39 g/100 g DM; Table 2). The main source of fat in the pasta was egg mass, the addition of which was the same in all samples (Tables 1 and 2). The addition of eggs improves the nutritional value of the obtained pasta. Although apple pomace contains more fat (2.90 g/100 g DM) than wheat flour (about 1.0 g/100 g DM), which was substituted, it was likely that gluten present in wheat flour bound egg fat better than apple pomace, which was why during the cooking process fat could be partially washed out [25]. According to available sources, the degree of fat binding depends mostly on the physicochemical properties of the raw material and next on the production process parameters [54], so it could be the reason for the reduced fat content in the final product. It was found that, regardless of the proportion of applied apple pomace, the content of reducing sugars in pasta was at the same level (about 1.07 g/100 g DM) and about 27% higher than the control. Although the source of reduced sugars in pasta was apple pomace, the process of cooking the pasta probably contributed to the washing out of the above ingredients from the products. Only strongly bound reducing sugars remained, which resulted in similar results, mostly not correlated with the applied proportion of apple pomace. However, the content of mineral components ranged from 0.50 g/100 g DM to 0.81 g/100 g DM, and the largest amount was observed in pasta with 50% apple pomace addition (0.81 g/100 g DM; Table 2). It is likely that the cooking of pasta contributed to a partial leaching of the above-mentioned ingredients from products, which resulted in different results that were not especially correlated with the applied apple pomace.

In a study by Gałkowska et al. [28] on noodles with the addition of 5 and 10% blackcurrant pomace, the amount of protein in these noodles did not change compared to the control. In contrast, the fat content increased twofold and the ash content by about 40% in pasta supplemented with blackcurrant when compared to the control [28]. In the study by Nur Azura et al. [31] on yellow alkaline noodles with mango peel (10–30%), it was noted that the content of ash, fat and protein in the noodles did not change after the use of the above-mentioned additive compared to the control. The amount of crude fibre increased 2–14-fold, and the amount of carbohydrate decreased by 5–25% after the introduction of mango peel to pasta relative to the control. In research by Isa et al. [32], in which pasta with mango powder (2–6%) was obtained, it was found that the ash content increased by an average of 33%, protein by 3% and dietary fibre 7-fold compared to the control. The amount of fat decreased by 50% in the mango peel powder pasta compared to the control.

Total polyphenol and flavonoid contents in pasta with the addition of apple pomace is given in Table 2. It was found that the total amount of polyphenols (TPC) and flavonoids, after apple pomace addition, increased from 120% to 410% and from 90% to 774% as compared to the control. It was also found that this increase was proportional to the level of pomace added, and the largest increase was observed for 50% replacement of wheat flour by apple pomace (Table 2).

Ajila et al. [55] reported a 3.9-fold increase in the total polyphenol content in wheat pasta with mango peel powder (MPP) in the range up to 20% when compared to the control [55]. In the study by Tolve et al. [29] concerning the influence of the addition of grape pomace on the quality of durum wheat pasta, it was observed that the amount of polyphenols after the use of 5 and 10% grape pomace addition increased, respectively, seven- and twelvefold compared to the control.

Taking into account the profiles of flavonoids, phenolic acids, flavan-3-ols and dihydrochalcones which were identified by the UPLC-PDA-MS/MS method, it was found that the control sample, which consisted mainly of wheat flour, contained only two phenolic compounds (di-p-coumaroylspermidine and feruloylquinic acid), and the addition of apple pomace caused a significant increase in the amount of phenolic compounds (Figure 2, Table 3). The content of phenolic acids increased in pasta with the addition of apple pomace, although this increase was smaller than expected. It was probably related to the production stages (Table 3). It can be suggested that during the low-temperature extrusion process, there was a partial release of phenolic acids from the pomace fibre fraction. Pasta drying could also decarboxylate phenolic acids to 4-vinyl guaiacol, while the cooking process additionally resulted in their degradation, leaching and dissolving them in water [56]. Michalska et al. [57] clearly observed that polyphenol losses were strongly affected by the type and parameters of drying.

Table 3. Quality and quantity of phenolic compounds in pasta with apple pomace.

Compound	Control	P 10%	P 20%	P 30%	P 50%
Flavonols (mg/100 g DM)					
isorhamnetin-3-O-galactoside	0.00 ± 0.00 a	0.06 ± 0.00 b	0.19 ± 0.00 c	0.34 ± 0.07 d	0.51 ± 0.11 e
isorhamnetin-3-O-glucoside	0.00 ± 0.00 a	0.09 ± 0.00 b	0.27 ± 0.02 c	0.38 ± 0.00 d	0.45 ± 0.09 d
luteolin 6-C-hexoside-O-hexoside	0.00 ± 0.00 a *	0.00 ± 0.00 a	0.00 ± 0.00 a	0.00 ± 0.00 a	0.00 ± 0.00 a
luteolin O- hexoside-C-hexoside	0.00 ± 0.00 a	0.00 ± 0.00 a	0.00 ± 0.00 a	0.00 ± 0.00 a	0.00 ± 0.00 a
quercetin-O-rutinoside	0.00 ± 0.00 a	0.19 ± 0.00 b	0.51 ± 0.20 c	0.73 ± 0.12 d	1.05 ± 0.10 e
quercetin-3-O-galactoside	0.00 ± 0.00 a	1.91 ± 0.15 b	5.22 ± 0.13 c	7.32 ± 0.17 d	8.97 ± 0.25 e
quercetin-3-O-glucoside	0.00 ± 0.00 a	0.17 ± 0.02 b	1.24 ± 0.00 c	2.35 ± 0.09 d	3.41 ± 0.12 e
quercetin-3-O-arabinoside	0.00 ± 0.00 a	0.58 ± 0.11 b	1.70 ± 0.00 c	3.03 ± 0.30 d	4.19 ± 0.13 e
quercetin-3-O-xyloside	0.00 ± 0.00 a	1.91 ± 0.10 b	4.26 ± 0.27 c	5.18 ± 0.23 d	6.28 ± 0.51 e
quercetin-3-O-rhamnoside	0.00 ± 0.00 a	1.90 ± 0.00 b	4.80 ± 0.00 c	5.84 ± 0.57 d	6.72 ± 0.14 e
Phenolic acids (mg/100 g DM)					
chlorogenic acid	0.00 ± 0.00 a	1.23 ± 0.00 b	3.48 ± 0.00 c	4.15 ± 0.09 d	5.27 ± 0.13 e
cryptochlorogenic acid	0.00 ± 0.00 a	0.08 ± 0.00 b	0.19 ± 0.00 c	0.37 ± 0.06 d	0.61 ± 0.05 e
p-coumaroylquinic acid	0.00 ± 0.00 a	0.10 ± 0.11 a	0.26 ± 0.05 ab	0.32 ± 0.01 c	0.45 ± 0.09 d
caffeoyl dihydroxyphenyllactaoyl-tartaric acid	0.00 ± 0.00 a	0.21 ± 0.02 b	0.56 ± 0.07 c	0.64 ± 0.06 c	0.70 ± 0.10 c
1-O-p-coumaroylglycerol	0.00 ± 0.00 a	0.00 ± 0.00 a	0.20 ± 0.08 b	0.31 ± 0.02 b	0.40 ± 0.05 c
p-coumaroylspermidine	0.00 ± 0.00 a	0.27 ± 0.02 c	0.10 ± 0.04 b	0.00 ± 0.00 a	0.00 ± 0.00 a
di-p-coumaroylspermidine	0.30 ± 0.00 b	0.14 ± 0.03 a	0.10 ± 0.00 a	0.00 ± 0.00 a	0.00 ± 0.00 a
Feruloylquinic acid	0.09 ± 0.00 a	0.00 ± 0.00 a	0.21 ± 0.08 b	0.00 ± 0.00 a	0.00 ± 0.00 a
Flavon-3-ol (mg/100 g DM)					
(+) catechin	0.00 ± 0.00 a	0.22 ± 0.01 b	0.20 ± 0.00 b	0.28 ± 0.03 c	0.37 ± 0.04 d
procyanidin B2	0.00 ± 0.00 a	0.21 ± 0.00 b	0.71 ± 0.17 c	0.98 ± 0.00 d	1.16 ± 0.12 e
(-)epicatechin	0.00 ± 0.00 a	0.22 ± 0.01 b	0.37 ± 0.05 c	0.53 ± 0.00 d	0.92 ± 0.00 e
Dihydrochalcone (mg/100 g DM)					
phloretin-2-O-xylosylglucoside	0.00 ± 0.00 a	0.11 ± 0.00 b	0.29 ± 0.01 c	0.42 ± 0.04 d	0.69 ± 0.05 e
phloretin 2-O-glucoside (phloridzin)	0.00 ± 0.00 a	1.74 ± 0.00 b	4.02 ± 0.23 c	5.12 ± 0.05 d	6.10 ± 0.07 e

* Different letters in the row represent the statistically significant difference of average values ($\alpha = 0.05$).

Pasta samples with apple pomace were characterized by a significantly higher content of quercetin derivatives as well as flavon-3-ols and dihydrochalcones. It was associated with applied addition because, as already mentioned, apple pomace is an excellent source of the above-mentioned biologically active compounds, whereas the control pasta does not contain such compounds (Table 3). Nevertheless, it should be remembered that the particular stage of pasta production can contribute to the losses of these compounds, similarly to phenolic acids. Nevertheless, apple pomace applied as an additive enriched pasta with antioxidants (Table 3). Ajila et al. [55] reached a similar conclusion when pasta was fortified with additional mango peel [55]. Similar conclusions were made by Tolve et al. [29], who studied the effect of adding grape pomace on the polyphenol content in pasta. Gaita et al. [58], examining the effect of grape pomace (3–9%) on the content of polyphenols, noted that the higher the level of the additive, the higher the polyphenol content in pasta with the above-mentioned additive compared to the control. Gaita et al. [58] observed an increase in polyphenol content in the range of 31–97% in pasta with the addition of 3 to 9% grape pomace compared to the control.

Data related to dietary fibre (DF) in pasta with apple pomace addition are presented in Table 4. It was found that the amount of insoluble and soluble DF increased, respectively, from 3.6- to 17-fold and from 1- to 4.4-fold in pasta made with apple pomace in comparison to the control. At the same time, it was found that this increase was proportional to the increase in the pomace content, and the largest increase was observed for the pasta with 50% apple pomace content. Pasta supplemented with apple pomace was also characterized by a high total fibre content (11–36.73 g/100 g DM; Table 4).

Table 4. Dietary fibre content, texture and max cut energy in wheat pasta with apple pomace.

Sample	Dietary Fibre g/100 g DM			Hardness (N)	Max Cut Energy (J)
	Insoluble Fraction	Soluble Fraction	Total Fibre		
Control *	1.37 ± 0.01 a *	2.17 ± 0.05 a	3.54 ± 0.06 a	2.08 ± 0.21 c *	3.03 ± 0.56 c
P 10%	6.26 ± 0.08 b	4.75 ± 0.03 b	11.01 ± 0.05 b	2.14 ± 0.17 cd	2.41 ± 0.38 ab
P 20%	10.88 ± 0.10 c	6.02 ± 0.04 c	16.90 ± 0.06 c	2.01 ± 0.24 c	2.50 ± 0.28 b
P 30%	15.97 ± 0.05 d	8.12 ± 0.08 d	24.08 ± 0.03 d	1.60 ± 0.26 b	2.30 ± 0.54 ab
P 50%	24.96 ± 0.07 e	11.77 ± 0.10 e	36.73 ± 0.17 e	1.44 ± 0.15 a	2.04 ± 0.45 a

* Different letters in the column represent the statistically significant difference of average values (α = 0.05).

In the research performed by Ajila et al. [55], the effects of the addition of MPP to pasta on the content of soluble and water-insoluble fractions and total DF fraction were evaluated [55]. In pasta with MPP addition (0–20%), an increase in the content of DF fractions was observed as follows: from 24% to 57% (water-soluble fraction), from 87% to 144% (water-insoluble fraction) and from 61% to 107% (total fibre) compared to the control. In the study by Gałkowska et al. [28], the amount of DF increased 2.5- to 5-fold after the application of blackcurrant pomace in pasta relative to the control. Pasta with blackcurrant pomace was characterized by a two-fold increase in the soluble fraction of DF, and the insoluble fraction increased 4- to 7-fold when compared to the control. In research by Padalino et al. [30] on the supplementation of pasta with tomato pomace, they observed that the use of 10 and 15% tomato pomace resulted in the increases in the insoluble DF fraction content from 35 to 60%, soluble fraction from 23 to 43% and total DF from 31 to 54% as compared to the control. In the study by Kultys and Moczkowska-Wyrwisz [33] for pasta with carrot pomace from 10 to 30% and beetroot-apple pomace (10–30%), an increase in total dietary fibre was observed, on average by 85% in pasta with carrot pomace and by 80% in pasta with beetroot-apple pomace compared to the control. The insoluble fraction of DF increased in pasta with carrot pomace in the range of 49 to 108%, and in pasta with beetroot-apple pomace it increased in the range of 73 to 150% compared to the control. The soluble DF fraction was up to three and a half times higher in pasta with carrot pomace, and in beetroot-apple pomace it was twice as high as the control. It can be concluded that the content of DF and its fractions in our study corresponded to the proportion of apple pomace in pasta (Table 4) because apple pomace was a rich source of DF (64.21 g/100 g DM), which is confirmed by the studies of the above-cited authors [28,33,55].

In addition to the nutritional and health-promoting properties of pasta, this product should also be characterized by the desired culinary features (especially water absorption) and textural properties (hardness), i.e., the characteristics of pasta that were revealed after hydrothermal treatment (cooking). These properties are primarily determined by the pasta recipe and processing technology. It should be remembered that the introduction of an additional enriching ingredient to the basic pasta recipe can cause some disturbances in the starch-gluten network microstructure and, consequently, can lead to modification of the culinary and sensory properties (including textural) of the finished product [25,33].

The hardness and maximum cutting energy of pasta supplemented with apple pomace were lower than those of the control (Table 4). Pasta with the largest percentage of apple pomace (50%) was characterized by the lowest cutting energy. This addition caused a decrease in hardness of the final product by approximately 30% as compared to the control, so these pastas were softer and more susceptible to damage (Table 4). Many factors could have an impact on these results, first of all the starch capacity for pasting, which according to Dexter & Matsuo [59] affected the hardness of cooked pasta. Secondly, DF, which was a main component of apple pomace, had a decisive impact on the hardness of the finished products [55]. However, a lot depends on the structure, form and composition of DF present in the product. For example, inulin (polysaccharide) reduces the hardness of finished pasta [60]; arabinoxylans, which are water soluble, do not affect the texture of cooked pasta [61]; and bran fibre increases the hardness of the pasta [62].

In their research, Ajila et al. [55] found that the increase in hardness from 19% to 67% compared to the control was directly proportional to the proportion of applied MPP [55]. In the studies of Padalino et al. [30] on the effect of tomato pomace on the hardness of pasta, a two-fold increase in hardness after adding this additive compared to the control was noted. A similar increase in hardness was noted by Xu et al. [27] by fortifying the pasta with 5 to 20% apple pomace. In the study by Kultys and Moczkowska-Wyrwisz [33], pasta with 10–20% carrot pomace and beetroot-apple pomace had slightly lower hardness. However, the 30% share of the above-mentioned additives resulted in a 20% decrease in hardness in the case of carrot pomace and 25% in the case of beetroot-apple pomace of these enriched pastas compared to the control. In the research of Nur Azura et al. [31] on noodles with mango peel powder (10–30%), no change in hardness was noted compared to the control. In the case of the research presented here, it should be taken into consideration that during the dough formation, egg albumins formed disulphide bonds (S-S) with gluten proteins, participating in the formation of the gluten network. The fibre that was introduced with the apple pomace interfered with the protein–starch matrix during dough development. Fibre became another ingredient competing for water with proteins and starch. Assuming that fibre had a greater water absorption capacity than starch, it interfered with the transformation of protein structure as well as the incorporation of starch granules into the gluten network [25–27]. As a consequence, the structure of the pasta after drying can be less homogeneous and less compact, i.e., less hard, hence the lower observed hardness of the pasta supplemented with apple pomace as compared to the control (Table 4). As previously mentioned, the hardness of the product was affected by the ingredients presented in the recipe. Fruit pomaces were a source of various types of DF with different compositions, molecular weights and degrees of polymerization or methoxylation as in the case of pectins [60], which mainly determined the hardness of the finished product, and this has been confirmed by other authors [27,30,55].

Water absorption is the ability of a product to bind water, in this case during cooking. Pasta with good parameters should be characterized by high water absorption. The water absorption of the pasta is directly proportional to the cooking time of the pasta. The longer the cooking time, the more water is bound in the product [33]. In this study, all samples were prepared according to the established optimal cooking time. The absorption of water by pasta is affected not only by the content of starch and pectins, which have, among others, hydrophilic properties [63], but also the structure of starch, the number and location of hydroxyl groups in pectin and the degree of their methylation. This parameter is influenced by the internal structure of starch and pectins in dried pasta, as well as the presence of polyphenols, as well as the composition of dietary fibre [25,33]. All these factors affected the interactions between the product ingredients and water, and finally the water absorption capacity [64].

Only the percentage of pomace at the level of 50% increases water absorption by approximately 26% (Figure 3) compared to other samples. It should be emphasized that along with the increase in the percentage of apple pomace, the content of pectins in the finished products also increased. However, the degree of methylation of these pectins was also an important factor, because highly methylated pectins (which include apple pomace pectins [65]) were more hydrophobic than pectin molecules with a low degree of methylation, which could have affected the results [64]. Additionally, as reported by Sivam et al. [66], polyphenols, of which apple pomace is a valuable source, could combine with proteins, starch and other polysaccharides through hydrogen bonds, leading to increased intermolecular interactions, and thus changing their hydrophilic properties, which affected the water absorption of products. Mineral components, which were also provided by apple pomace, increase the water absorption of finished products [67]. This was confirmed by the high level of ash in pasta supplemented with 50% pomace (0.81 g/100 g DM; Table 2), which contributed to, among other effects, up to 26% increase in water adsorption in these final products (Figure 3). Taking into account the recipe of pasta with apple pomace obtained in this research, it should be emphasized that the comparable water absorption of this pasta with the control sample (except for pasta with 50% apple pomace) may be due to the fact that the product was obtained by means of low-temperature extrusion during which free flour fats and egg monoacylglycerols interacted with the starch polymer amylose, thus limiting its swelling.

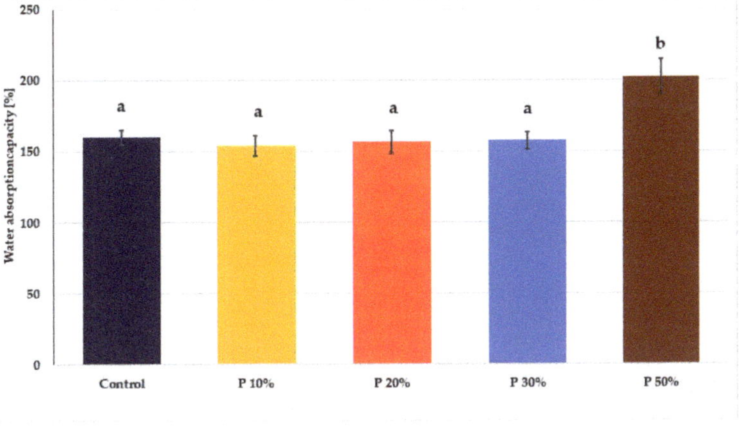

Figure 3. Water absorption capacity of wheat pasta with added apple pomace. Different letters over the bars mean statistically different average values ($\alpha = 0.05$).

According to Gałkowska et al. [28], pasta fortified with blackcurrant pomace was characterized by lower water absorption than the control, and the greater the addition of blackcurrant pomace, the lower the water absorption by the pasta. Xu et al. [27] observed in pasta with apple pomace that the increase in the proportion of apple pomace was associated with a slight increase in water absorption by the product (4%) as compared to the control. However, in the study by Tolve et al. [29], in which pasta was fortified with grape pomace, it was found that with the increase in the percentage of this additive, the water absorption of the pasta decreased. The research by Kultys and Moczkowska-Wyrwisz [33] clearly showed that the recipe and the type of additive used influence the water absorption of pasta. In the studies mentioned above, the authors used the same percentage of two additives (10–30%), carrot pomace and beetroot-apple pomace, claiming that the water absorption of pasta increases when carrot pomace is used or does not change when beetroot-apple pomace is used compared to the control. This can be explained by the different composition of dietary fibre in these two types of additives. In carrot pomace, the soluble fraction of fibre predominated, which guaranteed greater water absorption. It can therefore be said that in our study, higher water absorption (by 26%) by pasta with 50% apple pomace compared to the control may be associated not only with a higher content of ash, but also with the highest content of soluble fraction of dietary fibre in this pasta (Table 4).

Sun-Waterhouse et al. [64] observed that the proportion of elderberry juice in wheat pasta reduced its water absorption, except for pasta where there were also highly methylated pectins. In the sample with highly methylated pectins (exceeding hydrophobic properties) and elderberry juice, greater water absorption was observed than in pasta with only highly methylated pectins, most likely due to the presence of polyphenols in the added juice, which changed the conformation of gluten proteins and the structure of starch present in the flour, probably affecting the water absorption of the obtained pasta. Sun-Waterhouse et al. [64] also found that the addition of blackcurrant juice to wheat pasta reduced the volume of cooked pasta compared to samples without such addition.

4. Conclusions

It was found that apple pomace enriched wheat pasta with polyphenolic compounds (from 120% to 410% as compared to the control) and afforded an almost 8-fold increase in flavonoid content. Apple pomace also enriched wheat pasta with phenol acids, quercetin derivatives, flavon-3-ols and dihydrochalcones. The addition of apple pomace to pasta resulted in the growth of another health-promoting ingredient, dietary fibre. The content of dietary fibre was three to ten times higher in pasta with apple pomace relative to the control. Pasta with apple pomace was characterized by a higher content of soluble and insoluble fractions of dietary fibre, from 1 to 4.5 and from 3.5 to 17 times higher, respectively, compared to the control. In chemical composition, decreases in protein and fat content were observed, which were directly proportional to the percentage of apple pomace added. An increase in mineral components was also observed, particularly visible for the highest percentage of added pomace.

The hardness and maximum cutting energy for pasta decreased as the percentage of apple pomace increased in comparison to the reference sample. The addition of apple pomace to pasta did not negatively affect the water absorption of this product.

It was found that the smallest percentage (10%) of apple pomace in pasta resulted in the enrichment of wheat pasta with health-promoting compounds. This addition of apple pomace provided a 2-fold increase in polyphenols and flavonoids, and a significant increase in quercetin derivatives (quercetin-3-xyloside and quercetin-3-rhamnoside), chlorogenic and phloridzin. Such a small addition also results in a fivefold increase in the insoluble fraction, a twofold increase in the soluble fraction and a threefold increase in total dietary fibre. It also ensures the appropriate hardness and water absorption of pasta, which are important cooking properties.

Author Contributions: Conceptualization, D.G. and M.K.; methodology, D.G., E.I. and J.S.; validation, D.G., E.I. and J.S.; formal analysis, D.G, M.K., E.I. and J.S.; investigation, D.G, M.K., E.I., J.S. and S.K.; resources, D.G., M.K. and S.K.; data curation, D.G. and M.K.; writing—original draft preparation, D.G., M.K. and S.K. writing—review and editing, D.G., M.K. and S.K.; visualization, M.K.; supervision, D.G. and S.K.; project administration, D.G. and S.K.; funding acquisition, D.G. and M.K. All authors have read and agreed to the published version of the manuscript.

Funding: This research received no external funding.

Institutional Review Board Statement: Not applicable.

Informed Consent Statement: Not applicable.

Data Availability Statement: The data is included in this article.

Acknowledgments: This work was supported the by the Ministry of Science and Higher Education of the Republic of Poland.

Conflicts of Interest: The authors declare no conflict of interest.

References

1. Antonic, B.; Jancikova, S.; Dordevic, D.; Tremlova, B. Apple Pomace as Food Fortification Ingredient: A Systematic Review and Meta-Analysis. *J. Food Sci.* **2020**, *85*, 2977–2985. [CrossRef] [PubMed]
2. Santos, D.; Lopes da Silva, J.A.; Pintado, M. Fruit and Vegetable By-Products' Flours as Ingredients: A Review on Production Process, Health Benefits and Technological Functionalities. *LWT* **2022**, *154*, 112707. [CrossRef]
3. Tarko, T.; Duda-Chodak, A.; Bebak, A. Biological Activity of Selected Fruit and Vegetable Pomaces. *Zywnosc. Nauka. Technol. Jakosc/Food. Sci. Technol. Qual.* **2012**, *19*, 55–65. [CrossRef]
4. Tarko, T.; Duda-Chodak, A.; Tuszyński, T. Simulation of Phenolic Compounds Transformations and Interactions in an In Vitro Model of the Human Alimentary Tract. *Food Sci. Technol. Int.* **2009**, *15*, 235–241. [CrossRef]
5. Fernández-Martín, F.; Guerra, M.A.; López, E.; Solas, M.T.; Carballo, J.; Jiménez-Colmenero, F. Characteristics of Pressurised Pork Meat Batters as Affected by Addition of Plasma Proteins, Apple Fibre and Potato Starch. *J. Sci. Food Agric.* **2000**, *80*, 1230–1236. [CrossRef]
6. Nuñez de Gonzalez, M.T.; Boleman, R.M.; Miller, R.K.; Keeton, J.T.; Rhee, K.S. Antioxidant Properties of Dried Plum Ingredients in Raw and Precooked Pork Sausage. *J. Food Sci.* **2008**, *73*, H63–H71. [CrossRef]

7. Rodríguez, L.A.; Toro, M.E.; Vazquez, F.; Correa-Daneri, M.L.; Gouiric, S.C.; Vallejo, M.D. Bioethanol Production from Grape and Sugar Beet Pomaces by Solid-State Fermentation. *Int. J. Hydrog. Energy* **2010**, *11*, 5914–5917. [CrossRef]
8. Magyar, M.; da Costa Sousa, L.; Jin, M.; Sarks, C.; Balan, V. Conversion of Apple Pomace Waste to Ethanol at Industrial Relevant Conditions. *Appl. Microbiol. Biotechnol.* **2016**, *100*, 7349–7358. [CrossRef]
9. Lyu, F.; Luiz, S.F.; Azeredo, D.R.P.; Cruz, A.G.; Ajlouni, S.; Ranadheera, C.S. Apple Pomace as a Functional and Healthy Ingredient in Food Products: A Review. *Processes* **2020**, *8*, 319. [CrossRef]
10. Constenla, D.; Ponce, A.G.; Lozano, J.E. Effect of Pomace Drying on Apple Pectin. *LWT-Food Sci. Technol.* **2002**, *35*, 216–221. [CrossRef]
11. Ćetković, G.; Čanadanović-Brunet, J.; Djilas, S.; Savatović, S.; Mandić, A.; Tumbas, V. Assessment of Polyphenolic Content and in Vitro Antiradical Characteristics of Apple Pomace. *Food Chem.* **2008**, *2*, 340–347. [CrossRef] [PubMed]
12. Kammerer, D.R.; Kammerer, J.; Valet, R.; Carle, R. Recovery of Polyphenols from the By-Products of Plant Food Processing and Application as Valuable Food Ingredients. *Food Res. Int.* **2014**, *65*, 2–12. [CrossRef]
13. Rezk, B.M.; Haenen, G.R.M.M.; van der Vijgh, W.J.F.; Bast, A. The Antioxidant Activity of Phloretin: The Disclosure of a New Antioxidant Pharmacophore in Flavonoids. *Biochem. Biophys. Res. Commun.* **2002**, *295*, 9–13. [CrossRef] [PubMed]
14. He, X.; Liu, R.H. Triterpenoids Isolated from Apple Peels Have Potent Antiproliferative Activity and May Be Partially Responsible for Apple's Anticancer Activity. *J. Agric. Food Chem.* **2007**, *55*, 4366–4370. [CrossRef] [PubMed]
15. Bellion, P.; Digles, J.; Will, F.; Dietrich, H.; Baum, M.; Eisenbrand, G.; Janzowski, C. Polyphenolic Apple Extracts: Effects of Raw Material and Production Method on Antioxidant Effectiveness and Reduction of DNA Damage in Caco-2 Cells. *J. Agric. Food Chem.* **2010**, *58*, 6636–6642. [CrossRef] [PubMed]
16. Wijngaard, H.H.; Brunton, N. The Optimisation of Solid–Liquid Extraction of Antioxidants from Apple Pomace by Response Surface Methodology. *J. Food Eng.* **2010**, *96*, 134–140. [CrossRef]
17. Balasuriya, N.; Rupasinghe, H.P.V. Antihypertensive Properties of Flavonoid-Rich Apple Peel Extract. *Food Chem.* **2012**, *135*, 2320–2325. [CrossRef]
18. Rodríguez-Muela, C.; Rodríguez, H.E.; Arzola, C.; Díaz-Plascencia, D.; Ramírez-Godínez, J.A.; Flores-Mariñelarena, A.; Mancillas-Flores, P.F.; Corral, G. Antioxidant Activity in Plasma and Rumen Papillae Development in Lambs Fed Fermented Apple Pomace. *J. Anim. Sci.* **2015**, *93*, 2357–2362. [CrossRef]
19. Suárez, B.; Álvarez, Á.L.; García, Y.D.; del Barrio, G.; Lobo, A.P.; Parra, F. Phenolic Profiles, Antioxidant Activity and in Vitro Antiviral Properties of Apple Pomace. *Food Chem.* **2010**, *120*, 339–342. [CrossRef]
20. Jung, J.; Cavender, G.; Zhao, Y. Impingement Drying for Preparing Dried Apple Pomace Flour and Its Fortification in Bakery and Meat Products. *J. Food Sci. Technol.* **2015**, *52*, 5568–5578. [CrossRef]
21. Mir, S.A.; Bosco, S.J.D.; Shah, M.A.; Santhalakshmy, S.; Mir, M.M. Effect of Apple Pomace on Quality Characteristics of Brown Rice Based Cracker. *J. Saudi Soc. Agric. Sci.* **2017**, *1*, 25–32. [CrossRef]
22. Kırbaş, Z.; Kumcuoglu, S.; Tavman, S. Effects of Apple, Orange and Carrot Pomace Powders on Gluten-Free Batter Rheology and Cake Properties. *J. Food Sci. Technol.* **2019**, *56*, 914–926. [CrossRef]
23. Drożdż, W.; Tomaszewska-Ciosk, E.; Zdybel, E.; Boruczkowska, H.; Boruczkowski, T.; Regiec, P. Effect of Apple and Rosehip Pomaces on Colour, Total Phenolics and Antioxidant Activity of Corn Extruded Snacks. *Pol. J. Chem. Technol.* **2014**, *16*, 7–11. [CrossRef]
24. Reis, S.F.; Rai, D.K.; Abu-Ghannam, N. Apple Pomace as a Potential Ingredient for the Development of New Functional Foods. *Int. J. Food Sci. Technol.* **2014**, *49*, 1743–1750. [CrossRef]
25. Sobota, A.; Rzedzicki, Z.; Zarzycki, P.; Wirkijowska, A.; Sykut-Domańska, E. Zmiany zawartości tłuszczu wolnego w czasie procesu produkcji makaronu jajecznego. *Żywność Nauka-Technol.-Jakość* **2015**, *102*, 152–164. [CrossRef]
26. Bustos, M.C.; Perez, G.T.; Leon, A.E. Structure and Quality of Pasta Enriched with Functional Ingredients. *RSC Adv.* **2015**, *5*, 30780–30792. [CrossRef]
27. Xu, J.; Bock, J.E.; Stone, D. Quality and Textural Analysis of Noodles Enriched with Apple Pomace. *J. Food Process. Preserv.* **2020**, *44*, e14579. [CrossRef]
28. Gałkowska, D.; Witczak, T.; Pycia, K. Quality Characteristics of Novel Pasta Enriched with Non-Extruded and Extruded Blackcurrant Pomace. *Molecules* **2022**, *27*, 8616. [CrossRef]
29. Tolve, R.; Pasini, G.; Vignale, F.; Favati, F.; Simonato, B. Effect of Grape Pomace Addition on the Technological, Sensory, and Nutritional Properties of Durum Wheat Pasta. *Foods* **2020**, *9*, 354. [CrossRef]
30. Padalino, L.; Conte, A.; Lecce, L.; Likyova, D.; Sicari, V.; Pellicanò, T.M.; Poiana, M.; Del Nobile, M.A. Functional Pasta with Tomato By-Product as a Source of Antioxidant Compounds and Dietary Fibre. *Czech J. Food Sci.* **2017**, *35*, 48–56. [CrossRef]
31. Nur Azura, Z.; Radhiah, S.; Wan Zunairah, W.I.; Nurul Shazini, R.; Nur Hanani, Z.A.; Ismail-Fitry, M.R. Physicochemical, Cooking Quality and Sensory Characterization of Yellow Alkaline Noodle: Impact of Mango Peel Powder Level. *Food Res.* **2019**, *4*, 70–76. [CrossRef]
32. Isa, N.S.M.; MohdMaidin, N.; Hazmah, Y.; Madzuki, I.N.; Hamid, M.A. Nutritional Composition of Dried Noodle Incorporated with Mango Peel Powder. *Biosci. Res.* **2021**, *18*, 68–80.
33. Kultys, E.; Moczkowska-Wyrwisz, M. Effect of Using Carrot Pomace and Beetroot-Apple Pomace on Physicochemical and Sensory Properties of Pasta. *LWT* **2022**, *168*, 113858. [CrossRef]

34. Hirawan, R.; Ser, W.Y.; Arntfield, S.D.; Beta, T. Antioxidant Properties of Commercial, Regular- and Whole-Wheat Spaghetti. *Food Chem.* **2010**, *1*, 258–264. [CrossRef]
35. Official Methods of Analysis of AOAC International-18th Edition, Revision 3. Available online: https://www.techstreet.com/standards/official-methods-of-analysis-of-aoac-international-18th-edition-revision-3?product_id=1678986 (accessed on 15 February 2021).
36. CCAM. *Approved Methods of the American Association of Cereal Chemists*; AACC: Washington, DC, USA, 2000; ISBN 978-1-891127-12-0.
37. Taylor, K.A.C.C. A Colorimetric Method for the Quantitation of Galacturonic Acid. *Appl. Biochem. Biotechnol.* **1993**, *43*, 51–54. [CrossRef]
38. Singleton, V.L.; Orthofer, R.; Lamuela-Raventós, R.M. [14] Analysis of Total Phenols and Other Oxidation Substrates and Antioxidants by Means of Folin-Ciocalteu Reagent. In *Methods in Enzymology*; Oxidants and Antioxidants Part A; Academic Press: Cambridge, MA, USA, 1999; Volume 299, pp. 152–178.
39. El Hariri, B.; Sallé, G.; Andary, C. Involvement of Flavonoids in the Resistance of Two Poplar Cultivars to Mistletoe (*Viscum Album* L.). *Protoplasma* **1991**, *162*, 20–26. [CrossRef]
40. Tudorica, C.M.; Kuri, V.; Brennan, C.S. Nutritional and Physicochemical Characteristics of Dietary Fiber Enriched Pasta. *J. Agric. Food Chem.* **2002**, *50*, 347–356. [CrossRef]
41. Wang, X.; Kristo, E.; LaPointe, G. The Effect of Apple Pomace on the Texture, Rheology and Microstructure of Set Type Yogurt. *Food Hydrocoll.* **2019**, *91*, 83–91. [CrossRef]
42. Jin, H.; Kim, H.-S.; Kim, S.-K.; Shin, M.-K.; Kim, J.-H.; Lee, J.-W. Production of Heteropolysaccharide-7 by Beijerinckia Indica from Agro-Industrial Byproducts. *Enzym. Microb. Technol.* **2002**, *6*, 822–827. [CrossRef]
43. Jannati, N.; Hojjatoleslamy, M.; Hosseini, E.; Mozafari, H.R.; Siavoshi, M. Effect of Apple Pomace Powder on Rheological Properties of Dough and Sangak Bread Texture. *Carpathian J. Food Sci. Technol.* **2018**, *10*, 77–84.
44. Ktenioudaki, A.; O'Shea, N.; Gallagher, E. Rheological Properties of Wheat Dough Supplemented with Functional By-Products of Food Processing: Brewer's Spent Grain and Apple Pomace. *J. Food Eng.* **2013**, *116*, 362–368. [CrossRef]
45. Pieszka, M.; Gogol, P.; Pietras, M.; Pieszka, M. Valuable Components of Dried Pomaces of Chokeberry, Black Currant, Strawberry, Apple and Carrot as a Source of Natural Antioxidants and Nutraceuticals in the Animal Diet. *Ann. Anim. Sci.* **2015**, *15*, 475–491. [CrossRef]
46. Leyva-Corral, J.; Quintero-Ramos, A.; Camacho-Dávila, A.; de Jesús Zazueta-Morales, J.; Aguilar-Palazuelos, E.; Ruiz-Gutiérrez, M.G.; Meléndez-Pizarro, C.O.; de Jesús Ruiz-Anchondo, T. Polyphenolic Compound Stability and Antioxidant Capacity of Apple Pomace in an Extruded Cereal. *LWT-Food Sci. Technol.* **2016**, *65*, 228–236. [CrossRef]
47. Kieliszek, M.; Piwowarek, K.; Kot, A.M.; Pobiega, K. The Aspects of Microbial Biomass Use in the Utilization of Selected Waste from the Agro-Food Industry. *Open Life Sci.* **2020**, *15*, 787–796. [CrossRef]
48. Ferretti, G.; Turco, I.; Bacchetti, T. Apple as a Source of Dietary Phytonutrients: Bioavailability and Evidence of Protective Effects against Human Cardiovascular Disease. *Food Nutr. Sci.* **2014**, *5*, 1234–1246. [CrossRef]
49. Persic, M.; Mikulic-Petkovsek, M.; Slatnar, A.; Veberic, R. Chemical Composition of Apple Fruit, Juice and Pomace and the Correlation between Phenolic Content, Enzymatic Activity and Browning. *LWT-Food Sci. Technol.* **2017**, *82*, 23–31. [CrossRef]
50. Waldbauer, K.; McKinnon, R.; Kopp, B. Apple Pomace as Potential Source of Natural Active Compounds. *Planta Med.* **2017**, *83*, 994–1010. [CrossRef]
51. Escarpa, A.; González, M.C. High-Performance Liquid Chromatography with Diode-Array Detection for the Determination of Phenolic Compounds in Peel and Pulp from Different Apple Varieties. *J. Chromatogr. A* **1998**, *823*, 331–337. [CrossRef]
52. Rabetafika, H.N.; Bchir, B.; Blecker, C.; Richel, A. Fractionation of Apple By-Products as Source of New Ingredients: Current Situation and Perspectives. *Trends Food Sci. Technol.* **2014**, *1*, 99–114. [CrossRef]
53. Fares, C.; Menga, V. Effects of Toasting on the Carbohydrate Profile and Antioxidant Properties of Chickpea (*Cicer Arietinum* L.) Flour Added to Durum Wheat Pasta. *Food Chem.* **2012**, *4*, 1140–1148. [CrossRef]
54. Pilli, T.D.; Derossi, A.; Talja, R.A.; Jouppila, K.; Severini, C. Study of Starch-Lipid Complexes in Model System and Real Food Produced Using Extrusion-Cooking Technology. *Innov. Food Sci. Emerg. Technol.* **2011**, *4*, 610–616. [CrossRef]
55. Ajila, C.M.; Aalami, M.; Leelavathi, K.; Rao, U.J.S.P. Mango Peel Powder: A Potential Source of Antioxidant and Dietary Fiber in Macaroni Preparations. *Innov. Food Sci. Emerg. Technol.* **2010**, *11*, 219–224. [CrossRef]
56. Maillard, M.-N.; Berset, C. Evolution of Antioxidant Activity during Kilning: Role of Insoluble Bound Phenolic Acids of Barley and Malt. *J. Agric. Food Chem.* **1995**, *43*, 1789–1793. [CrossRef]
57. Michalska, A.; Wojdyło, A.; Lech, K.; Łysiak, G.P.; Figiel, A. Effect of Different Drying Techniques on Physical Properties, Total Polyphenols and Antioxidant Capacity of Blackcurrant Pomace Powders. *LWT-Food Sci. Technol.* **2017**, *78*, 114–121. [CrossRef]
58. Gaita, C.; Alexa, E.; Moigradean, D.; Conforti, F.; Poiana, M.-A. Designing of High Value-Added Pasta Formulas by Incorporation of Grape Pomace Skins. *Rom. Biotechnol. Lett.* **2020**, *25*, 1607–1614. [CrossRef]
59. Dexter, J.E.; Matsuo, R.E. Changes in Spaghetti Protein Solubility during Cooking [Durum Wheats and a Hard Red Spring Wheat]. *Cereal Chem.* **1979**, *56*, 394–398.
60. Brennan, C.S.; Kuri, V.; Tudorica, C.M. Inulin-Enriched Pasta: Effects on Textural Properties and Starch Degradation. *Food Chem.* **2004**, *86*, 189–193. [CrossRef]
61. Edwards, N.M.; Biliaderis, C.G.; Dexter, J.E. Textural Characteristics of Wholewheat Pasta and Pasta Containing Non-Starch Polysaccharides. *J. Food Sci.* **1995**, *60*, 1321–1324. [CrossRef]

62. Sozer, N.; Dalgıç, A.C.; Kaya, A. Thermal, Textural and Cooking Properties of Spaghetti Enriched with Resistant Starch. *J. Food Eng.* **2007**, *2*, 476–484. [CrossRef]
63. Lawton, J.W. Effect of Starch Type on the Properties of Starch Containing Films. *Carbohydr. Polym.* **1996**, *29*, 203–208. [CrossRef]
64. Sun-Waterhouse, D.; Jin, D.; Waterhouse, G.I.N. Effect of Adding Elderberry Juice Concentrate on the Quality Attributes, Polyphenol Contents and Antioxidant Activity of Three Fibre-Enriched Pastas. *Food Res. Int.* **2013**, *54*, 781–789. [CrossRef]
65. Sato, M.F.; Vieira, R.G.; Zardo, D.M.; Falcão, L.D.; Nogueira, A.; Wosiacki, G. Apple Pomace from Eleven Cultivars: An Approach to Identify Sources of Bioactive Compounds. *Acta Sci. Agron.* **2010**, *32*, 29–35. [CrossRef]
66. Sivam, A.S.; Sun-Waterhouse, D.; Perera, C.O.; Waterhouse, G.I.N. Application of FT-IR and Raman Spectroscopy for the Study of Biopolymers in Breads Fortified with Fibre and Polyphenols. *Food Res. Int.* **2013**, *50*, 574–585. [CrossRef]
67. Vetrimani, R.; Sudha, M.L.; Haridas Rao, P. Effect of Extraction Rate of Wheat Flour on the Quality of Vermicelli. *Food Res. Int.* **2005**, *38*, 411–416. [CrossRef]

foods

MDPI

Article

Environmental and Yield Comparison of Quick Extraction Methods for Caffeine and Chlorogenic Acid from Spent Coffee Grounds

Ibtissam Bouhzam [1], Rosa Cantero [1], Mercè Balcells [2], María Margallo [3], Rubén Aldaco [3], Alba Bala [4], Pere Fullana-i-Palmer [4] and Rita Puig [1,*]

[1] Department of Computer Science and Industrial Engineering, University of Lleida (UdL), Pla de la Massa, 8, 08700 Igualada, Spain
[2] Department of Chemistry, University of Lleida (UdL), Rovira _Roure 191, 25198 Lleida, Spain
[3] Department of Chemical and Biomolecular Engineering, University of Cantabria, Av. de Los Castros s/n, 39005 Santander, Spain
[4] UNESCO Chair in Life Cycle and Climate Change ESCI-UPF, Pg. Pujades 1, 08003 Barcelona, Spain
* Correspondence: author: rita.puig@udl.cat

Abstract: This study aims to provide an overview of different extraction methods to obtain chlorogenic acid (CA) and caffeine (Caf) from spent coffee grounds (SCG). This overview shows that the quantity extracted is highly dependent on the type of SCG, so experiments using the same SCG are needed to compare different methods. Three easy and simple extraction methods will be tested at a laboratory scale and environmentally compared. All three experiments were of 1 min duration: first, using supramolecular solvent; second, with water and vortex; and third, with water assisted by ultrasound. Water extraction assisted by ultrasound at room temperature yielded the greatest quantity of chlorogenic acid and caffeine, with 1.15 mg CA/g and 0.972 mg Caf/g, respectively. Extraction using supra-solvent leads to a lower content of CA in the supra-phase since it has more affinity for the water-based inferior phase. An environmental assessment using life cycle assessment has been carried out to compare water and supra extraction methods for the manufacture of two different commercial products: a face cream and an eye contour serum. Results show that the type of solvent and the amount of active substance extracted have a great influence on the environmental results. The results presented here are important for companies willing to obtain these active substances at an industrial scale.

Keywords: life cycle assessment; supra solvents method; water extraction; spent coffee grounds circularity

Citation: Bouhzam, I.; Cantero, R.; Balcells, M.; Margallo, M.; Aldaco, R.; Bala, A.; Fullana-i-Palmer, P.; Puig, R. Environmental and Yield Comparison of Quick Extraction Methods for Caffeine and Chlorogenic Acid from Spent Coffee Grounds. *Foods* **2023**, *12*, 779. https://doi.org/10.3390/foods12040779

Academic Editors: Stanisław Kowalski and Dorota Gumul

Received: 23 December 2022
Revised: 27 January 2023
Accepted: 3 February 2023
Published: 10 February 2023

1. Introduction

Food waste is one of the important issues facing the world [1]. The impacts and the real value of food waste have not been fully understood. Most researchers have agreed that food waste is linked to economic, environmental, and social problems [2]. In 2020, 20% of global food production was wasted [3], and it is estimated that 8% to 10% of global greenhouse gas emissions are attributed to food waste [4]. Therefore, there is an urgent need for sustainable methods to reduce food waste [5–7].

Spent coffee grounds (SCG) are among the most common organic waste generated all around the world. Coffee is the most widely consumed product in the world after water; it is also ranked the second-traded commodity worldwide after petroleum [8]. According to the International Coffee Organization, the world coffee consumption for the year 2020/2021 reached around 166 million bags of 60 kg, of which Europe accounts for almost a third, generating a great number of residues every year [9]. The process of making coffee involves the formation of various by-products, including the coffee husk, pulp, silverskin, and spent coffee grounds (SCG). Spent coffee grounds (SCG), for example, are produced in large amounts during the process of instant coffee preparation (at home, in restaurants, bars, and other businesses in the food industry) and mostly discharged into the environment [10] or burned [8,11]. It has been calculated that one ton of green coffee beans converted into coffee beverages produces 650 kg of discarded coffee grounds [11].

Nonetheless, there has been a surging interest in recycling these residues in recent years, particularly in the last years when the number of papers published on the theme "waste coffee grounds" has increased because they include various components that could be useful in the food, cosmetics, and pharmaceutical industries, such as caffeine, carbohydrates, lipids, and phenolic acids such as chlorogenic acid (CA) [12–14].

Chlorogenic acid and caffeine have received considerable attention due to their interesting properties. CA's health benefits for humans are being supported by an increasing body of scientific evidence. CA appears to work as a protective agent, in particular, preventing or lowering oxidative stress of cell structures and functions while also boosting health. CA has also been studied for its potential to improve blood pressure and glucose regulation [13,15,16]. Caffeine, which is the most important alkaloid of the coffee species, presents various biological activities, such as stimulation of the central nervous system, myocardial stimulation, and peripheral vasoconstriction. Some studies suggest that caffeine is effective in reducing weight through thermogenesis and fat oxidation, effective in reducing the effects of ultraviolet radiation and damage induced by free radicals on the skin, and also as an adjunct to the treatment of hair loss [17].

Chlorogenic acid, an important part of the phenolic compounds, and caffeine are present in spent coffee grounds [14]. Thus this residue has been suggested as a sustainable alternative to obtain these substances.

Some methods have been tested at a laboratory scale for their recovery using different solvents and techniques, such as liquid–solid extraction [18–20], Soxhlet extraction [21,22], autohydrolysis [23], ultrasound [21,24–26], microwave-assisted extraction [27,28], subcritical water extraction [29], pressurized liquid [12], hydrothermal treatment [8], and supramolecular solvent extraction [30].

Supramolecular solvents (SUPRAS) have been reported as an efficient alternative method to organic solvents, effective in solubilizing a variety of solutes, and a readily available method, making it a very attractive process for extraction applications. SUPRAS are nanostructured liquids generated from aqueous or hydro-organic colloidal dispersions of amphiphilic substances through spontaneous processes of self-assembly and coacervation [31].

Our final aim is to contribute to the promotion of circularity for the processing of spent coffee grounds (SCG) by providing comparable information on different extraction methods. The purpose of this study is to compare, for the first time, supramolecular solvent extraction with conventional methods to extract chlorogenic acid and caffeine. The comparison will include both extraction yields and environmental performance. In addition, an overview of the different reported extraction methods will be provided.

2. Literature Review

Different methods to obtain chlorogenic acids and caffeine from spent coffee grounds can be found in the literature (Tables 1 and 2).

Chlorogenic acid is one of the phenolic compounds most commonly quantified in the literature studies. According to Andrade et al. [15], chlorogenic acid was the phenolic component found in higher concentrations in the extracts of coffee grounds. Gallic acid, p-hydroxybenzoic acid, protocatechuic acid, vanillic acid, and tannic acid were also found, but in lower amounts. Chlorogenic acids include three main groups: caffeoylquinic acids (CQAs: 3-CQA, 4-CQA, and 5-CQA), dicaffeoylquinic acids (diCQAs: 3,4-diCQA, 3,5-diCQA, and 4,5-diCQA), and feruloylquinic acids (FQAs: 3- FQA, 4- FQA, and 5-FQA) [32].

In only a few studies, researchers have evaluated the content of chlorogenic acids as 5-O-Caffeoylquinic acid (abbreviated as CA) or total chlorogenic acids as tCA (those including the different isomers). Most of them have extensively reported the total phenolic compounds as gallic acid equivalents (GAE). The total phenolic content can be measured using the Folin–Ciocalteu method, and results are expressed as gallic acid equivalents (GAE).

The Folin–Ciocalteu method, developed by Singleton et al. [33], was first employed to determine "wine tannin" level, then gained widespread acceptance and is being used on a variety of plant materials and foods [34]. The principle of the method relies on the reduction of the Folin–Ciocalteu reagent in the presence of an antioxidant. Thus, the reductive capacity increases with the concentration of the phenolic compounds [35]. Calibration curves using various chemical standards should be used when calculating the total phenolic content. The most commonly used standard is gallic acid, which has been widely used in a variety of fruits, vegetables, red wines, and plant extracts, including berries, herbs, cereals, tree materials, plant sprouts, seeds, and spent coffee grounds [18,19,23,34].

Table 1 presents studies conducted by different authors, which focus on extracting caffeine and chlorogenic acid through different methods. However, the results are difficult to compare because the coffee sample, the solvent/solid ratio, and the way to measure the results are different. Nevertheless, some conclusions from the literature review will be noted. Results are usually quantified in three different ways: gallic acid equivalents (GAE), which measure the total phenolic compounds; chlorogenic acid (CA); and caffeine, which specifically quantifies those analytes.

The most common solvent used in the literature was ethanol in different proportions, with high variation in the solvent/SCG ratio (from 5 to 83 mL/g). The recovery rate of the total phenolic compounds ranged from 17 to 273 mg GAE/g, with an extraction time varying from 5 min to 6 h (Table 1), except for the supramolecular method, which only takes 1 min. The highest content was obtained by the Soxhlet method using ethanol (273.34 mg GAE/g). Water also successfully accomplished the extraction of high amounts of phenolic compounds; around 86.23 mg GAE/g was extracted with subcritical water extraction (SWE), also known as pressurized hot water extraction. SWE has emerged as a viable method for extracting bioactive and nutritional components because it has several advantages, including reduced organic solvent use, excellent extraction efficiency, and an environmentally friendly technology [29]. Autohydrolysis, pressurized liquid, and hydrothermal pretreatment have also been demonstrated to be effective methods for recovering phenolic compounds from spent coffee grounds [12,23].

Chlorogenic acid can be measured as CA (using 5-O-Caffeoylquinic acid as standard) or as total chlorogenic acid (tCA) when all the peaks corresponding to the different isomers of chlorogenic acid are integrated. Results found in the literature (expressed as mg of analyte per gram of dry SCG) have a wide range of variation, from 0.02 mg CA/g to 5.97 mg tCA/g (Tables 1 and 2).

Caffeine (1, 3, 7-trimethylxanthine) is also the most desired compound to isolate from spent coffee grounds. The maximum contents reported in the literature ranged from 3.32 to 11.50 mg Caf/g in mixtures of Arabica and Robusta varieties (Tables 1 and 2). The recovery rate may vary in function of the type of coffee variety, as said before. According to Campos-Vega et al. [36], the caffeine content in Robusta is almost twice (1.4–2.9%) compared with Arabica (0.9–1.6%) or a Robusta mix (60 Arabica/40 Robusta) (1.7%).

Several studies [12,18,23,29,37] (see Table 1) have tried to optimize condition parameters (time, temperature, and solvent ratio) for the same extraction method and solvent using statistical tools, such as response surface methodology or chemometrics in general.

Only a few studies, those represented in Table 2, have compared extraction methods and solvents using the same spent coffee ground.

Andrade et al. [15] have tested CO_2 supercritical fluid extraction (SFE CO_2), which is known as a decaffeination process to extract caffeine. The results are expressed in terms of concentration (mg per g of extract). Using this method, Andrade et al. [15] achieved an extraction rate of 41.3 mg/g of extract at 333.15 K and 300 bar. Regarding low-pressure methods, ultrasound gave a higher concentration (38.2 mg/g extracts) than Soxhlet (25.9 mg/g extracts) (Table 2). The solvent dichloromethane recovered the most caffeine (38.2 mg/g extracts), followed by ethanol (25.7 mg/g extracts), and then hexane (0.734 mg/g extracts) with ultrasound extraction [15] (Table 2). According to Shalmashi et al. [38], caffeine's solubility decreases in the following order: chloroform, dichloromethane, acetone, ethyl acetate, water, methanol, ethanol, and carbon tetrachloride. Hence, the polarity of the solvent has a significant impact on the extraction level of caffeine.

Table 1. Extraction methods and conditions reported in the literature.

Extraction Type	Solvent	Time	Temperature	Ratio Solvent/Solid	Results	Ref.
Hydrothermal pretreatment	Water	20 min	120 °C	20 mL/g	32.92 mg GAE/g	[8]
Pressurized liquid	Ethanol 25–75%	5–20 min	195 °C	-	22.91 mg GAE/g 9.66 mg Caf/g	[12]
Solid/liquid	Ethanol 70%	2 h	50 °C	40 mL/g	17.09 mg GAE/g (Espresso)	[18]
Solid/liquid	Ethanol 70%	2 h	50 °C	40 mL/g	19.98 mg GAE/g (Capsules)	[18]
Solid/liquid	Ethanol 60%	30 min	60 °C	50 mL/g	28.26 mg GAE/g	[19]
Soxhlet	Ethanol 50%	-	Ambient	5 mL/g	273.34 mg GAE/g	[22]
Autohydrolysis	Water	50 min	200 °C	15 mL/g	40.36 mg GAE/g	[23]
Solid/liquid	Ethanol 60%	8 h	25 °C	20 mL/g	0.54 mg CA/g	[24]
Soxhlet	Ethanol 60%	6 h	-	25 mL/g	0.63 mg CA/g	[24]
Ultrasound	Ethanol 60%	1 h	50 °C	20 mL/g	0.93 mg CA/g	[24]
Ultrasound	Ethanol	34 min	40 °C	17 mL/g	33.84 mg GAE/g 1.43 mg CA/g	[25]
Subcritical water	Water	55 min	177 °C	71 mL/g	86.23 mg GAE/g	[29]
Subcritical water	Water	30 min	170 °C	50 mL/g	1.41 mg CA/g	[29]
Supramolecular solvent	1-hexanol 24% Ethanol 30% Water 46%	1 min	Ambient	5.7 mL/g	4.3 mg CA/g 3.32 mg Caf/g	[30]
Solid/liquid	Ethanol 25%	15 min	60 °C	83 mL/g	0.02 mg CA/g (Marcilla coffee)	[37]
Solid/liquid	Ethanol 25%	15 min	60 °C	83 mL/g	0.8 mg CA/g (Arabica coffee)	[37]

CA: Chlorogenic acid (measured from one single chromatographic peak, the one of the standard 5-O-Caffeoylquinic acid); GAE: gallic acid equivalent (measuring the total phenolic compounds); and Caf: caffeine (measured using the standard 1,3,7-trimethylxantine). The color is used to easily identify the results/methods described in the same reference and distinguish from the others. The extraction types belong to the same reference (and probably same SCG) when marked with the same color.

Vandeponseele et al. [11] extracted higher content of caffeine (4.32 mg Caf/g) using an efficient combination of ethanol (40%) with water than pure water (3.63 mg Caf/g).

Regarding chlorogenic acid, some investigations have compared ultrasound with high hydrostatic pressure on the extraction of phenolics (measured as chlorogenic acid, CA). The greatest recovery rate of chlorogenic acid was achieved using ultrasound, which gave a slightly greater yield than high hydrostatic pressure [39] (Table 2).

Concerning the influence of the solvent, Panusa et al. [19] have detected small differences between the use of pure water (5.67 mg tCA/g) and ethanol 60% (5.97 mg tCA/g) in SCGs collected from bars, as well as in coffee capsules (2.15 mg tCA/g for pure water) and (2.26 mg tCA/g for ethanol 60%), which also revealed that SCGs from bars are twice as rich in chlorogenic acid as capsules residue (see Table 2).

Table 2. Comparison of the extraction methods and solvents for the extraction of caffeine and chlorogenic acid using the same SCG.

Extraction Methods	Solvent	Time	Temperature	Ratio Solvent/Solid	Results	Ref.
Solid/liquid	Ethanol 40%	15 min	20 °C	25 mL/g	4.32 mg Caf/g	[11]
	Water 100%				3.63 mg Caf/g	
	Ethanol 100%				0.26 mg Caf/g	
Ultrasound extraction	Hexane	2 h	Ambient T	30 mL/g	0.734 mg Caf/g extract	[15]
	Dichloromethane				38.2 mg Caf/g extract	
	Ethanol				25.7 mg Caf/g extract	
Soxhlet	Hexane	6 h	Ambient T	30 mL/g	3.27 mg Caf/g extract	
	Dichloromethane				25.9 mg Caf/g extract	
	Ethanol				11.8 mg Caf/g extract	
SFE CO_2	CO_2	4.30 h	333.15 K (200 bar)	-	27.2 mg Caf/g extract	
	CO_2	4.30 h	333.15 K (300 bar)	-	41.3 mg Caf/g extract	
SFE CO_2 + ethanol	CO_2 + ethanol	4.30 h	333.15 K	-	23.4 mg Caf/g extract	
Solid/liquid	Water 100%	30 min	60 °C	50 mL/g	5.67 mg tCA/g 11.23 mg Caf/g (Arabica 80%, Robusta 20%)	[19]
	Ethanol 60%				5.97 mg tCA/g 11.50 mg Caf/g (Arabica 80%, Robusta 20%)	
Ultrasound	Methanol 80%	15 min	25 °C	10 mL/g	0.085 mg CA/g	[39]
High hydrostatic pressure					0.081 mg CA/g	

CA: Chlorogenic acid (measured from one single chromatographic peak, the one of the standard 5-O-Caffeoylquinic acid); tCA: total chlorogenic acid (measured integrating all the peaks corresponding to the different chlorogenic acid isomers); and Caf: caffeine (measured using the standard 1,3,7-trimethylxantine).). The color is used to easily identify the results/methods described in the same reference and distinguish from the others. The extraction types belong to the same reference (and probably same SCG) when marked with the same color.

Thus, the recovery rate of CA and caffeine extracted depends a great deal on the composition of SCGs, which is affected by the type of coffee bean, the roasting condition, and the coffee-making process [40–44].

Many studies have reported that the caffeine/CA ratio directly increases with the roasting intensity of the coffee bean [43]. In addition, it depends on the extraction method and conditions used (solvent type, solid/solvent ratio, temperature, pressure, and extraction time) [30]. This is why it is important to determine the most convenient solvent and extraction method for caffeine and CA extraction using the same sample of SCG.

3. Materials and Methods

In this section, the materials and methods for the three experimental extraction methods tested will be presented.

3.1. Chemicals and Reagents

1-hexanol (98%), ethanol (99.9%), and glacial acetic acid (99.8%) were purchased from Scharlab (Barcelona, Spain). Methanol (99.9%) was supplied by ITW reagents (Barcelona, Spain). Caffeine (1,3,7-trimethylxantine, HPLC grade) and 5-chlorogenic acid (5-O-Caffeoylq uinic acid, 5-CA, 98%) were supplied by Sigma-Aldrich (St. Louis, MO, USA).

All chemicals were analytical reagent grade. Ultrapure water was prepared using a Milli-Q System equipped with a 0.22 μm filter purchased from Merck Millipore (Burlington, MA, USA).

3.2. Equipment

Quantification of caffeine and 5-CA was carried out using a high-performance liquid chromatograph (HPLC) coupled to a Waters 2996 UV Detector (Milford, MA, USA). The stationary phase was an Ultra C8 column (5 μm particle size, 150 mm length, 4.6 mm i.d.) from Scharlab (Barcelona, Spain). All data were processed using the Empower Solutions Software (Orlando, FL, USA).

A Velp Scientifica F202A0176 vortex shaker (Monza et de la Brianza, Italy), a J.P. Selecta 7002575 centrifuge, and a 3000865 ultrasound bath (Barcelona, Spain) were used for sample preparation.

3.3. Spent Coffee Grounds

Spent coffee grounds were provided by two restaurants (SCG1 and SCG2) located in the municipality of Igualada in the province of Barcelona, Spain. Both samples were derived from mixtures of Arabica and Robusta coffee varieties. The roasted coffee beans were from different producers (SCG1, "Novell gourmet Responsable", and SCG2, "Cafes MamaSame", with SCG1 being less roasted than SCG2). Both SCGs were obtained after the preparation of coffee using an espresso machine. The mixtures were manually homogenized, naturally dried, and stored in capsules at room temperature for further extractions. Experiments were performed in triplicate.

3.4. Extraction with Supra Solvents

The so-called supra solvents are a mixture of water with two additional solvents. In the present case, 1-hexanol, ethanol, and water were used. In the first step, a colloidal suspension of ethanol and 1-hexanol was formed. In the second step (coacervation), water was added, giving rise to a separation phenomenon of liquid phases: the superior (supra) and the inferior phases [31].

The extraction experiments were carried out according to the principle of the SUPRAS method mentioned above. A sample of 0.7 g of dry SCG was mixed with ethanol (1.2 mL), 1-hexanol (0.96 mL), and water until a total volume of 4 mL was obtained. The whole solution was vortex-shaken for 1 min at 3000 rpm and then centrifuged for 30 min at 4200 rpm. The volume of SUPRAS produced was calculated from the cylindrical volume formula $\pi r^2 h$, where r is the radius of the tube and h is the height of the supra phase of the extract. The height was measured using a digital caliper. The resulting supra solution was filtrated with a syringe filter (0.45 μm) before HPLC analysis. Experiments were performed in triplicate.

3.5. Extraction with Water

The extraction method with water was performed by adding 0.7 g of SCG to 4 mL of ultrapure water. The tube with the mixture was vortex-shaken at 3000 rpm or immersed in an ultrasound bath (50/60 Hz, 195 W) for 1 min, then centrifuged for 30 min at 4200 rpm and filtrated with a syringe filter (0.45 μm) before HPLC analysis. Experiments were performed in triplicate.

3.6. Analysis of Caffeine and Chlorogenic Acid by HPLC-UV

The mobile phase was prepared with ultrapure water and methanol, which were both acidified with 0.1% (*v*/*v*) of HPLC-grade acetic acid, and under a constant flow of 0.6 mL/min. The elution gradient started with 18% methanol and 82% ultrapure water and ended with 30% methanol and 70% ultrapure water for 15 min. The sample injection volume was 20 μL at 36 °C, and the detection was performed with a UV/Visible diode at a wavelength of 278 nm (for both chlorogenic acid and caffeine detection). Quantitative analysis was conducted by external calibration using standard solutions of 5-O-Caffeoylquinic acid and caffeine prepared in ultrapure water in concentration ranges from 1 mg/L to 200 mg/L and 100 mg/L to 600 mg/L, respectively.

4. Experimental Results and Discussion

All the results are expressed as the arithmetic mean of three replicates and their coefficients of variation. The coefficient of variation is the standard deviation divided by the arithmetic mean. The standard deviation measures the dispersion of the three values obtained with respect to the mean (higher dispersion of values and higher standard deviation).

4.1. Extraction with Supra Solvents Using Vortex

The capacity of SUPRAS to extract caffeine (Caf) and chlorogenic acid (CA) was previously reported in the literature, using 1-hexanol as an amphiphile and ethanol:water as a hydro-organic media [30]. The ratio of 1-hexanol/ethanol (24/30, *v*/*v*) has been reported to give the best results. Therefore, all the experiments were performed using this ratio.

The chromatogram of the extract obtained (see Figure 1) revealed different chlorogenic acid isomers (peaks 1, 2 and 3) at retention times close to one of the standards (5-O-Caffeoylquinic acid, 5-CA). Spiking experiments with the chlorogenic acid standard (5-CA, 98%) were carried out to determine the peaks corresponding to the standard. The spiking consisted of replacing water with a certain volume of the chlorogenic acid standard solution. The comparison of the chromatograms with and without spiking showed that the chlorogenic acid standard has isomerized (due to its interaction with SCG) and increased the three peaks of the spiked sample instead of appearing as one single peak. The area of the first peak (peak 1) increased by 75%, while the second and third peaks (peaks 2 and 3) had a relative rise of roughly 29% and 31%, respectively. Therefore, all peaks were included in the measurement of CA content.

Neither the extraction of chlorogenic acid nor that of caffeine is selective (see Figure 1), which means that the extract obtained will have other components transferred from the coffee grounds.

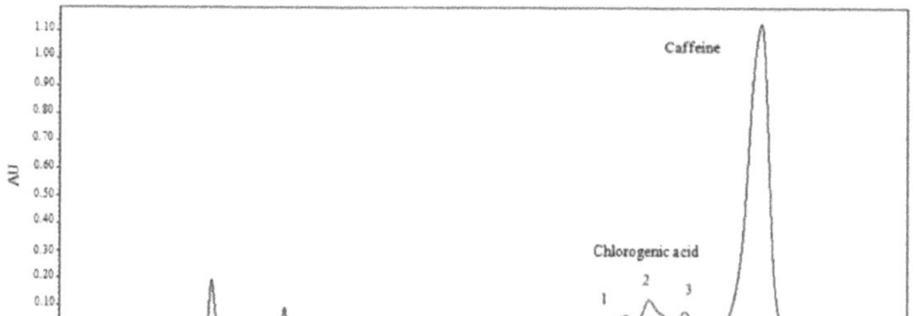

Figure 1. HPLC chromatogram of the supra extract for spent coffee grounds 1 (SCG1) subjected to the supramolecular method.

The results obtained are presented in Figure 2. Results are expressed as mg of bioactive substance (chlorogenic acid and caffeine) per g of dry SCG.

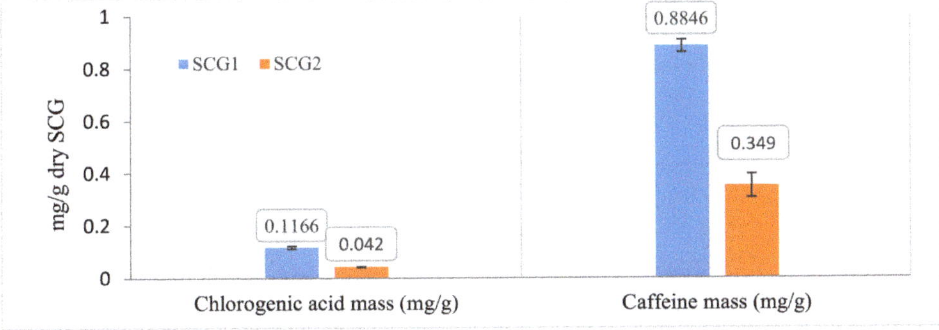

Figure 2. Chlorogenic acid and caffeine extracted with the supramolecular method (average of three replicates).

The experimental measurements have shown that SCG1 is richer in chlorogenic acid and caffeine than SCG2. Chlorogenic acid and caffeine obtained for SCG1 were 0.116 mg CA/g and 0.884 mg Caf/g, respectively.

Torres-Valenzuela et al. [30] have reported higher extraction values of chlorogenic acids and caffeine (4.3 mg CA/g and 3.32 mg Caf/g) using the same extraction method of SCG obtained from drip filter brewing of a Colombian 100% Arabica variety. The reasons for these differences may be mainly attributed to the type of coffee variety and coffee preparation method. As said before, coffee variety, genetic characteristics, agricultural processes, storage conditions, brewing time, and roasting degree can affect the content of CA and caffeine in SCG and their recovery rate [11,41,45,46].

In addition, a series of experiments were carried out to measure the affinity of the supra phase for chlorogenic acid compared to the inferior phase. The experiments were performed in triplicate without the use of coffee grounds but with the addition of approximately 1 mL of a 100 mg/L chlorogenic acid solution. The results demonstrate that only 34% of the total added chlorogenic acid was recovered in the supra phase, and 66% remained in the lower phase (inferior phase). This proved that the phase with the highest affinity for chlorogenic acid is the inferior phase.

Therefore, a quantification of the contents at the inferior phase (infra) was also measured to be able to compare the extraction content in both phases.

As can be seen in Figure 3, a significant amount of chlorogenic acid is still present at the inferior phase, while for caffeine, the main quantity is found at the supra phase (although around 28% still remained at the inferior phase).

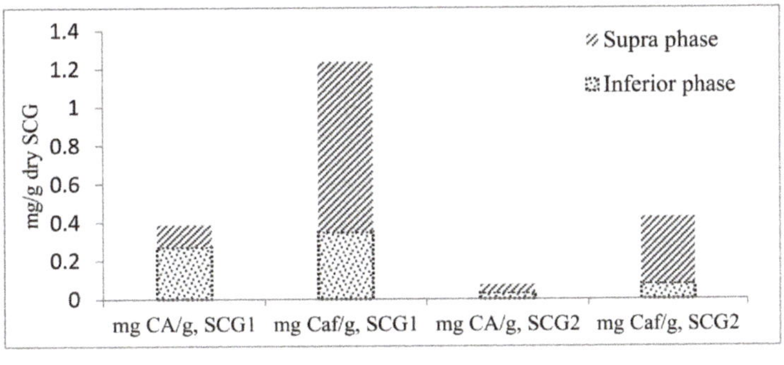

Figure 3. Chlorogenic acid and caffeine mass at the superior and inferior phases (average of three replicates) with the supramolecular solvent method.

In terms of concentration, the supramolecular method extracted 48.7 mg/L of chlorogenic acid and 370 mg/L of caffeine at the superior phase from SCG1, while 277 mg/L of chlorogenic acid and 355 mg/L of caffeine concentrations were found at the inferior phase. Concentration results are important to know when direct use of the extract (without subsequent evaporation of the solvent) is possible for further industrial applications.

4.2. Extraction with Water Using Vortex

The influence of the solvent on the extraction results for chlorogenic acid and caffeine has been investigated using ultrapure water. Water is much cheaper and far more widely accessible than other solvents, such as dichloromethane and ethanol.

A comparison of spiked and non-spiked water extractions was also carried out. The objective of these experiments was to ensure which of the chlorogenic acid peaks corresponds to that of the standard. In the extractions with water, the first peak is practically negligible, while the others (peaks 2 and 3) present significant results in relation to the area (see Figure 4).

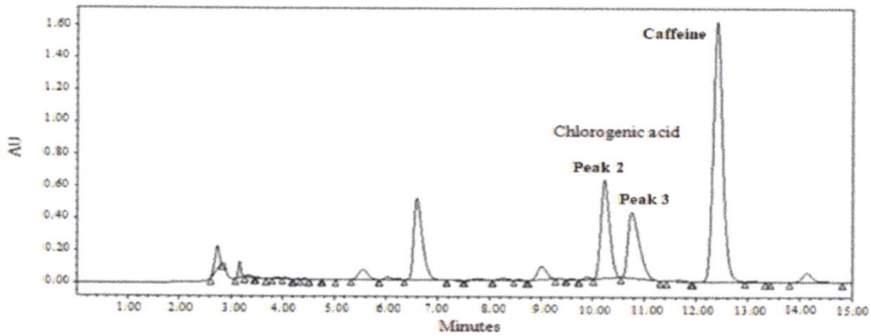

Figure 4. HPLC chromatogram of the water extract for spent coffee grounds 1 (SCG1) subjected to the water extraction method.

The effect of spiking promotes a simultaneous increase in both peaks of chlorogenic acid. The second peak of the spiked sample increased by a factor of 1.6, whereas the third peak increased by a factor of 1.2.

Therefore, to determine the amount of chlorogenic acid extracted in the coffee grounds samples, the sum of the two peak areas (peaks 2 and 3) has been used.

Results also revealed that SCG1 is richer in chlorogenic acid and caffeine than SCG2 (see Figure 5). Chlorogenic acid and caffeine obtained for SCG1 were 0.827 mg CA/g and 0.766 mg Caf/g, respectively.

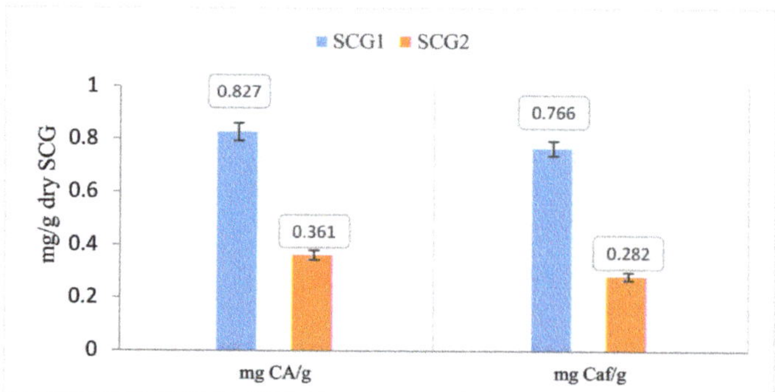

Figure 5. Extraction rate of chlorogenic acid mass and caffeine mass (average of three replicates) with the water extraction method.

In terms of concentration in the extract obtained, water recovered 223 mg/L of chlorogenic acid and 207 mg/L of caffeine from SCG1.

4.3. Ultrasound-Assisted Water Extraction

Ultrasound-assisted water extraction at different temperatures (room temperature and 50 °C) was evaluated on chlorogenic acid and caffeine extraction.

The experiments did not show a variation in chlorogenic acid and caffeine extracted at 50 °C compared to room temperature for both spent coffee grounds. A slight positive effect was detected on SCG2 because the chlorogenic acid and caffeine masses were increased by about 11% and 12%, respectively, whereas for SCG1, a slight decrease was observed (about 11–14%). The best results were obtained at room temperature (1.15 mg CA/g and 0.972 mg Caf/g) (see Figure 6).

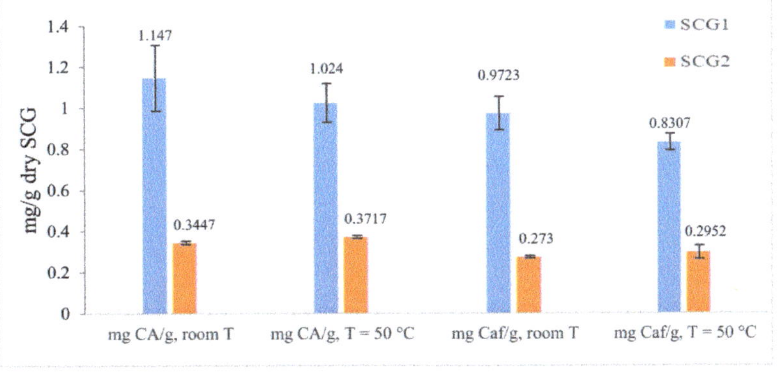

Figure 6. Extraction rate of chlorogenic acid mass (average of three replicates) using the ultrasound method at ambient temperature and at 50 °C.

The effect of temperature on chlorogenic acid and caffeine is also discussed in the literature. Vandeponseele et al. [11] reported a small increase (+12%) of the caffeine mass (mg/g SCG) between 20 and 80 °C when using a mixture of H_2O/EtOH (60/40 *v*/*v*) as extraction solvent, while Butiuk et al. [47] have found that a temperature between 30 and 80 °C did not affect the extraction content of chlorogenic acid from yerba mate stems when performed with water as the extraction solvent. However, when the temperature exceeds 80 °C, the amount of CA extracted decreases. Hence, we can infer that CA becomes unstable above a temperature of 80 °C.

In terms of concentration, ultrasound at ambient temperature extracted the most chlorogenic acid and caffeine from SCG1 with 332 mg/L and 282 mg/L, respectively, whereas ultrasound at 50 °C obtained only 284 mg/L and 231 mg/L, respectively.

4.4. Longer Extraction Time

The influence of the extraction time between 1 min and 5 min on the caffeine and chlorogenic acid results was investigated for two extraction methods: vortex-shaken with water and using supra solvents.

The effect of the extraction time widely varies depending on the extraction method. For SCG1, an increase in chlorogenic acid mass of 25.1% and caffeine mass of 23.6% between 1 min and 5 min was associated with the extraction with supra, while there was an increase of 70% in the chlorogenic acid mass and 41.9% in the caffeine mass for the extraction with water, as can be seen in Figure 7 for the chlorogenic acid results and Figure 8 for caffeine.

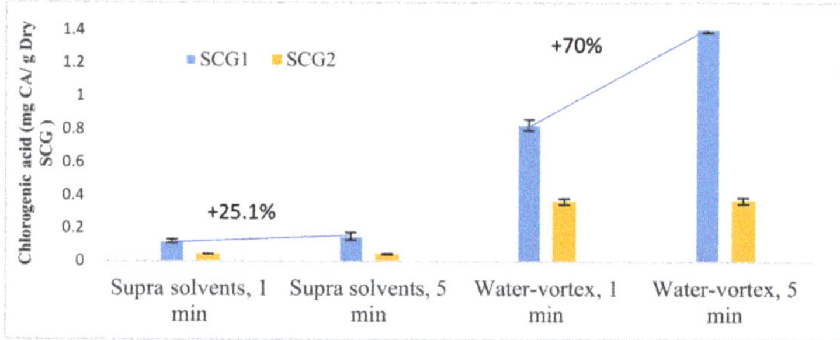

Figure 7. Influence of time for supra and water extraction of chlorogenic acid from spent coffee grounds.

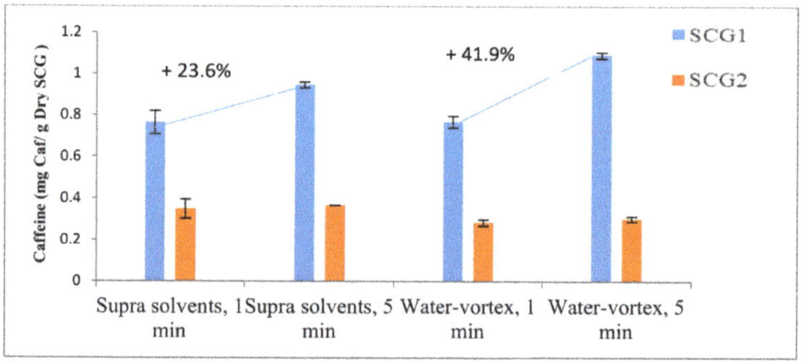

Figure 8. Influence of time for supra and water extraction of caffeine from spent coffee grounds.

For SCG2, the experiments revealed a very small increase in the chlorogenic acid (+0.3% for supra solvents or +1.8% for water) and the caffeine mass (+5% or +6%, respectively) between 1 min and 5 min.

According to Ballesteros-Gómez et al. [31], a few extraction minutes (5–10 min) are sufficient for liquid samples, while a longer extraction time is needed in the case of solid samples to accelerate and increase the mass transfer between the sample and the solvent. Thus, longer extraction times can probably be used in the present case if the aim is to extract as much chlorogenic acid and caffeine as possible.

4.5. Comparison of the Different Extraction Methods Tested

The results comparing the different methods to extract chlorogenic acid and caffeine are presented in Figures 9 and 10, respectively.

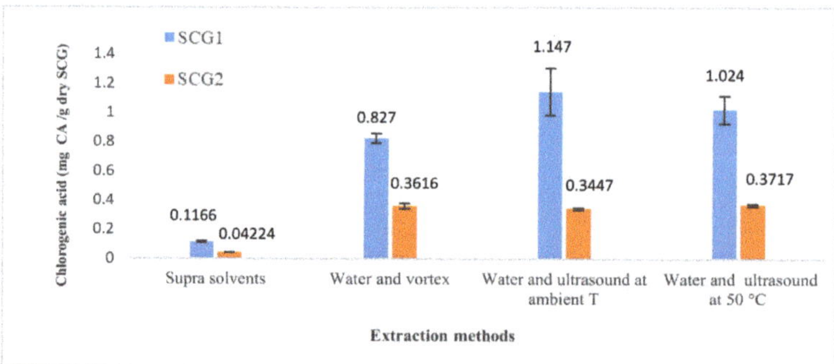

Figure 9. Chlorogenic acid mass extracted as a function of the different extraction methods.

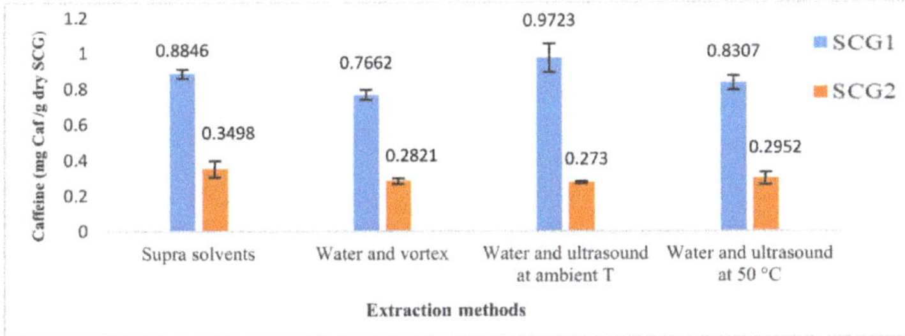

Figure 10. Caffeine mass extracted as a function of the different extraction methods.

A comparison of the results indicates significant differences between the bioactive substances extracted (chlorogenic acid and caffeine) and the type of coffee. On the other hand, remarkable differences were also detected among all the extraction methods examined (Figures 9 and 10).

SCG1 showed the highest extraction values, which may be attributed to the characteristics of the coffee variety, the proportions in the coffee blend used, the manufacturing process (roasting) and the coffee preparation method. As said before, Robusta has about twice the caffeine content as Arabica and is slightly richer in chlorogenic acid [19]. In the present case, SCG1 was less roasted than SCG2, and both are mixtures with unknown proportions of Arabica and Robusta.

The comparison of methods that are discussed below refers just to SCG1, the one showing better results.

The method that led to the best results was the extraction with ultrasound at room temperature, while extraction of supra solvents was, surprisingly, not the best. Nevertheless, ultrasounds at room temperature present a higher coefficient of variation, which means that the results of the triplicate have more dispersion (14% and 8.3% for chlorogenic acid and caffeine, respectively).

For chlorogenic acid, water extraction assisted by ultrasound at room temperature yielded the greatest amount with 1.15 mg CA/g, followed by ultrasound at 50 °C and then extraction with water using a vortex. This best result was seven times higher than the one obtained by supramolecular solvent extraction.

For caffeine, ultrasound at room temperature also demonstrated the best results with 0.972 mg Caf/g, followed by supramolecular solvent extraction (0.885 mg Caf/g).

Supramolecular solvent extraction did not succeed at extracting higher contents of bioactive components, especially chlorogenic acid, in comparison with the conventional methods. This can be attributed either to the distribution of analytes between supra and infra phases or the thermodynamic difficulty in obtaining "supra" solvents (which require self-assembly and coacervation processes that are spontaneous and sequential).

5. Environmental Comparison: Methodology and Results

The extraction methods to be compared will be the supramolecular solvent method (of greater novelty) and the water (vortex-shaken) method (of easier industrial scaling). The extraction yields obtained for chlorogenic acid (CA) were 0.827 mg CA/g SCG and 0.116 mg CA/g SCG using water and supramolecular extraction methods, respectively, while for caffeine (Caf), yields were 0.766 mg Caf/g SCG and 0.885 mg Caf/g SCG, respectively (see Section 4.5).

In order to perform an environmental comparison using the life cycle assessment (LCA) methodology, we need to compare products/processes performing identical functions. In the present study, the extraction methods were compared for the manufacture of the same commercial product. Two different commercial products have been chosen for comparison: one of them is a cosmetic face cream containing chlorogenic acid (CA) because the antioxidant property (from polyphenols) is the one required in the product, and the other is an eye contour serum containing caffeine due to its anti-inflammatory property.

The LCA methodology, as specified by ISO 14,044 [48], was used for the environmental comparison. The systems were modeled using the software Gabi v10.6.135, and the impact categories used for the environmental assessment were the ones suggested by the EF 3.0 method. Thirteen impact categories were evaluated, and all of them were classified into three types: damage to health, damage to ecosystems, and resource consumption.

5.1. Environmental Comparison for the Manufacture of a Face Cosmetic Cream

5.1.1. Chlorogenic Acid Recovery: Scope and Inventory Data

As said before, in cosmetic cream, the antioxidant property of chlorogenic acid (and other polyphenols) is the property required.

Thus, the Functional Unit (FU) chosen for this comparison was "125 mg of chlorogenic acid needed for 100 g of a face cosmetic cream".

The amount of chlorogenic acid in the cream was obtained by the relation between CA and the total phenolic compounds, as the cream manufacturing laboratory reported a 4% content of total phenolic compounds in the cream. Thus, the proportion of chlorogenic acid related to the total polyphenols in an SCG extract was estimated by Xu et al. [29], where the authors reported that chlorogenic acid represents 3.12% of the total phenolic compounds found in a specific SCG. A content of 125 mg of chlorogenic acid per 100 g of cream was calculated using this percentage as the FU of the present study.

System boundaries of the LCA comparison, as well as the inventory data, are illustrated in Figure 11. The system boundaries of the study are from the extraction process to the concentration of the extract just before its incorporation into the face cream. The extraction process consists of the extraction of CA (and caffeine) from SCG, followed by filtration to separate the extract from the solid. The concentration process involved the complete evaporation of ethanol and 1-hexanol in System 2 or the partial evaporation of water (taking into account the percentage of water, 70.7%, required in the composition of the cream) in System 1. The evaporated 1-hexanol and ethanol were reused as inputs for the extraction process. A reusing efficiency of 95% was estimated for this study, and the other 5% of the spent solvents were treated in an incineration plant.

The stages studied were those that varied in the two systems and excluded the ones that intervened in the same way (such as the energy used in the vortex shaker, centrifuge, and filtration).

After extraction, the remaining wet SCG and the inferior phase (in System 2) were treated in a composting plant and a wastewater treatment facility, respectively.

The inventory data was obtained through experiments on a laboratory scale where a 0.7 g sample of dry spent coffee grounds was mixed with 4 mL of solvent. The solvent is water for System 1 and a mixture of 1.2 mL ethanol, 0.96 mL 1-hexanol, and 1.84 mL water for System 2.

After the extraction and for both methods, 60% humidity for SCG was considered (so the amount of wet SCG is 1.75 g from 0.7 g of dried SCG). The mass of the extract for System 1 consisted of the remaining water calculated according to the mass balance, while for System 2, the mass of the extract was calculated based on the volume of the supra phase obtained (1.67 mL from 0.7 g dry SCG), assuming this phase contains all of the 1-hexanol (0.96 mL) and part of the ethanol. The volume of ethanol is the total volume of the extract minus the volume of 1-hexanol. Subsequently, the mass of the different solvents (water and ethanol) in the inferior phase (going to wastewater treatment) was determined by applying the mass balance.

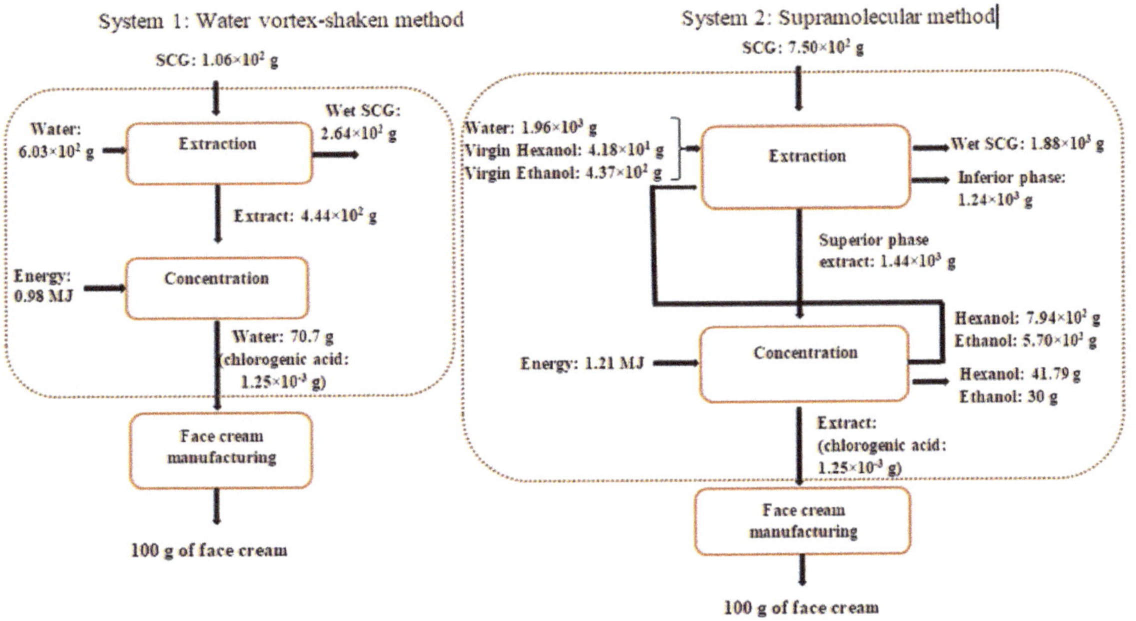

Figure 11. System boundaries of the LCA study to obtain a face cream.

According to the experiments carried out in the laboratory, 0.827 mg CA/g SCG and 0.116 mg CA/g SCG were obtained using water and supramolecular extraction methods, respectively. These results were used to determine the different inputs and outputs to obtain 125 mg CA/FU.

The thermal energy required for each system was theoretically calculated by taking into account the energy required to heat the solvent until boiling point plus the energy required to evaporate this solvent.

5.1.2. Chlorogenic Acid Recovery: Impact Assessment and Discussion of the Results

Results in Figure 12 reveal that water extraction is the eco-friendliest method to obtain 125 mg of CA (the amount needed to produce 100 g of face cosmetic cream), with lower impacts compared to the supramolecular extraction method. Impacts on System 1 are especially linked to the use of thermal energy in the concentration stage (except for land use, LU, and water consumption, Water), while, for the supramolecular method, the impacts are mainly attributed to the lower extraction yield of CA and to the production of the solvents consumed in the extraction stage, since 1-hexanol and ethanol have higher impacts than water.

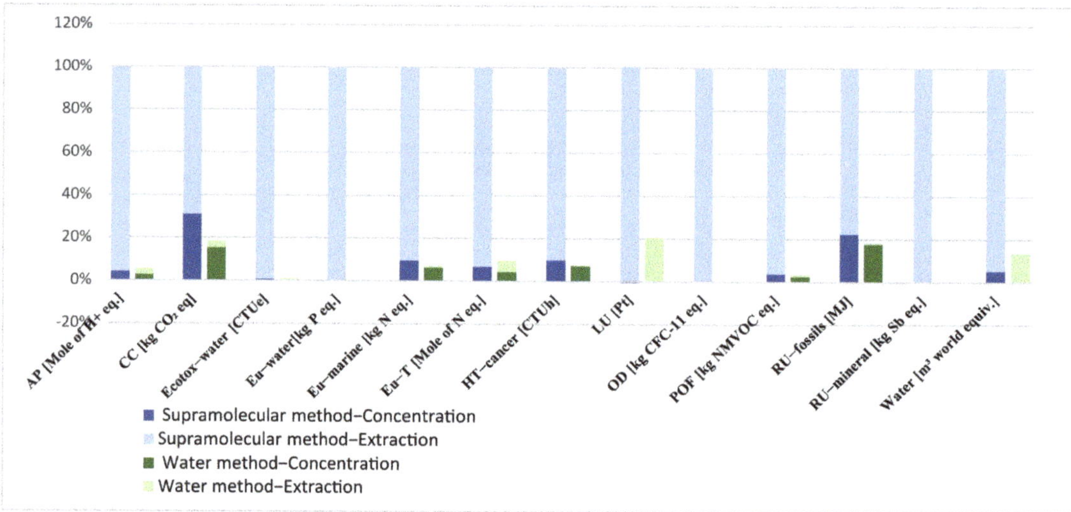

Figure 12. Environmental impacts of chlorogenic acid extraction using water and supra methods. AP: acidification; CC: climate change; Ecotox−water: freshwater ecotoxicity; Eu−water: freshwater eutrophication; Eu−marine: marine eutrophication; Eu−T: terrestrial eutrophication; HT−cancer: human toxicity, cancer; LU: land use; OD: ozone depletion; POF: photochemical ozone formation; RU−fossils: resource use, fossils; RU−mineral: resource use, mineral and metals; and Water: water use.

5.2. *Environmental Comparison for the Production of an Eye Contour Serum*

5.2.1. Caffeine Recovery: Scope and Inventory Data

In the case of an eye contour serum, the anti-inflammatory property of the caffeine is appreciated. Thus, the Functional Unit (FU) chosen for the comparison was "5 g of caffeine needed for 100 g of an eye contour serum", as the serum contains 5% of caffeine (content specified in commercial products).

System boundaries of the LCA comparison, as well as the inventory data, are shown in Figure 13. The system boundaries of the study are the same as before, from the extraction process to the concentration of the extract, just before its incorporation into the serum. The extraction process consists of the extraction of caffeine from SCG, followed by separation to isolate the extract from the solid. The concentration process involves the complete evaporation of ethanol and 1-hexanol in System 2 and partial evaporation of water (taking into account the 70% content of water in the serum).

The same hypothesis as before was taken, where 95% of 1-hexanol and ethanol is reused and 5% not reused is incinerated, wet SCG contains 60% humidity and is composted, and finally, wastewater is treated in a wastewater treatment plant.

Inventory data were obtained in the same way explained in Section 5.1.1 using the yield of caffeine recovery obtained in the laboratory (see Section 4.5): a total of 0.766 mg Caf/g SCG and 0.885 mg Caf/g SCG for water and supramolecular extraction methods, respectively.

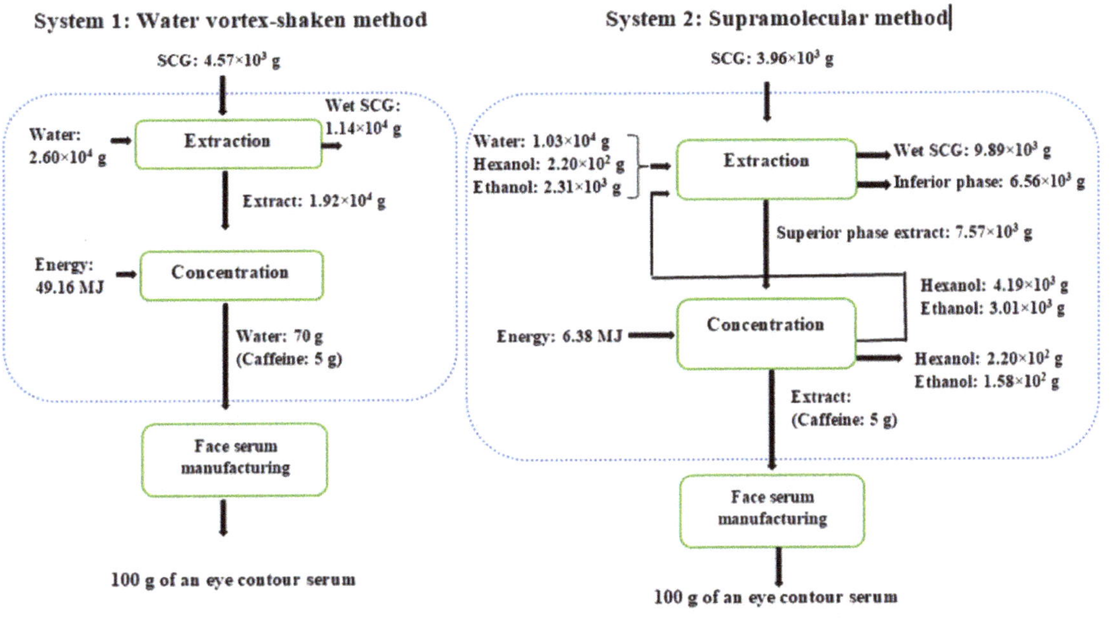

Figure 13. System boundaries of the LCA study to obtain an eye contour serum.

5.2.2. Caffeine Recovery: Impact Assessment and Discussion of the Results

Results in Figure 14 revealed that water extraction is still environmentally better to obtain 5 g of caffeine (the amount needed to produce 100 g of eye serum), with generally lower impacts compared with the supramolecular extraction method (except for climate change, CC; land use, LU; fossil resources use, RU—fossils; and water use, Water).

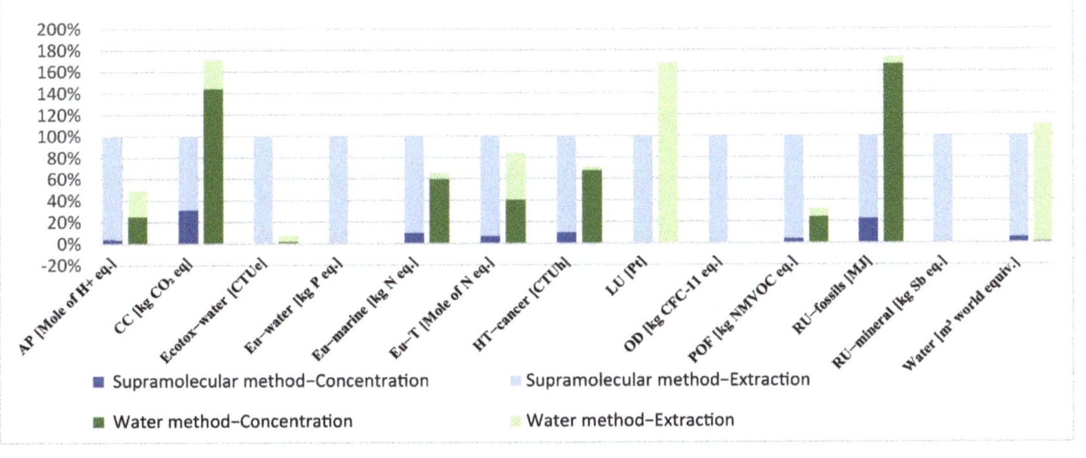

Figure 14. Environmental impacts of caffeine extraction using water and supra methods. AP: acidification; CC: climate change; Ecotox—water: freshwater ecotoxicity; Eu—water: freshwater eutrophication; Eu—marine: marine eutrophication; Eu—T: terrestrial eutrophication; HT—cancer: human toxicity, cancer; LU: land use; OD: ozone depletion; POF: photochemical ozone formation; RU—fossils: resource use, fossils; RU—mineral: resource use, mineral and metals; and Water: water use.

6. Conclusions and Recommendations

A review of the literature on chlorogenic acid and caffeine extraction methods from spent coffee grounds (SCGs) was conducted. Results obtained by different studies are difficult to compare due to the differences in coffee variety, roasting degree, and coffee preparation method. Most authors have optimized a specific extraction method in terms of time, temperature, and solvent/solid ratio. Authors usually do not compare different extraction methods and solvents using the same SCG. In order to generate comparative information, in the present paper, four different extraction methods were compared using the shortest extraction time and lowest solvent/solid ratio (1 min and 5.7 mL/g) described in the literature. Supramolecular solvent extraction (vortex-shaken) and water (vortex-shaken) using ultrasound at two different temperatures were experimentally tested.

Two different SCGs were extracted: SCG1 ("Novell Gourmet Responsable") and SCG2 ("MamaSame"). Results revealed significant differences between the two brands, which are attributed to their origin, roasting degree, and coffee infusion preparation, with SCG1 (which is a lighter roast) being richer in chlorogenic acid and caffeine. Therefore, results are discussed for SCG1.

Ultrasound at room temperature using water as a solvent was demonstrated to be the most effective method to recover chlorogenic acid and caffeine with 1.15 mg CA/g and 0.972 mg Caf/g, respectively. While the supramolecular solvent is more selective for caffeine, obtaining a concentration of 370 mg Caf/L in the extract compared to 49 mg/L for CA.

Additionally, the influence of the extraction time, between 1 min and 5 min, on the extraction yield of chlorogenic acid and caffeine revealed that a longer extraction time (5 min) helps to extract higher content of chlorogenic acid and caffeine (a +70% and +42% increase, respectively).

When the supramolecular solvent method is environmentally compared with water extraction (vortex-shaken) using LCA methodology, the results show that water extraction is usually better for both chlorogenic acid and caffeine extraction. This is mainly due to the impact of the production of solvents (ethanol and 1-hexanol) used in the supra solvent method and to the lower amount of the analyte extracted (especially in the case of chlorogenic acid). It is worth mentioning that lengthening the extraction time (from 1 to 5 min) increases the water extraction yield much more than the supramolecular yield, and, in addition, water extraction would be easier to implement at an industrial scale. The results show that when the water extraction yield is similar to the yield obtained with other solvents, water will usually be the preferred solvent from an environmental perspective. The supramolecular method will certainly be efficient to use and environmentally preferable when extracting other types of analytes having less affinity for water.

The results presented here are intended to promote the circularity of SCG by obtaining valuable substances, such as caffeine and chlorogenic acid. There is no standard method to extract those analytes from SCG; therefore, a large number of alternatives are presented in the literature. Surprisingly, water, which delivers better results than those new alternatives, is not discussed in the literature, which, until now, has not presented comparative studies of different extraction methods. Our proposal is that when a new extraction alternative is reported, comparative studies are needed to evaluate both extraction yields and environmental performance in comparison with older methods.

Author Contributions: Conceptualization: R.P.; Methodology: R.C. and R.P.; Formal analysis and investigation: I.B. and R.C.; Resources and data curation: A.B.; Writing—original draft preparation: I.B. and R.C.; Writing—review and editing: I.B., R.P., M.B. and P.F.-i.-P.; Supervision: R.C. and R.P.; Funding acquisition: R.A., M.M. and R.P.; Project administration: M.B.; Conception of the Project's original idea: R.A. and P.F.-i.-P. All authors have read and agreed to the published version of the manuscript.

Funding: The authors wish to acknowledge the financial support from the Spanish Ministry of Science and Innovation to the project KAIROS-BIOCIR (PID2019-104925RB-C32). The first author also appreciates the support (2021FI SDUR 00130) from the Secretariat for Universities and Research of the Ministry of Business and Knowledge of the Government of Catalonia and the European Social Fund.

Data Availability Statement: Data is contained within the article. Nevertheless, if more details are needed they will be available from the corresponding author.

Conflicts of Interest: The authors declare no conflict of interest.

References

1. Lang, L.; Wang, Y.; Chen, X.; Zhang, Z.; Yang, N.; Xue, B.; Han, W. Awareness of Food Waste Recycling in Restaurants: Evidence from China. *Resour. Conserv. Recycl.* **2020**, *161*, 104949. [CrossRef]
2. Ishangulyyev, R.; Kim, S.; Lee, S.H. Understanding Food Loss and Waste—Why Are We Losing and Wasting Food? *Foods* **2019**, *8*, 297. [CrossRef] [PubMed]
3. European Commission Communication from the Commission to the European Parliament, the Council, the European Economic and Social Committee and the Committee of the Regions. Available online: https://eur-lex.europa.eu/legal-content/EN/TXT/?qid=1551965345008&uri=CELEX:32018L0852 (accessed on 12 May 2022).
4. Food and Agriculture Organization of the United Nations. *Food Waste Index Report 2021*; FAO: Rome, Italy, 2021.
5. Wu, T.-W.; Zhang, H.; Peng, W.; Lü, F.; He, P.-J. Applications of Convolutional Neural Networks for Intelligent Waste Identification and Recycling: A Review. *Resour. Conserv. Recycl.* **2023**, *190*, 106813. [CrossRef]
6. Ferdous, W.; Manalo, A.; Siddique, R.; Mendis, P.; Zhuge, Y.; Wong, H.S.; Lokuge, W.; Aravinthan, T.; Schubel, P. Recycling of Landfill Wastes (Tyres, Plastics and Glass) in Construction—A Review on Global Waste Generation, Performance, Application and Future Opportunities. *Resour. Conserv. Recycl.* **2021**, *173*, 105745. [CrossRef]
7. De Sadeleer, I.; Brattebø, H.; Callewaert, P. Waste Prevention, Energy Recovery or Recycling—Directions for Household Food Waste Management in Light of Circular Economy Policy. *Resour. Conserv. Recycl.* **2020**, *160*, 104908. [CrossRef]
8. Conde, T.; Mussatto, S.I. Isolation of Polyphenols from Spent Coffee Grounds and Silverskin by Mild Hydrothermal Pretreatment. *Prep. Biochem. Biotechnol.* **2016**, *46*, 406–409. [CrossRef]
9. International Coffee Organization. Available online: https://www.ico.org/ (accessed on 31 May 2022).
10. Gouws, S.; Muller, M. Valorization of Products from Grounded-Coffee Beans. *Sci. Rep.* **2021**, *11*, 20445. [CrossRef]
11. Vandeponseele, A.; Draye, M.; Piot, C.; Chatel, G. Study of Influential Parameters of the Caffeine Extraction from Spent Coffee Grounds: From Brewing Coffee Method to the Waste Treatment Conditions. *Clean Technol.* **2021**, *3*, 335–350. [CrossRef]
12. Shang, Y.F.; Xu, J.L.; Lee, W.J.; Um, B.H. Antioxidative Polyphenolics Obtained from Spent Coffee Grounds by Pressurized Liquid Extraction. *South African J. Bot.* **2017**, *109*, 75–80. [CrossRef]
13. Silva, M.d.O.; Honfoga, J.N.B.; Medeiros, L.L.d.; Madruga, M.S.; Bezerra, T.K.A. Obtaining Bioactive Compounds from the Coffee Husk (*Coffea Arabica* L.) Using Different Extraction Methods. *Molecules* **2020**, *26*, 46. [CrossRef]
14. Perta-Crisan, S.; Ursachi, C.; Munteanu, F.-D. Trends in Valorisation of Spent Coffee Grounds: A Review. *Sci. Tech. Bull. Ser. Chem. Food Sci. Eng.* **2019**, *16*, 31–42.
15. Andrade, K.S.; Gonalvez, R.T.; Maraschin, M.; Ribeiro-Do-Valle, R.M.; Martínez, J.; Ferreira, S.R.S. Supercritical Fluid Extraction from Spent Coffee Grounds and Coffee Husks: Antioxidant Activity and Effect of Operational Variables on Extract Composition. *Talanta* **2012**, *88*, 544–552. [CrossRef]
16. Gemechu, F.G. Embracing Nutritional Qualities, Biological Activities and Technological Properties of Coffee Byproducts in Functional Food Formulation. *Trends Food Sci. Technol.* **2020**, *104*, 235–261. [CrossRef]
17. Cruz Velasquez, S.M. *Evaluación de La Actividad Antioxidante y Detección de Marcadores Químicos En Extractos de Hojas y Granos de Siete Variedades de Café Comercializadas En Guatemala*; Guatemala University San Carlos Guatemala, Facultad de Ciencias Químicas y Farmacia: Guatemala City, Guatemala, 2018.
18. Zuorro, A.; Lavecchia, R. Spent Coffee Grounds as a Valuable Source of Phenolic Compounds and Bioenergy. *J. Clean. Prod.* **2012**, *34*, 49–56. [CrossRef]
19. Panusa, A.; Zuorro, A.; Lavecchia, R.; Marrosu, G.; Petrucci, R. Recovery of Natural Antioxidants from Spent Coffee Grounds. *J. Agric. Food Chem.* **2013**, *61*, 4162–4168. [CrossRef]
20. Mussatto, S.I.; Ballesteros, L.F.; Martins, S.; Teixeira, J.A. Extraction of Antioxidant Phenolic Compounds from Spent Coffee Grounds. *Sep. Purif. Technol.* **2011**, *83*, 173–179. [CrossRef]
21. Andrade, K.S. Avaliação Das Técnicas de Extração e Do Potencial Antioxidante Dos Extratos Obtidos a Partir de Casca e de Borra de Café (*Coffea Arabica* L.). Master's Thesis, Universidade Federal de Santa Catarina, Centro Tecnológico, Florianópolis, Brazil, 2012.
22. Acevedo, F.; Rubilar, M.; Scheuermann, E.; Cancino, B.; Uquiche, E.; Garcés, M.; Inostroza, K.; Shene, C. Spent Coffee Grounds as a Renewable Source of Bioactive Compounds. *J. Biobased Mater. Bioenergy* **2013**, *7*, 420–428. [CrossRef]
23. Ballesteros, L.F.; Ramirez, M.J.; Orrego, C.E.; Teixeira, J.A.; Mussatto, S.I. Optimization of Autohydrolysis Conditions to Extract Antioxidant Phenolic Compounds from Spent Coffee Grounds. *J. Food Eng.* **2017**, *199*, 1–8. [CrossRef]
24. Caballero-Galván, A.S.; Restrepo-Serna, D.L.; Ortiz-Sánchez, M.; Cardona-Alzate, C.A. Analysis of Extraction Kinetics of Bioactive Compounds from Spent Coffee Grounds (Coffea Arábica). *Waste Biomass Valorization* **2018**, *9*, 2381–2389. [CrossRef]

25. Al-Dhabi, N.A.; Ponmurugan, K.; Maran Jeganathan, P. Development and Validation of Ultrasound-Assisted Solid-Liquid Extraction of Phenolic Compounds from Waste Spent Coffee Grounds. *Ultrason. Sonochem.* **2017**, *34*, 206–213. [CrossRef]
26. Arauzo, P.J.; Lucian, M.; Du, L.; Olszewski, M.P.; Fiori, L.; Kruse, A. Improving the Recovery of Phenolic Compounds from Spent Coffee Grounds by Using Hydrothermal Delignification Coupled with Ultrasound Assisted Extraction. *Biomass Bioenergy* **2020**, *139*, 105616. [CrossRef]
27. Pavlović, M.D.; Buntić, A.V.; Šiler-Marinković, S.S.; Dimitrijević-Branković, S.I. Ethanol Influenced Fast Microwave-Assisted Extraction for Natural Antioxidants Obtaining from Spent Filter Coffee. *Sep. Purif. Technol.* **2013**, *118*, 503–510. [CrossRef]
28. Pettinato, M.; Casazza, A.A.; Ferrari, P.F.; Palombo, D.; Perego, P. Eco-Sustainable Recovery of Antioxidants from Spent Coffee Grounds by Microwave-Assisted Extraction: Process Optimization, Kinetic Modeling and Biological Validation. *Food Bioprod. Process.* **2019**, *114*, 31–42. [CrossRef]
29. Xu, H.; Wang, W.; Liu, X.; Yuan, F.; Gao, Y. Antioxidative Phenolics Obtained from Spent Coffee Grounds (Coffea Arabica L.) by Subcritical Water Extraction. *Ind. Crops Prod.* **2015**, *76*, 946–954. [CrossRef]
30. Torres-Valenzuela, L.S.; Ballesteros-Gómez, A.; Sanin, A.; Rubio, S. Valorization of Spent Coffee Grounds by vecular Solvent Extraction. *Sep. Purif. Technol.* **2019**, *228*, 115759. [CrossRef]
31. Ballesteros-Gómez, A.; Sicilia, M.D.; Rubio, S. Supramolecular Solvents in the Extraction of Organic Compounds. A Review. *Anal. Chim. Acta* **2010**, *677*, 108–130. [CrossRef]
32. Clifford, M.N.; Wight, J. The Measurement of Feruloylquinic Acids and Caffeoylquinic Acids in Coffee Beans. Development of the Technique and Its Preliminary Application to Green Coffee Beans. *J. Sci. Food Agric.* **1976**, *27*, 73–84. [CrossRef]
33. Singleton, V.L.; Rossi, J.A. Colorimetry of Total Phenolics with Phosphomolybdic-Phosphotungstic Acid Reagents. *Am. J. Enol. Vitic.* **1965**, *16*, 144–158.
34. Kyoung Chun, O.; Kim, D.O. Consideration on Equivalent Chemicals in Total Phenolic Assay of Chlorogenic Acid-Rich Plums. *Food Res. Int.* **2004**, *37*, 337–342. [CrossRef]
35. Malta, L.G.; Liu, R.H. Analyses of Total Phenolics, Total Flavonoids, and Total Antioxidant Activities in Foods and Dietary Supplements. In *Encyclopedia of Agriculture and Food Systems*; Academic Press: Cambridge, MA, USA, 2014; pp. 305–314. [CrossRef]
36. Campos-Vega, R.; Loarca-Piña, G.; Vergara-Castañeda, H.A.; Dave Oomah, B. Spent Coffee Grounds: A Review on Current Research and Future Prospects. *Trends Food Sci. Technol.* **2015**, *45*, 24–36. [CrossRef]
37. Ramón-Gonçalves, M.; Gómez-Mejía, E.; Rosales-Conrado, N.; León-González, M.E.; Madrid, Y. Extraction, Identification and Quantification of Polyphenols from Spent Coffee Grounds by Chromatographic Methods and Chemometric Analyses. *Waste Manag.* **2019**, *96*, 15–24. [CrossRef]
38. Shalmashi, A.; Golmohammad, F. Solubility of Caffeine in Water, Ethyl Acetate, Ethanol, Carbon Tetrachloride, Methanol, Chloroform, Dichloromethane, and Acetone between 298 and 323 K. *Lat. Am. Appl. Res.* **2010**, *40*, 283–285.
39. Okur, I.; Soyler, B.; Sezer, P.; Oztop, M.H.; Alpas, H. Improving the Recovery of Phenolic Compounds from Spent Coffee Grounds (SCG) by Environmentally Friendly Extraction Techniques. *Molecules* **2021**, *26*, 613. [CrossRef] [PubMed]
40. Iziar, A.L.; Mena, P.; Calani, L.; Cid, C.; Del Rio, D.; Lean, M.E.; Crozier, A. Variations in Caffeine and Chlorogenic Acid Contents of Coffees: What Are We Drinking? *Food Funct.* **2014**, *5*, 1718–1726. [CrossRef]
41. Mills, C.E.; Oruna-Concha, M.J.; Mottram, D.S.; Gibson, G.R.; Spencer, J.P.E. The Effect of Processing on Chlorogenic Acid Content of Commercially Available Coffee. *Food Chem.* **2013**, *141*, 3335–3340. [CrossRef] [PubMed]
42. Vignoli, J.A.; Bassoli, D.G.; Benassi, M.T. Antioxidant Activity, Polyphenols, Caffeine and Melanoidins in Soluble Coffee: The Influence of Processing Conditions and Raw Material. *Food Chem.* **2011**, *124*, 863–868. [CrossRef]
43. Jeon, J.S.; Kim, H.T.; Jeong, I.H.; Hong, S.R.; Oh, M.S.; Yoon, M.H.; Shim, J.H.; Jeong, J.H.; Abd El-Aty, A.M. Contents of Chlorogenic Acids and Caffeine in Various Coffee-Related Products. *J. Adv. Res.* **2019**, *17*, 85–94. [CrossRef]
44. Kulapichitr, F.; Borompichaichartkul, C.; Fang, M.; Suppavorasatit, I.; Cadwallader, K.R. Effect of Post-Harvest Drying Process on Chlorogenic Acids, Antioxidant Activities and CIE-Lab Color of Thai Arabica Green Coffee Beans. *Food Chem.* **2022**, *366*, 130504. [CrossRef]
45. McCusker, R.R.; Goldberger, B.A.; Cone, E.J. Caffeine Content of Specialty Coffees. *J. Anal. Toxicol.* **2003**, *27*, 520–522. [CrossRef]
46. Caprioli, G.; Cortese, M.; Sagratini, G.; Vittori, S. The Influence of Different Types of Preparation (Espresso and Brew) on Coffee Aroma and Main Bioactive Constituents. *Food Sci. Nutr.* **2015**, *66*, 505–513. [CrossRef]
47. Butiuk, A.P.; Maidana, S.A.; Adachi, O.; Akakabe, Y.; Martos, M.A.; Hours, R.A. Optimization and Modeling of the Chlorogenic Acid Extraction from a Residue of Yerba Mate Processing. *J. Appl. Res. Med. Aromat. Plants* **2021**, *25*, 100329. [CrossRef]
48. *ISO 14040:2006;* Environmental Management: Life Cycle Assessment; Principles and Framework. International Organization for Standardization: Geneva, Switzerland, 2006.

foods

MDPI

Article

Physicochemical and Sensory Properties and Shelf Life of Block-Type Processed Cheeses Fortified with Date Seeds (*Phoenix dactylifera* L.) as a Functional Food

Nashi K. Alqahtani [1], Tareq M. Alnemr [1,2], Abdullah M. Alqattan [3], Salah M. Aleid [1] and Hosam M. Habib [4,*]

1 Department of Food and Nutrition Sciences, College of Agricultural and Food Sciences, King Faisal University, P.O. Box 400, Al-Ahsa 31982, Saudi Arabia
2 Dairy Science and Technology Department, Faculty of Agriculture, Alexandria University, Alexandria P.O. Box 21545, Egypt
3 Research & Development Department, Almarai Company, P.O. Box 8524, Riyadh 12411, Saudi Arabia
4 Research & Innovation Hub, Alamein International University (AIU), Alameen City 5060310, Egypt
* Correspondence: hosaml@hotmail.com; Tel.: +20-1003991550

Abstract: Processed cheese has rapidly been established as a commercial product in recent years. A new ingredient, a byproduct from date fruit seed (DFS), was obtained and tested as a fortified fiber from food industrial waste in block-type processed cheese. This is the first inclusive investigation to report such a test. Different concentrations of DFS (0%, 5%, 10%, 15%, and 20%) were added to block-type processed cheese as a partial substitution for butter. The current investigation was undertaken to estimate the impact of the partial substitution of butter by DFS and its effect on the product's quality in terms of its shelf life and physicochemical, microstructure, color, and sensory properties. Quality was assessed over a 150-day storage period. The results indicate that adding DFS to cheese increased its nutritional value due to the addition of fiber. Additionally, the texture profile of cheese was decreased in terms of hardness, adhesion, springiness, and cohesiveness. The overall structure of cheeses became less compact and had a more open cheese network, which increased with increasing DFS% and duration of storage. Moreover, DFS exhibited the darkest color with increasing ratios of supplementary DFS and duration of storage. Based on the results found in the present investigation, it was concluded that an acceptable quality of block-type processed cheese could be achieved using DFS fiber at 5% and 10% levels of fortification.

Keywords: date fruit seeds; block-type processed cheeses; microstructure properties; texture properties; sensory properties; shelf life; byproduct; fortified fiber; substitution; food waste

Citation: Alqahtani, N.K.; Alnemr, T.M.; Alqattan, A.M.; Aleid, S.M.; Habib, H.M. Physicochemical and Sensory Properties and Shelf Life of Block-Type Processed Cheeses Fortified with Date Seeds (*Phoenix dactylifera* L.) as a Functional Food. *Foods* **2023**, *12*, 679. https://doi.org/10.3390/foods12030679

Academic Editors: Antonio Cilla, Stanisław Kowalski and Dorota Gumul

Received: 2 January 2023
Revised: 25 January 2023
Accepted: 30 January 2023
Published: 3 February 2023

1. Introduction

The valorization of byproducts and industrial food waste has become the main focus of the investigation to improve the sustainability of functional food [1–4]. Food fortification increases the amount of essential micronutrients in foods, improves their nutritional quality, and provides health benefits with little or no risk [5,6]. Fortification acts as a vehicle to add a nutrient (enhancer, restorative, or additive) that is absent or present in low amounts in the food matrix [7,8]. Fortification prevents nutrient deficiencies and associated issues to balance the overall nutrient profile and restore nutrients lost during processing, increasing the appeal for consumers looking to supplement their diets. Food fortification is also a public health strategy to increase nutrient intake in the population [9,10]. Food fortification adds many primary or secondary beneficial compounds to the food chain [11–13]. Enrichment can prevent or minimize the risk of micronutrient deficiencies in a given population or group. Therefore, it is a process for improving nutritional status and food intake [6,14]. The effect of fortification on health depends on certain parameters, such as the level of fortification bioavailability and the amount of fortified food consumed [15,16].

The challenge of producing good-tasting functional foods within the dairy industry rises when using food waste. Numerous authors have conveyed organoleptic issues related to the fortification of byproducts in dairy foods, mostly due to bitter, rheological, and textural properties or the salty off-flavors that are typical of date fruit seeds' phytonutrients [17,18]. Furthermore, there is a deficiency of information from the consumer's perspective with regard to consuming food byproducts as ingredients in other foods. The understandable concerns about food safety and quality that may arise, as well as the importance of sustainability as a driver of food choice, should be studied [4,19–21].

The Statistics Division of the Food and Agriculture Organization of the United Nations (FAOSTAT, 2020) [22] reported that the global production of date palm fruit (*Phoenix dactylifera* L.) is around 9.5 million tons yearly. Date palm fruit processing results in the production of massive quantities of date seeds as a waste byproduct, establishing 6.11–11.50% of date fruit weight depending on the variety, grade, and quality [23,24]. In the past, date fruit seeds were utilized as soil fertilizers and animal feed [25]. Nevertheless, they have increasingly been utilized in functional foods such as beef burgers, bread, muffins, cookies, processed cheese, coffee, ketchup, gluten-free cookies and biscuits, cooking oil, cakes, jams, and bio-oil, and have also been utilized in dietary supplementation, pharmacology, and cosmetics [26–35]. Figure 1 shows products that are fortified with DFS. Furthermore, date fruit seed (DFS) was shown to be principally rich in fiber and considerable amounts of vitamins, minerals, lipids, carotenoids, and protein [36–39]. Likewise, DFS is also considered to be a source of flavonoids and polyphenols, depending on the variety. The polyphenolic and flavonoid content of DFS is approximately 51 g/kg, which is higher when equated with other polyphenol-rich foods such as flaxseeds, grapes, and green tea [40–42]. Nevertheless, the polyphenolic compounds present in DFS have been shown to have anti-inflammatory and antioxidant properties that could be used to prevent and treat obesity, cancer, cardiovascular diseases, and neurodegenerative conditions [23,43–45].

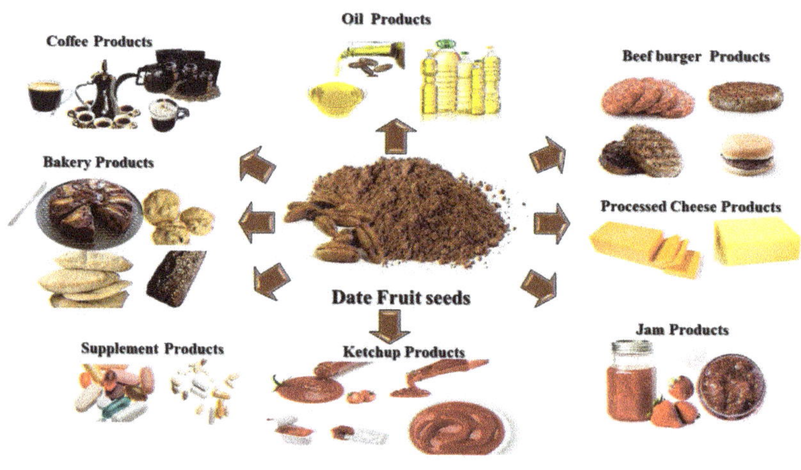

Figure 1. Products fortified with DFS.

Due to a lack of studies on the integral use of DFS for prolonging the shelf life of fortified block-type processed cheeses, the objective of this study was to evaluate the microbiological characteristics, chemical composition, and physical and functional potential of DFS. Furthermore, the production of functional block-type processed cheeses enriched with DFS was studied in order to assess the impact of these fortifications on the microstructure, shelf life, and physicochemical, texture, and sensory characteristics of DFS block-type processed cheeses. This allowed for the identification of the most ideal and functional form of DFS to be consumed in order to unlock its maximum potential with acceptable sensorial properties for consumers.

2. Materials and Methods

2.1. Materials

2.1.1. Plant Material and Preparation

The DFS was used as a powdered food ingredient. Date Khalas variety fruit seeds, at the Tamar stage at the end of the harvested season, were obtained from the local date processing industry of Al-Ahasa Eastern Province, Saudi Arabia. The DFS was soaked and washed in normal water to eliminate any date flesh and then dried at 50 °C under a vacuum. The dried DFS was ground using a heavy-duty grinder and a stainless-steel hammermill (Guangzhou Mingyue, Nancun, China) grinding mill. Finally, a 0.30-mm sieve was used to obtain the DFS. A final step of drying for 1 h was repeated to remove the moisture from the powder. Finally, the DFS was vacuum-packaged and kept frozen until it was used.

2.1.2. Study Materials

Standard cheddar cheese (Fat/DM: 55.8 and protein 25.1%), frozen cheddar cheese (Fat/DM: 54.2 and protein 24.2%), and milk fat butter (fat: 81.7%, SNF: 1.4) were obtained from Fonterra Inc., Auckland, New Zealand. Emulsifying salt, Joha C Special, was obtained from BK Giulini, Ladenburg, Germany. Antibacterial agent, Nisin (E234), was obtained from Danisco, Copenhagen, Denmark. Antifungal agent, potassium sorbate, was obtained from Nutrinova, Sulzbach, Germany. Plate count agar medium, Baird–Parker agar medium, sulfite–cycloserine agar medium, egg yolk polymyxin agar medium, violet red bile agar medium, nutrient agar medium, and notato dextrose agar medium were purchased from HiMedia Laboratories Pvt. Ltd., Mumbai, India. Sulfuric acid 98%, H_3BO_3 (boric acid; reagent-grade, ≥98%; pellets (anhydrous)), phenolphthalein reagent (3,3-Bis(4-hydroxyphenyl)-1(3*H*)-isobenzofuranone), glutaraldehyde solution (Grade I, 25% in H_2O, specially purified for use as an electron microscopy fixative), phosphate-buffered solution, osmium tetroxide (ReagentPlus®, 99.8%), ethyl ether (analytical-grade), ethanol (absolute for analysis EMSURE® ACS, ISO, Reag), and petroleum ether (ACS reagent) were purchased from Millipore Sigma Chemical Co. (St. Louis, MO, USA).

2.2. Characterization of DFS and Final Products

2.2.1. Proximate Analysis Composition

The proximate analysis composition of protein, moisture, ash, fat, and fiber was determined for DFS and the final products from day 0 to 5 months (monthly).

Total Protein

Total protein was determined according to the method described previously [46], using Kjeldahl Semi-automized Foss model 2300 (Foss Tecator, Hoganas, Sweden).

Moisture

All samples were analyzed for moisture content in a moisture balance with a halogen lamp heating element (MS-70, A&D Instruments Company Ltd., Abingdon, UK.), and subjected to step heating to 130 °C; persistent weight was gained at +/−0.02% precision [47,48].

Ash

Ash contents were determined by carbon elimination of 1 g of sample, which was ignited and incinerated in a muffle furnace at 550 °C for 2 h. The flask was removed from heat and left to cool. Two milliliters of H_2O_2 were added and the flask was put back in the muffle furnace for further incineration over 1 h. The total ash was expressed as a percentage of dry weight [49].

Total Fat

Total fat was determined according to the method described previously [50], using the Soxhlet method (Ankom XT15 Extractor, Ankom Technology, Macedon, NY, USA; Soxtec™)

Total Fiber

Total fiber content was measured using a modified version of the method described in [51]. Briefly, 1 g of sample and 10 mL of distilled water were added and the mixture was maintained at 100 °C for 10 min. Then, 40 mL of absolute ethanol was added and the mixture was agitated and left in ice water for 30 min. The mixture was centrifuged (Universal 320 R, Andreas Hettich GmbH & Co. KG, Tuttlingen, Germany) at 1500× *g* for 10 min at 4 °C and the residue was added to 50 mL of 85% ethanol, mixed, and centrifuged again. This step was repeated using 50 mL of absolute ethanol. Finally, the residue was dried at 80 °C for 24 h and weighed.

2.2.2. Physicochemical Analysis

The pH and acidity were assessed in the final products as previously described [33]. Briefly, the pH of 20 g of cheese in 20 mL of distilled water was measured using a pH meter (Crison Instrument, Barcelona, Spain). The visual color was measured with a Hunter colorimeter model D2s A-2 (Hunter Assoc. Lab., Inc., Reston, VA, USA), in terms of a* (redness and greenness), L* (lightness), and b* (yellowness and blueness). The instrument (45°/0° geometry, 10° observer) was calibrated with a standard white and black tile followed by the measurement of the final products at different storage times [52].

2.2.3. Microbiological Analysis

Microbiological counts of *Escherichia coli*, *Staphylococcus aureus*, *Clostridium perfringens*, *Bacillus cereus*, yeast and mold, and coliform in the DFS (before cooking) and block-type processed cheese samples were determined monthly, from 0 to 5 months. For this determination, 10 g of each sample of block-type processed cheese was mixed with 90 mL of sterilized Ringer solution and the diluted samples were consistently spread into a sterile Petri dish containing each culture medium using a spatula. Potato dextrose agar medium was used for yeast counts and the subsequent incubation of molds for 5 days in aerobic conditions at 25 ± 2 °C. Furthermore, the presence of *E. coli* was detected using violet red bile agar medium at 37 °C for 24 h. Violet red bile agar medium was also used to distinguish coliforms, incubated at 37 °C for 24 h. Baird–Parker agar was used to distinguish *Staphylococcus aureus*, incubated at 37 °C for 48 h. Sulfite–cycloserine agar, incubated anaerobically at 37 °C for 18–22 h, was used to distinguish *Clostridium perfringens*, and egg yolk polymyxin agar was used to distinguish *Bacillus cereus*, incubated at 30 °C for 48 h [33]. The results were counted directly as colony-forming units (CFU/g) [53].

2.3. *Cheese-Making Procedure*

The ingredients used for the manufacture of fortified block-type processed cheeses were prepared according to the formulations presented in Table 1. Totals of 5%, 10%, 15%, and 20% (*w*/*w*) of the cheese mixture butter were replaced with DFS.

Table 1. Chemical ingredients and experimental design of block-type processed cheeses.

Ingredients (g), (*w*/*w*)	Cheese Codes				
	Control	5% DFS *	10% DFS *	15% DFS *	20% DFS *
Butter	103.50	98.35	93.15	88.00	83.00
DFS	0.00	5.175	10.35	15.53	20.70
Standard cheddar cheese	540	540	540	540	540
Frozen cheddar cheese	300	300	300	300	300
Emulsifying salt (phosphate salts, Joha C)	27.30	27.30	27.30	27.30	27.30
Potassium sorbate	1.17	1.17	1.17	1.17	1.17
Nisin	0.117	0.117	0.117	0.117	0.117
Water	200	200	200	200	200
Total	1172.1	1172.1	1172.1	1172.1	1172.1

* The % of DFS was calculated as a partial replacement for butter compared with the amount of control butter.

The procedure for the production of fortified block-type processed cheeses is presented in the flowchart in Figure 2.

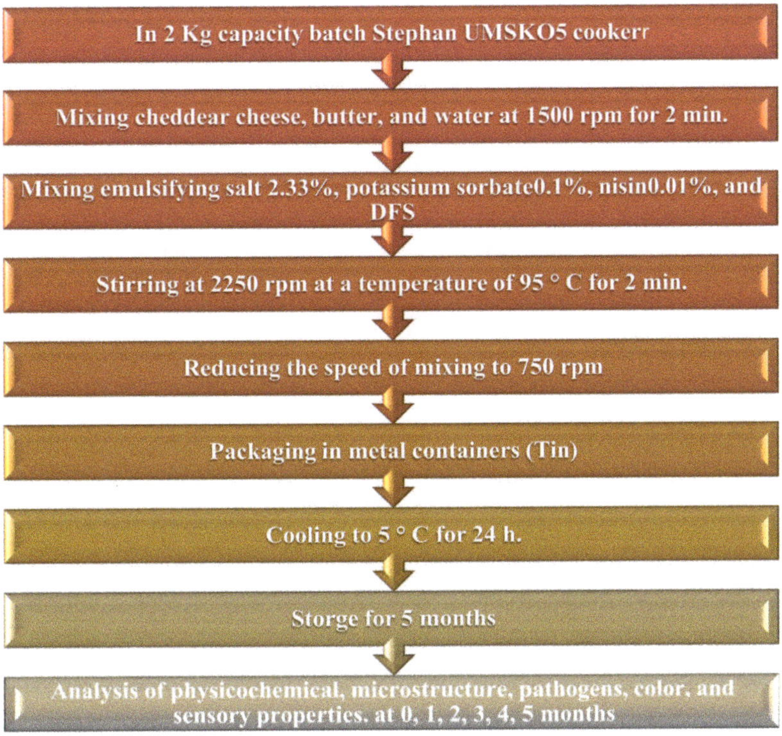

Figure 2. Flowchart of the effect of each ingredient in the manufacturing process of block-type processed cheeses [54].

2.4. Texture Analysis

The texture was measured using the Brookfield Texture Analyzer (Model CT3 4500, AMETEK Brookfield, Middleboro, MA, USA). Pre-test speed, test speed, trigger type, and deformation were 1.5 mm/s, 2 mm/s, 2.5 g, and 5 mm, respectively. The following parameters were assessed by texture profile analysis (TPA), as reported previously by [33]. Hardness was selected as the force that is essential to attain a given deformation; fracture ability is the force at which the material fractures; springiness or elasticity is defined as the rate at which a deformed material returns to its un-deformed state after the deforming force is removed; and cohesiveness is defined as the quantity simulating the strength of the internal bond assembly in the body of the cheese product. The cheese samples were analyzed at a temperature of 25 ± 1 °C, with a minimum of five replicates.

2.5. Microstructural Analysis

To investigate the microstructural properties of the fortified block-type processed cheese samples, scanning electron microscopy (SEM) (JEOL—Japan Electron Optics Laboratory—JSM-6380 LA, JEOL USA, Inc., Peabody, MA, USA) was performed at a voltage of 30.0 kV, as previously reported by [55], with a few modifications. Briefly, to stabilize the samples, pieces of block-type processed cheeses (4 × 4 × 4 mm) were cut from the center portions of the cheese blocks and placed in a 4% (v/v) solution of glutaraldehyde in 0.1 M phosphate buffer, pH 7.2, at 4 °C for 4 h. The cheese samples were washed four times in phosphate buffer for 15 min, then stabilized by osmium tetroxide 2% as a secondary stabilizer for 4 h. Then, the samples were washed three times in cacodylate buffer, followed by dehydration using a series of ethanol solutions (from 20% to 100%). After dehydration, the samples were dried using CO_2 at the critical point and then directly mounted on a copper stub and coated with gold up to a thickness of 400 Å in a sputter-coating unit. The resolution was 3.0 nm (30 kV, WD 8 mm, SEI), the accelerating voltage was from 0.5 to 30 kV (53 steps), and the magnification was ×5 to 300,000 (149 steps). Samples from each treatment were taken both initially and after 5 months for analysis.

2.6. Sensory Analysis

The sensory evaluation of fortified cheese samples was achieved at the beginning of storage time and every month for 5 months. The sensory properties of cheese samples were assessed using a five-point hedonic scale (1: dislike extremely, 5: like very much). This scale consisted of the test parameters of appearance, firmness, stickiness, breakdown, gumminess, smoothness, chewiness, flavor, and overall acceptance. The sensory attributes were measured by 20 experienced staff members of the Food and Nutrition Science Department, King Faisal University, who had previously participated in the hedonic sensory assessment test. The panelist selection criteria were as follows: (1) between 25 and 60 years of age, (2) not allergic to dairy products, (3) non-smoker, and (4) available to participate in the sensory analysis during testing time [56].

2.7. Statistical Analysis

All analyses were performed in triplicate (n = 3). Statistical analyses were performed using SPSS for Windows (version 25; SPSS Inc., Chicago, IL, USA). The differences in mean values among sample varieties were determined ($p < 0.05$) using a one-way analysis of variance (ANOVA). Means separation was performed using Tukey multiple range tests. Values for the mean ± standard deviation are also presented.

3. Results and Discussion

3.1. Characterization of DFS

Total fiber, protein, and fat compounds were the major components of DFS. Table 2 illustrates the chemical composition of DFS. The moisture, TS (total solids), protein, fat, fiber, and ash were 9.39 ± 0.10, 90.61 ± 3.01, 4.86 ± 0.58, 18.44 ± 1.72, 65.95 ± 0.70, and 1.36 ± 0.01, respectively. From the chemical composition of DFS, it can be concluded that the protein and fiber content influence the functional properties and agreeable texture of the fortified cheese samples during storage, and can act as good emulsifying agents [26,57]. The results in terms of the chemical composition were in the same range and similar to results that were reported previously [26,32,33,37,58].

Table 2. Chemical and microbiological characterization of DFS.

Contents	Chemical Compositions g/100 g	Microbiological Analysis CFU/gm	
Moisture%	9.39 ± 0.10	Coliform	Nd
TS%	90.61 ± 3.01	Yeast and mold	<12
Protein%	4.86 ± 0.58 (5.36% DM)	*Escherichia coli*	Nd
Fat%	18.44 ± 1.72 (20.35% DM)	*Staphylococcus aureus*	Nd
Total Fiber%	65.95 ± 0.70 (71.99% DM)	*Clostridium perfringens*	Nd
Ash%	1.36 ± 0.01(1.50% DM)	*Bacillus cereus*	Nd

Nd: not detected; DM: dry matter.

3.2. Characterization of Cheeses Fortified with DFS during Storage

The average chemical composition values, moisture, pH, acidity, ash, fat, and protein of the five fortified block-type processed cheese samples are shown in Figure 3a–f. Figure 3a shows the values of moisture content; it is obvious that the values decreased after fortification with DFS. The values were 49.63 ± 0.39% for the control treatment, while, for treatments with 5%, 10%, 15%, and 20% DFS, the values were 48.34 ± 0.17, 47.99 ± 1.26,

47.66 ± 1.25, and 48.40 ± 0.5%, respectively. Moisture values were decreased ($p \leq 0.05$) compared with the control treatment. Cheese fortified with 10% and 15% DFS as a partial substitution for butter had 3.3 and 3.97% less moisture than the control treatment, respectively, whereas cheese fortified with 5% and 20% DFS possessed 2.6 and 2.48% less moisture than the control treatment, respectively.

During storage, the cheese samples showed gradual decreases ($p \leq 0.05$) in their moisture content as the storage period progressed; however, in the first month, the decreases were not significant ($p \geq 0.05$). The decrease in moisture values in the control sample after five months of storage was 6.61%, while, for 5%, 10%, 15%, and 20% DFS (w/w), these values were 4.06, 2.19, 1.15, and 3.47%, respectively, after five months of storage. The change in the moisture contents can be explained due to the increase in the amount of fiber and polyphenols with increasing DFS. Dönmez et al. [59] reported that fiber and polyphenols could interact with proteins, resulting in the formation of polyphenol–protein complexes, which decreases the release of moisture contents in a concentration-dependent manner. On the other hand, the excess polyphenol concentrations increased the release of moisture.

Likewise, Figure 3b shows that the effect of the incorporation of DFS on the pH values was not significant ($p \geq 0.05$), although a slight decrease was shown. The values were 5.17 ± 0.02 for the control treatment, while, for fortified treatment with 5%, 10%, 15%, and 20% DFS as a partial substitution for butter, these values were 5.17 ± 0.02, 5.16 ± 0.01, 5.18 ± 0.03, and 5.16 ± 0.00, respectively. During storage, the cheese samples showed slight but significant gradual decreases ($p \leq 0.05$) in their pH values due to the hydrolysis of emulsifying salts and their interaction with proteins [60]. Figure 3c shows the observations of the acidity values.

Furthermore, Figure 3d shows the effect of fortification with DFS as a partial substitution for butter on the ash values in cheese samples, which were slightly increased, although not significantly ($p \geq 0.05$). The values were 4.23 ± 0.04 for the control treatment, while, for fortified treatments with 5%, 10%, 15%, and 20% DFS (w/w), these values were 4.36 ± 0.11, 4.26 ± 0.07, 4.47 ± 0.13, and 4.33 ± 0.05, respectively. During storage, the cheese samples showed a slight but significant gradual increase ($p \leq 0.05$) in their ash values due to the slight decrease in the moisture content.

The fat% is represented in Figure 3e, showing that the fat values in the cheese samples were slightly decreased, although not significantly ($p \geq 0.05$) for 5%, 10%, and 15% fortification with DFS, while this change was significant for 20% DFS.

The fat value of the control cheese sample was 28.00 ± 0.16%, while, for 5%, 10%, 15%, and 20% DFS, these values were 27.63 ± 1.0%, 27.33 ± 0.57%, 27.0 ± 1.00%, and 26.16 ± 0.28%, respectively. During storage, the cheese samples showed a slight but not-significant gradual increase ($p \geq 0.05$) in their fat values, except for 20% DFS, for which this change was significant. The increase in fat values was due to the slight decrease in the moisture content.

Likewise, Figure 3f shows that the protein values in the cheese samples were slightly decreased, although not significantly ($p \geq 0.05$). The control value was 16.62 ± 0.28%, while, for 5%, 10%, 15%, and 20% DFS, these values were 16.25 ± 0.31%, 16.53 ± 0.21%, 16.53 ± 0.10%, and 16.10 ± 0.34%, respectively. During storage, the cheese samples showed a slight but not significant gradual increase ($p \geq 0.05$) in their fat values. Our data exhibit the same trend as that found in previous studies, which used rice bran, oat, inulin, and bulger for the fortification of processed cheese [60–63].

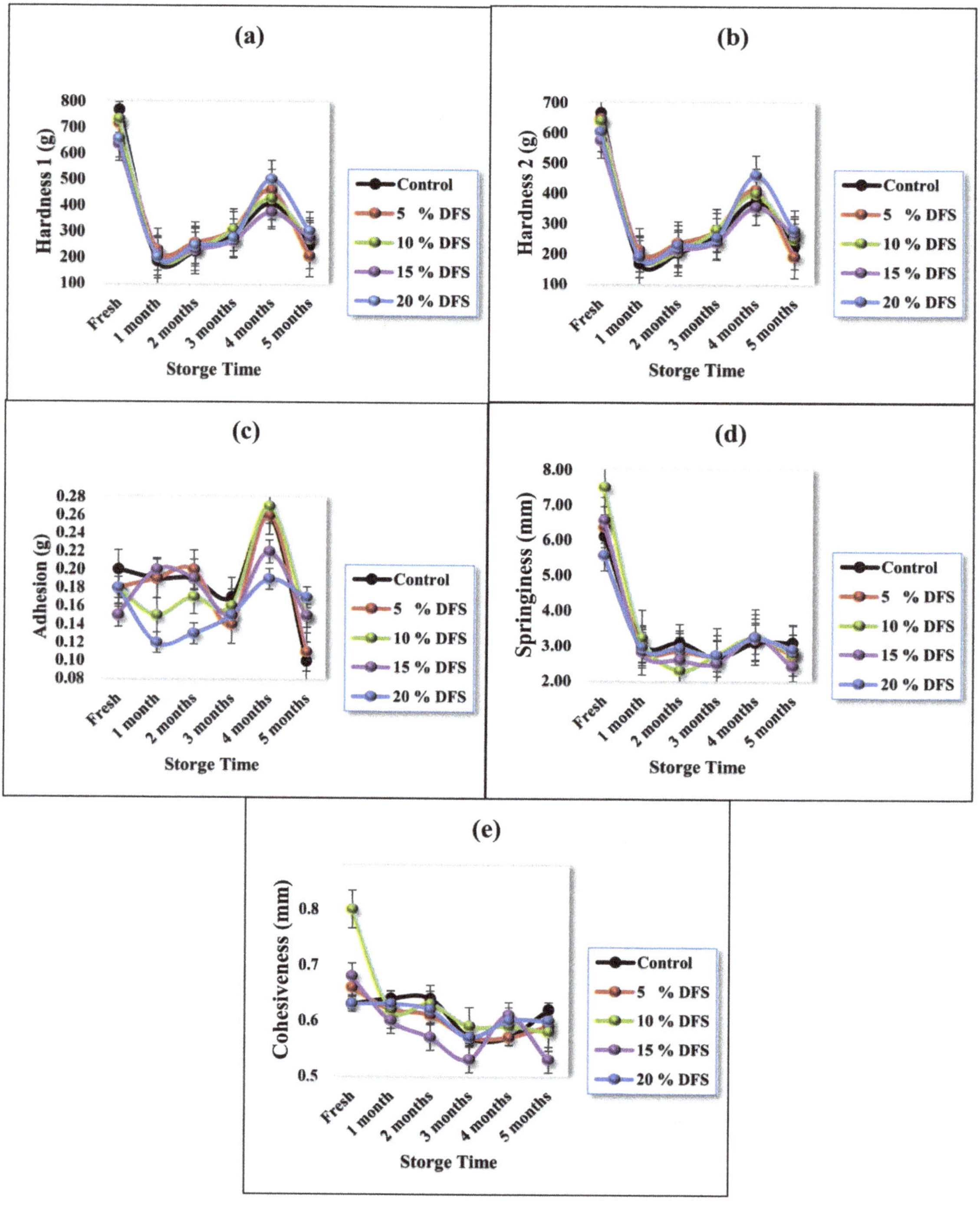

Figure 4. (**a**–**e**) Texture profile analysis of processed cheese with various ratios of DFS as a partial substitution for butter, in hardness 1, hardness 2, adhesion, springiness, and cohesiveness.

3.5. Microstructure Properties

The microstructure properties of block-type processed cheese fortified with 5%, 10%, 15%, and 20% DFS were examined in terms of their internal microstructure by scanning electron microscopy (magnitude 500×; at 20 kV), when fresh and after five months of storage, as illustrated in Figure 5. For fresh cheese, the control cheese had a compact structure and was homogenously uniform compared with the cheeses fortified with 5%, 10%, 15%, and 20% DFS, while the fortified cheeses showed a less compact and inhomogeneous structure which increased with an increase in the % of DFS as a partial substitution for butter. Nevertheless, a filamentous-like structure could be observed through the cheese network as one of the structural characteristics of cheeses fortified with DFS. This structure appeared to have low electronic density, which might be attributed to the fiber content of DFS-fortified cheeses. After five months of storage, the overall structure of the cheeses became more open with a less compact cheese network, which increased with increasing DFS%. Furthermore, the surface of the protein matrix of fortified cheeses by DFS appeared to be coarse and the matrix itself was less compact and denser, which increased with an increase in the DFS%. These changes could probably be attributed to the protein-binding ability and emulsification capacity of added DFS. This description of the microstructure is in agreement with that previously reported for block-type processed cheese [63].

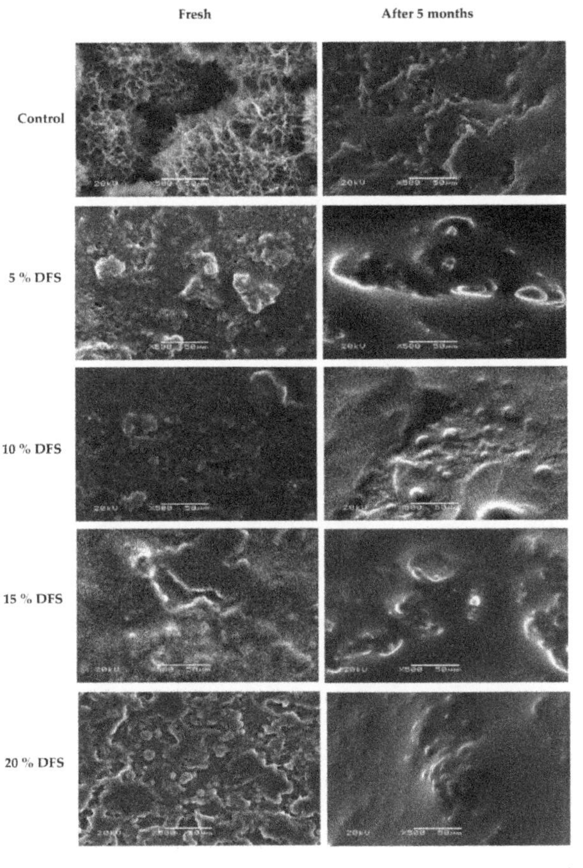

Figure 5. Scanning electron micrographs of the effect of DFS on the overall structure of block-type processed cheese when fresh and after five months of storage at room temperature. Magnitude 500×; 20 kV. Bar = 50 μm.

The microstructure properties of processed cheese are affected the most by the chemical composition: fat, moisture content, pH, degree of proteolysis in the raw cheese, type of cheese, and the amount of emulsifier used. Nevertheless, processing conditions such as processing temperature, stirring speed, cooling temperature, cooling rate, storage temperature, and processing time also had an effect [63].

3.6. Colorimetric Measurements

Color is one of the most significant visual features in dairy products, and color differences affect the storage, shelf life, and color deterioration of processed cheeses. All differences in the L*, a*, and b* color values of the samples of fortified cheese and the controls are taken into account in Table 3. DFS fortification led to significant changes in the color of the cheese; it can be seen that the lightness (L*) values of the cheese were significantly decreased in all cheese batches ($p \leq 0.05$) compared with the control. The mean L* value for the control sample was 80.96 ± 0.66, while those at 5%, 10%, 15%, and 20% DFS were 72.90 ± 0.85, 68.96 ± 2.54, 65.86 ± 4.73, and 64.45 ± 3.01, respectively. During storage, cheese samples showed a non-significant ($p \geq 0.05$) gradual decrease, except after four months of storage, which showed a significant ($p \leq 0.05$) increase.

Table 3. Effect of different DFS% on color parameters during storage.

	Storage	Control	5% DFS **	10% DFS **	15% DFS **	20% DFS **
L* Value	Fresh	80.96 ± 0.66 ^{Ba}	72.90 ± 0.85 ^{Cb}	68.96 ± 2.54 ^{Bcb}	65.86 ± 4.73 ^{Bc}	64.45 ± 3.01 ^{ABc}
	1 month	80.18 ± 1.14 ^{Ba}	73.98 ± 1.41 ^{BCb}	69.95 ± 1.50 ^{Bc}	66.40 ± 2.47 ^{Bd}	64.13 ± 1.72 ^{Bd}
	2 months	79.44 ± 0.32 ^{Ba}	73.11 ± 1.02 ^{Cb}	70.22 ± 0.95 ^{Bb}	64.92 ± 4.53 ^{Bc}	64.43 ± 1.85 ^{Bc}
	3 months	80.76 ± 1.34 ^{Ba}	73.08 ± 0.84 ^{Cb}	69.93 ± 0.73 ^{Bc}	66.66 ± 1.99 ^{Bd}	64.06 ± 1.18 ^{Be}
	4 months	85.37 ± 0.10 ^{Aa}	76.39 ± 0.99 ^{Ab}	75.74 ± 2.15 ^{Ab}	72.50 ± 0.66 ^{Abc}	69.43 ± 4.21 ^{Ac}
	5 months	80.16 ± 0.70 ^{Ba}	74.00 ± 0.59 ^{Bb}	68.36 ± 1.05 ^{Bc}	63.76 ± 0.58 ^{Bd}	63.04 ± 2.30 ^{Bd}
a* Value	Fresh	−3.35 ± 0.04 ^{Cd}	0.53 ± 0.10 ^{Bc}	2.34 ± 0.32 ^{Bb}	3.66 ± 0.82 ^{Aa}	4.24 ± 0.23 ^{Aa}
	1 month	−6.79 ± 0.16 ^{Ee}	−2.99 ± 0.32 ^{Dd}	−1.08 ± 0.28 ^{Dc}	0.32 ± 0.46 ^{Bb}	1.21 ± 0.28 ^{Ca}
	2 months	−6.49 ± 0.30 ^{Dd}	−2.70 ± 0.17 ^{DCc}	−1.03 ± 0.17 ^{Db}	0.85 ± 1.11 ^{Ba}	1.28 ± 0.26 ^{Ca}
	3 months	−6.62 ± 0.06 ^{DEe}	−2.77 ± 0.11 ^{DCd}	−0.90 ± 0.21 ^{Dc}	0.49 ± 0.47 ^{Bb}	1.40 ± 0.16 ^{Ca}
	4 months	−7.39 ± 0.18 ^{Fe}	−2.47 ± 0.29 ^{Cd}	−0.29 ± 0.43 ^{Cc}	1.47 ± 0.10 ^{Bb}	2.64 ± 0.29 ^{Ba}
	5 months	−2.74 ± 0.11 ^{Be}	0.75 ± 0.23 ^{Bd}	2.71 ± 0.06 ^{Bc}	4.26 ± 0.03 ^{Ab}	4.57 ± 0.18 ^{Aa}
b* Value	Fresh	22.66 ± 1.00 ^{Ca}	18.22 ± 0.43 ^{Cb}	16.35 ± 0.20 ^{Cc}	16.65 ± 0.52 ^{Cc}	14.31 ± 0.35 ^{Cc}
	1 month	25.20 ± 0.62 ^{Ba}	20.94 ± 0.04 ^{Bb}	18.60 ± 0.42 ^{Bc}	17.66 ± 0.16 ^{Bd}	16.70 ± 0.34 ^{Be}
	2 months	25.39 ± 0.54 ^{Ba}	20.84 ± 0.11 ^{Bb}	18.72 ± 0.01 ^{Bc}	17.52 ± 0.15 ^{Bd}	16.51 ± 0.44 ^{Be}
	3 months	24.87 ± 0.54 ^{Ba}	20.69 ± 0.64 ^{Bb}	18.74 ± 0.36 ^{Bc}	17.77 ± 0.35 ^{Bd}	16.53 ± 0.30 ^{Be}
	4 months	23.40 ± 0.38 ^{Aa}	22.74 ± 0.63 ^{Ab}	22.26 ± 0.77 ^{Ac}	20.65 ± 0.24 ^{Ad}	19.56 ± 1.39 ^{Ad}
	5 months	21.58 ± 0.15 ^{Da}	17.99 ± 0.02 ^{Cb}	16.46 ± 0.26 ^{Cc}	15.43 ± 0.23 ^{Cd}	14.22 ± 0.33 ^{Ce}

The values are expressed as mean ± standard deviation (N = 3); significant differences between means in a row are indicated by different superscript uppercase letters ($p < 0.05$), and the means within a column indicated by superscript lowercase letters differ ($p < 0.05$). ** The % of DFS was calculated as a partial replacement for butter in comparison with the control amount of butter.

Fresh processed cheese samples fortified with DFS exhibited the darkest color, whereas, with an increasing ratio of fortification of DFS (Figure 6), the whiteness values of the processed cheese samples decreased. During the storage period, the processed cheese became darker. This could be due to the Maillard browning reaction, which occurs during storage between the lactose and proteins in cheese [62].

On the other hand, it can be seen that the redness (a*) value of the cheeses was significantly increased in all cheese batches ($p \leq 0.05$) with increasing DFS%, compared with the control. Storage showed a significant ($p \leq 0.05$) gradual decrease, except after four- and five-month storage, which showed a significant ($p \leq 0.05$) increase.

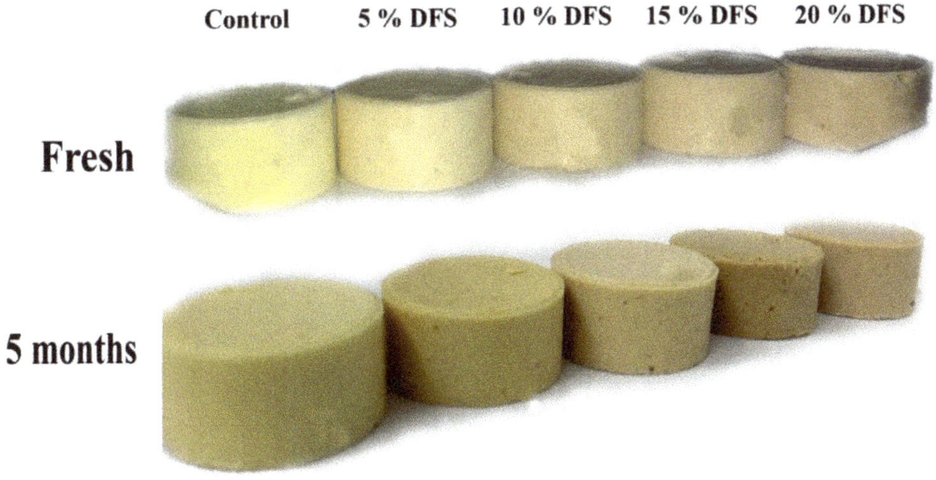

Figure 6. The visual appearance of block-type processed cheese fortified with DFS as a partial substitution for butter.

Nevertheless, the yellowness (b*) values of cheese were significantly decreased in all cheese batches ($p \leq 0.05$) with increasing DFS% as a partial substitution for butter, compared with the control. Storage showed a significant ($p \leq 0.05$) gradual increase, except after five-month storage, which showed a significant ($p \leq 0.05$) decrease.

3.7. Sensorial Attributes

Since this newly-developed block-type processed cheese based on DFS fiber is destined to be consumed by humans, the acceptance of sensory attributes is an important aspect. In general, higher concentrations of DFS% and extended storage had adverse effects on the sensorial attributes of cheeses. Figures 7a–h and 8 show that the appearance, firmness, stickiness, breakdown, gumminess, smoothness, chewiness, flavor, and overall acceptance gradually decreased, although not significantly ($p \geq 0.05$), with increased DFS, compared with the control. However, at 15 and 20% DFS, the decrease was significant ($p \leq 0.05$), and the breakdown in particular was decreased significantly ($p \leq 0.05$) at 20% DFS. During the storage period, there were increases and decreases in the values; however, these were not significant, except for those at 15 and 20% DFS after four and five months. The data are summarized in Figure 8 as overall acceptance.

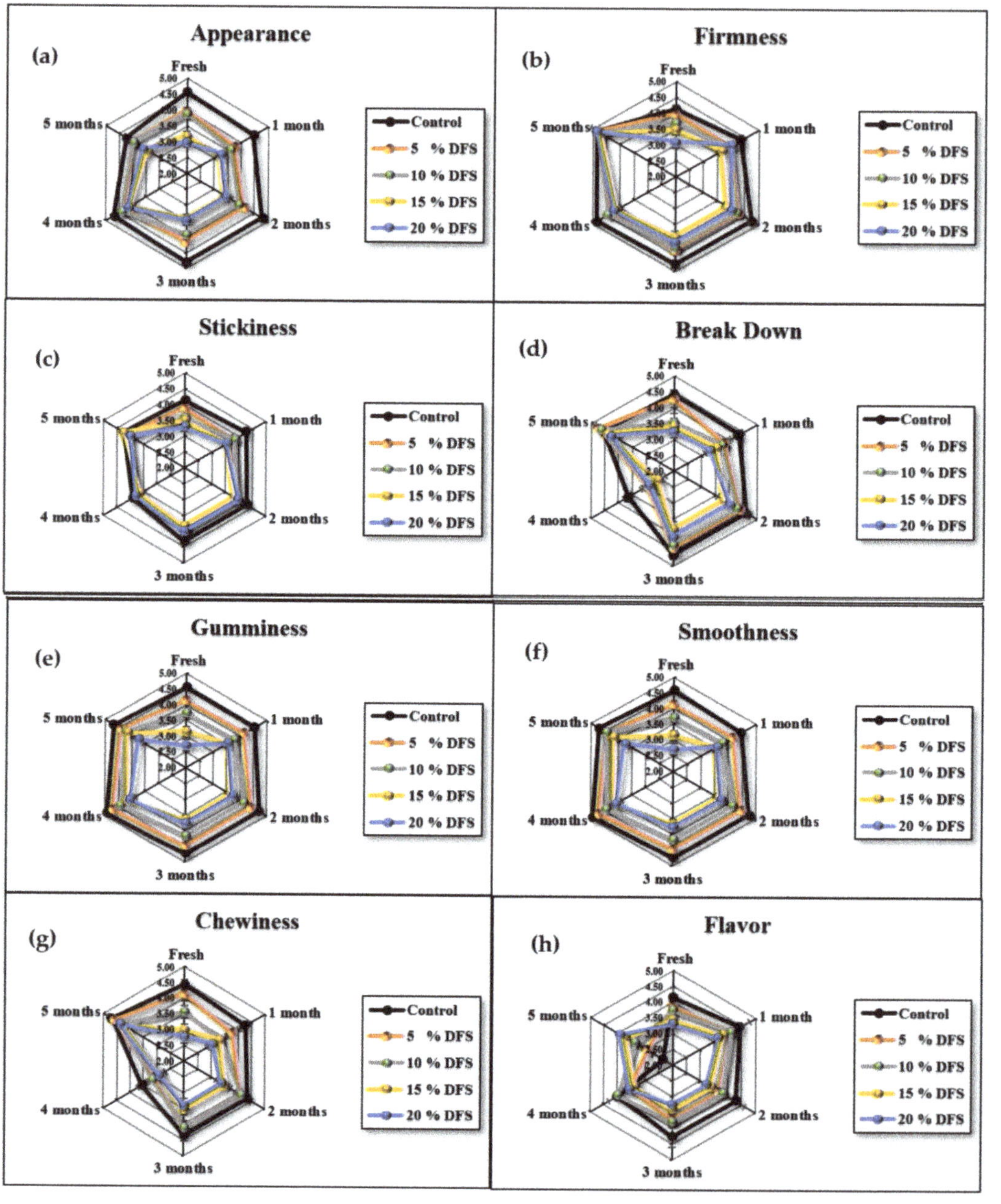

Figure 7. (**a**–**h**), Appearance, firmness, stickiness, breakdown, gumminess, smoothness, chewiness, and flavor of cheeses fortified with 5%, 10%, 15%, and 20% DFS as a partial substitution for butter during storage.

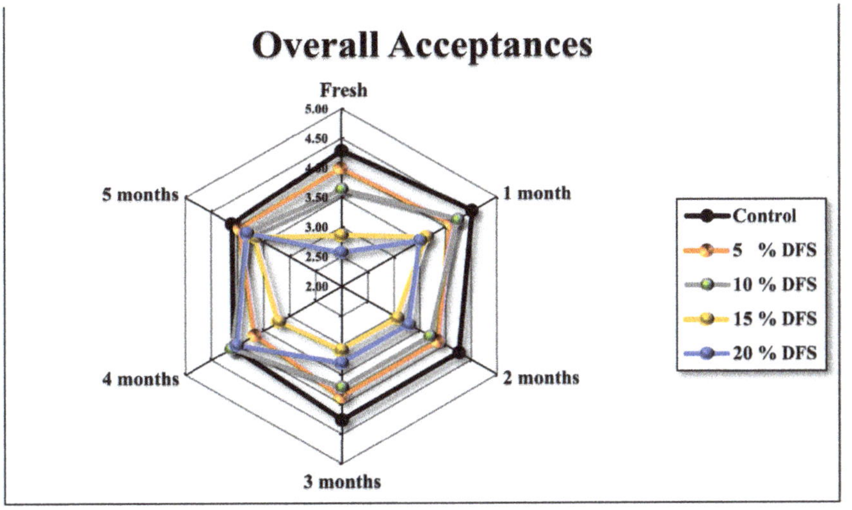

Figure 8. Overall acceptances of cheeses fortified with 5%, 10%, 15%, and 20% DFS as a partial substitution for butter during storage.

4. Conclusions

In this study, we present evidence that DFS is a rich source of fiber and that it can be useful in block-type processed cheese as a partial substitution for butter. Therefore, the primary goal of the current investigation conducted on DFS was to explore its effect on the product's quality in terms of its shelf life and physicochemical, microstructure, color, and sensory properties. Quality was assessed over a 150-day storage period. The overall structure of cheeses became more open with a less compact cheese network, which increased with increasing DFS% as a partial substitution for butter and the duration of storage. It can be concluded that an acceptable quality of block-type processed cheese could be obtained using DFS fiber at the 5% and 10% levels of fortification as a partial substitution for butter. These conclusions should support the use of DFS in functional foods and nutraceutical products. Further studies are required to assess the techno-functional properties of the final products, such as water and oil binding capacity, foaming, and emulsifying activities, in order to confirm the quality of these novel foods. Furthermore, the biological activity should be investigated, inspiring both clinical and in vivo trials of DFS.

Author Contributions: Conceptualization, T.M.A. and A.M.A.; methodology, A.M.A.; software, H.M.H. and A.M.A.; validation, H.M.H. and T.M.A.; formal analysis, S.M.A.; investigation, N.K.A., T.M.A. and H.M.H.; resources, N.K.A. and A.M.A.; data curation, H.M.H. and T.M.A.; writing—original draft preparation, H.M.H. and T.M.A.; writing—review and editing, T.M.A., S.M.A. and N.K.A.; visualization, N.K.A. and A.M.A.; supervision, T.M.A. and S.M.A.; project administration, T.M.A. and A.M.A.; funding acquisition, T.M.A. and N.K.A. All authors have read and agreed to the published version of the manuscript.

Funding: This research was supported by a project (DPRC-05-2022) financed by Date Palm Research Center of Excellence, King Faisal University, Saudi Arabia.

Institutional Review Board Statement: Not applicable.

Informed Consent Statement: Not applicable.

Data Availability Statement: Data is contained within the article.

Acknowledgments: The authors acknowledge Almarai Company for Dairy Products, Riyadh, KSA, the Department of Food and Nutrition Sciences, the College of Agricultural and Food Sciences, and the Deanship of Scientific Research at King Faisal University for their support.

Conflicts of Interest: The authors declare no conflict of interest.

References

1. Kandemir, K.; Piskin, E.; Xiao, J.; Tomas, M.; Capanoglu, E. Fruit Juice Industry Wastes as a Source of Bioactives. *J. Agric. Food Chem.* **2022**, *70*, 6805–6832. [CrossRef] [PubMed]
2. Rao, M.; Bast, A.; de Boer, A. Valorized Food Processing By-Products in the EU: Finding the Balance between Safety, Nutrition, and Sustainability. *Sustainability* **2021**, *13*, 4428. [CrossRef]
3. Caponio, F.; Piga, A.; Poiana, M. Valorization of Food Processing By-Products. *Foods* **2022**, *11*, 3246. [CrossRef]
4. Iriondo-DeHond, M.; Miguel, E.; del Castillo, M. Food Byproducts as Sustainable Ingredients for Innovative and Healthy Dairy Foods. *Nutrients* **2018**, *10*, 1358. [CrossRef] [PubMed]
5. Picciotti, U.; Massaro, A.; Galiano, A.; Garganese, F. Cheese Fortification: Review and Possible Improvements. *Food Rev. Int.* **2022**, *38*, 474–500. [CrossRef]
6. Olson, R.; Gavin-Smith, B.; Ferraboschi, C.; Kraemer, K. Food Fortification: The Advantages, Disadvantages and Lessons from Sight and Life Programs. *Nutrients* **2021**, *13*, 1118. [CrossRef]
7. Ottaway, P.B. *Food Fortification and Supplementation Technological, Safety and Regulatory Aspects*, 1st ed.; Elsevier: Cambridge, UK, 2008.
8. Kumar, D.S. *Herbal Bioactives and Food Fortification: Extraction and Formulation*; CRC Press: Boca Raton, FL, USA, 2015.
9. Dwyer, J.T.; Wiemer, K.L.; Dary, O.; Keen, C.L.; King, J.C.; Miller, K.B.; Philbert, M.A.; Tarasuk, V.; Taylor, C.L.; Gaine, P.C.; et al. Fortification and Health: Challenges and Opportunities. *Adv. Nutr.* **2015**, *6*, 124–131. [CrossRef]
10. Chadare, F.J.; Idohou, R.; Nago, E.; Affonfere, M.; Agossadou, J.; Fassinou, T.K.; Kénou, C.; Honfo, S.; Azokpota, P.; Linnemann, A.R.; et al. Conventional and Food-to-food Fortification: An Appraisal of Past Practices and Lessons Learned. *Food Sci. Nutr.* **2019**, *7*, 2781–2795. [CrossRef]
11. Das, J.K.; Salam, R.A.; bin Mahmood, S.; Moin, A.; Kumar, R.; Mukhtar, K.; Lassi, Z.S.; Bhutta, Z.A. Food Fortification with Multiple Micronutrients: Impact on Health Outcomes in General Population. *Cochrane Database Syst. Rev.* **2019**, *2020*, CD011400. [CrossRef]
12. Thielecke, F.; Lecerf, J.-M.; Nugent, A.P. Processing in the Food Chain: Do Cereals Have to Be Processed to Add Value to the Human Diet? *Nutr. Res. Rev.* **2021**, *34*, 159–173. [CrossRef]
13. Sajid Arshad, M.; Khalid, W.; Shabir Ahmad, R.; Kamran Khan, M.; Haseeb Ahmad, M.; Safdar, S.; Kousar, S.; Munir, H.; Shabbir, U.; Zafarullah, M.; et al. Functional Foods and Human Health: An Overview. In *Functional Foods-Phytochemicals and Health Promoting Potential*; IntechOpen: London, UK, 2021.
14. Koç, E.; Karayiğit, B. Assessment of Biofortification Approaches Used to Improve Micronutrient-Dense Plants That Are a Sustainable Solution to Combat Hidden Hunger. *J. Soil Sci. Plant Nutr.* **2022**, *22*, 475–500. [CrossRef]
15. Romina Alina, V.; Crina Carmen, M.; Sevastita, M.; Andruţa, M.; Vlad, M.; Ramona, S.; Georgiana, P.; Mihaela, M. Food Fortification through Innovative Technologies. In *Food Engineering*; IntechOpen: London, UK, 2019.
16. Kaur, N.; Agarwal, A.; Sabharwal, M. Food Fortification Strategies to Deliver Nutrients for the Management of Iron Deficiency Anaemia. *Curr. Res. Food Sci.* **2022**, *5*, 2094–2107. [CrossRef]
17. Bianchi, F.; Tolve, R.; Rainero, G.; Bordiga, M.; Brennan, C.S.; Simonato, B. Technological, Nutritional and Sensory Properties of Pasta Fortified with Agro-industrial By-products: A Review. *Int. J. Food Sci. Technol.* **2021**, *56*, 4356–4366. [CrossRef]
18. FAO. *Thinking about the Future of Food Safety*; FAO: Roma, Italy, 2022; ISBN 978-92-5-135783-5.
19. Vermeir, I.; Weijters, B.; de Houwer, J.; Geuens, M.; Slabbinck, H.; Spruyt, A.; van Kerckhove, A.; van Lippevelde, W.; de Steur, H.; Verbeke, W. Environmentally Sustainable Food Consumption: A Review and Research Agenda from a Goal-Directed Perspective. *Front. Psychol.* **2020**, *11*, 1603. [CrossRef]
20. Petrescu, D.C.; Vermeir, I.; Petrescu-Mag, R.M. Consumer Understanding of Food Quality, Healthiness, and Environmental Impact: A Cross-National Perspective. *Int. J. Environ. Res. Public Health* **2019**, *17*, 169. [CrossRef]
21. FAO FAOSTAT. Food and Agriculture Organization of the United Nations. 2020. Available online: http://faostat.fao.org/ (accessed on 1 December 2022).
22. Habib, H.M.; El-Fakharany, E.M.; Souka, U.D.; Elsebaee, F.M.; El-Ziney, M.G.; Ibrahim, W.H. Polyphenol-Rich Date Palm Fruit Seed (*Phoenix dactylifera* L.) Extract Inhibits Labile Iron, Enzyme, and Cancer Cell Activities, and DNA and Protein Damage. *Nutrients* **2022**, *2022*, 3536. [CrossRef]
23. Kamal, H.; Habib, H.M.; Ali, A.; Show, P.L.; Koyande, A.K.; Kheadr, E.; Ibrahim, W.H. Food Waste Valorization Potential: Fiber, Sugar, and Color Profiles of 18 Date Seed Varieties (*Phoenix dactylifera* L.). *J. Saudi Soc. Agric. Sci.* **2022**, *in press*. [CrossRef]
24. Najjar, Z.; Stathopoulos, C.; Chockchaisawasdee, S. Utilization of Date By-Products in the Food Industry. *Emir. J. Food Agric.* **2020**, *32*, 808. [CrossRef]
25. Sayas-Barberá, E.; Martín-Sánchez, A.M.; Cherif, S.; Ben-Abda, J.; Pérez-Álvarez, J.Á. Effect of Date (*Phoenix dactylifera* L.) Pits on the Shelf Life of Beef Burgers. *Foods* **2020**, *9*, 102. [CrossRef]
26. Devaraj, A.; Mahalingam, G. Bioactive Molecules from Medicinal Plants as Functional Foods (Biscuits) for the Benefit of Human Health as Antidiabetic Potential. In *Bioactive Compounds in Nutraceutical and Functional Food for Good Human Health*; IntechOpen: London, UK, 2021.

27. Oladipupo Kareem, M.; Edathil, A.A.; Rambabu, K.; Bharath, G.; Banat, F.; Nirmala, G.S.; Sathiyanarayanan, K. Extraction, Characterization and Optimization of High Quality Bio-Oil Derived from Waste Date Seeds. *Chem. Eng. Commun.* **2021**, *208*, 801–811. [CrossRef]
28. Mohammadi, M.; Khorshidian, N.; Yousefi, M.; Khaneghah, A.M. Physicochemical, Rheological, and Sensory Properties of Gluten-Free Cookie Produced by Flour of Chestnut, Date Seed, and Modified Starch. *J. Food Qual.* **2022**, *2022*, 5159084. [CrossRef]
29. Sriharsha, C.H.; Swamy, R.; Padmavathi, T.V.N. Development and Quality Evaluation of Date Seed Powder Incorporated Cookies. *J. Exp. Agric. Int.* **2021**, *43*, 87–93. [CrossRef]
30. Djaoudene, O.; Mansinhos, I.; Gonçalves, S.; Jara-Palacios, M.J.; Bachir Bey, M.; Romano, A. Phenolic Profile, Antioxidant Activity and Enzyme Inhibitory Capacities of Fruit and Seed Extracts from Different Algerian Cultivars of Date (*Phoenix dactylifera* L.) Were Affected by in Vitro Simulated Gastrointestinal Digestion. *S. Afr. J. Bot.* **2021**, *137*, 133–148. [CrossRef]
31. Platat, C.; Habib, H.M.; Hashim, I.B.; Kamal, H.; AlMaqbali, F.; Souka, U.; Ibrahim, W.H. Production of Functional Pita Bread Using Date Seed Powder. *J. Food Sci. Technol.* **2015**, *52*, 6375–6384. [CrossRef] [PubMed]
32. Alqattan, A.M.; Alqahtani, N.K.; Aleid, S.M.; Alnemr, T.M. Effects of Date Pit Powder Inclusion on Chemical Composition, Microstructure, Rheological Properties, and Sensory Evaluation of Processed Cheese Block. *Am. J. Food Nutr.* **2020**, *8*, 69–77. [CrossRef]
33. Alqahtani, N. Effects of Replacing Pectin with Date Pits Powder in Strawberry Jam Formulation. *Basic Appl. Sci. Sci. J. King Faisal Univ.* **2020**, *8*, 69–77. [CrossRef]
34. Habib, H.M.; Kamal, H.; Ibrahim, W.H.; Dhaheri, A.S.A. Carotenoids, Fat Soluble Vitamins and Fatty Acid Profiles of 18 Varieties of Date Seed Oil. *Ind. Crops Prod.* **2013**, *42*, 567–572. [CrossRef]
35. Habib, H.M.; Ibrahim, W.H. Nutritional Quality Evaluation of Eighteen Date Pit Varieties. *Int. J. Food Sci. Nutr.* **2009**, *60*, 99–111. [CrossRef]
36. Mrabet, A.; Jiménez-Araujo, A.; Guillén-Bejarano, R.; Rodríguez-Arcos, R.; Sindic, M. Date Seeds: A Promising Source of Oil with Functional Properties. *Foods* **2020**, *9*, 787. [CrossRef]
37. Aljaloud, S.; Colleran, H.L.; Ibrahim, S.A. Nutritional Value of Date Fruits and Potential Use in Nutritional Bars for Athletes. *Food Nutr. Sci.* **2020**, *11*, 463–480. [CrossRef]
38. Habib, H.M.; Platat, C.; Meudec, E.; Cheynier, V.; Ibrahim, W.H. Polyphenolic Compounds in Date Fruit Seed (*Phoenix dactylifera*): Characterisation and Quantification by Using UPLC-DAD-ESI-MS. *J. Sci. Food Agric.* **2014**, *94*, 1084–1089. [CrossRef]
39. Habib, H.M.; El-Fakharany, E.M.; Kheadr, E.; Ibrahim, W.H. Grape Seed Proanthocyanidin Extract Inhibits DNA and Protein Damage and Labile Iron, Enzyme, and Cancer Cell Activities. *Sci. Rep.* **2022**, *12*, 12393. [CrossRef]
40. Hilary, S.; Tomás-Barberán, F.A.; Martinez-Blazquez, J.A.; Kizhakkayil, J.; Souka, U.; Al-Hammadi, S.; Habib, H.; Ibrahim, W.; Platat, C. Polyphenol Characterisation of *Phoenix dactylifera* L. (Date) Seeds Using HPLC-Mass Spectrometry and Its Bioaccessibility Using Simulated in-Vitro Digestion/Caco-2 Culture Model. *Food Chem.* **2020**, *311*, 125969. [CrossRef]
41. Habib, H.M.; Ibrahim, W.H. Effect of Date Seeds on Oxidative Damage and Antioxidant Status in Vivo. *J. Sci. Food Agric.* **2011**, *91*, 1674–1679. [CrossRef]
42. Habib, H.; Platat, C.; AlMaqbali, F.; Ibrahim, W. Date Seed (*Phoenix dactylifera*) Extract Reduces the Proliferation of Pancreatic Cancer Cells, DNA Damage and Superoxide-dependent Iron Release from Ferritin in Vitro (829.20). *FASEB J.* **2014**, *28*, 829. [CrossRef]
43. Al-Meqbaali, F.; Habib, H.; Othman, A.; Al-Marzooqi, S.; Al-Bawardi, A.; Pathan, J.Y.; Hilary, S.; Souka, U.; Al-Hammadi, S.; Ibrahim, W.; et al. The Antioxidant Activity of Date Seed: Preliminary Results of a Preclinical in Vivo Study. *Emir. J. Food Agric.* **2017**, *29*, 822. [CrossRef]
44. Guinee, T.P.; O'Kennedy, B.T. Reducing the Level of Added Disodium Phosphate Alters the Chemical and Physical Properties of Processed Cheese. *Dairy Sci. Technol.* **2012**, *92*, 469–486. [CrossRef]
45. Talbot-Walsh, G.; Selomulya, C. The Effect of Rennet Casein Hydration on Gel Strength and Matrix Stability of Block-Type Processed Cheese. *Food Struct.* **2021**, *28*, 100174. [CrossRef]
46. *ISO Standard No. 5534*; Cheese and Processed Cheese–Determination of the Total Solids Content (Reference Method). International Organization for Standardization: Geneva, Switzerland, 2004.
47. Nehdi, I.; Omri, S.; Khalil, M.I.; Al-Resayes, S.I. Characteristics and Chemical Composition of Date Palm (Phoenix Canariensis) Seeds and Seed Oil. *Ind. Crops Prod.* **2010**, *32*, 360–365. [CrossRef]
48. Nielsen, S.S. Heldman, D.R., Ed.; *Correction to: Food Analysis*, 5th ed.; Heldman Associates: Mason, OH, USA, 2021; pp. C1–C3.
49. Englyst, H.N.; Quigley, M.E.; Hudson, G.J.; Cummings, J.H. Determination of Dietary Fibre as Non-Starch Polysaccharides by Gas–Liquid Chromatography. *Analyst* **1992**, *117*, 1707–1714. [CrossRef]
50. Habib, H.M.; Theuri, S.W.; Kheadr, E.E.; Mohamed, F.E. Functional, Bioactive, Biochemical, and Physicochemical Properties of the Dolichos Lablab Bean. *Food Funct.* **2017**, *8*, 872–880. [CrossRef]
51. Nandy, S.K.; Venkatesh, K. Study of CFU for Individual Microorganisms in Mixed Cultures with a Known Ratio Using MBRT. *AMB Express* **2014**, *4*, 38. [CrossRef] [PubMed]
52. Guinee, T.P. Cheese: Pasteurized Processed Cheese Products. In *Reference Module in Food Science*; Elsevier: Amsterdam, The Netherlands, 2016.
53. Nasiri, M.; Tavakolipour, H.; Safaeian, S.; Mousavi Nadushan, R. Exploring the Potential of Modified Potato Starch and Seaweed Salt as Structuring Agents to Design Processed Cheeses with Desirable Properties. *Int. Dairy J.* **2022**, *133*, 105439. [CrossRef]

54. Delahunty, C.M.; Drake, M.A. Sensory Character of Cheese and Its Evaluation. *Cheese Chem.* **2004**, *1*, 455–487.
55. Akasha, I.; Campbell, L.; Lonchamp, J.; Euston, S.R. The Major Proteins of the Seed of the Fruit of the Date Palm (*Phoenix dactylifera* L.): Characterisation and Emulsifying Properties. *Food Chem.* **2016**, *197*, 799–806. [CrossRef] [PubMed]
56. Dönmez, Ö.; Mogol, B.A.; Gökmen, V. Syneresis and Rheological Behaviors of Set Yogurt Containing Green Tea and Green Coffee Powders. *J. Dairy Sci.* **2017**, *100*, 901–907. [CrossRef]
57. Mohamed, A.G.; Abbas, H.M.; Bayoumi, H.M.; Kassem, J.M.; Enab, A.K.; Mohamed, A.G. Processed Cheese Spreads Fortified with Oat. *J. Am. Sci.* **2011**, *7*, 631–637.
58. El-Assar, M.A.; Abou-Dawoo, S.A.; Sakr, S.S.; Younis, N.M. Low-Fat Processed Cheese Spread with Added Inulin: Its Physicochemical, Rheological and Sensory Characteristics. *Int. J. Dairy Sci.* **2018**, *14*, 12–20. [CrossRef]
59. Abd El-Gawad, M.A.M.; El-Shibiny, S.; Assem, F.M.; Seleet, F.L.; Dawood, S.A.A.; Elaaser, M. Preparation, Composition and Microbiological and Rheological Properties of Functional Processed Cheese Supplemented with Rice Bran. *J. Appl. Sci. Res.* **2013**, *9*, 4927–4934.
60. Aly, E.S.; el Saadany, K.; el Dakhakhny, E.; Kheadr, E. The use of Bulgur in the Production of Reduced-Fat and Substituted Block-Type Processed Cheeses. *J. Appl. Sci. Res.* **2017**, *13*, 17–30.
61. Szafrańska, J.O.; Muszyński, S.; Tomasevic, I.; Sołowiej, B.G. The Influence of Dietary Fibers on Physicochemical Properties of Acid Casein Processed Cheese Sauces Obtained with Whey Proteins and Coconut Oil or Anhydrous Milk Fat. *Foods* **2021**, *10*, 759. [CrossRef]
62. El-Bakry, M.; Mehta, B. *Processed Cheese Science and Technology*; Elsevier: Amsterdam, The Netherlands, 2022; ISBN 9780128214459.
63. El-Loly, M.M.; Farahat, E.S.A.; Mohamed, A.G. Novel Approach for Producing Processed Cheese Fortified with Date Syrup. *ACS Food Sci. Technol.* **2021**, *1*, 737–744. [CrossRef]

foods MDPI

Article

Quality Assessment of Waste from Olive Oil Production and Design of Biodegradable Packaging

Joanna Grzelczyk *, Joanna Oracz and Ilona Gałązka-Czarnecka

Institute of Food Technology and Analysis, Faculty of Biotechnology and Food Sciences, Lodz University of Technology, 90-537 Lodz, Poland
* Correspondence: joanna.grzelczyk@p.lodz.pl

Abstract: The use of olive pomace from olive oil production is still insufficient. The lingering olive pomace is harmful to the environment. On the other hand, the world is increasingly polluted with plastic or by-products from the production of biodegradable products. Considering these two problems, the aim of this work was to develop a mixture and create biodegradable disposable tableware characterized by high antioxidant activity. The disposable tableware was made by mixing olive pomace with teff flour or/and sorghum groats and lecithin. Baking was carried out at the temperature of 180 °C. The best variant of the mixture for the preparation of disposable tableware was olive pomace, teff flour, sorghum groats and lecithin. These vessels were the toughest, with low water absorption and had a high antioxidant potential due to the high content of polyphenols and omega acids. Protecting the cups and bowls with beeswax had a positive effect on reducing water absorption.

Keywords: pomace; olives; disposable tableware; biodegradable; antioxidants; polyphenols

Citation: Grzelczyk, J.; Oracz, J.; Gałązka-Czarnecka, I. Quality Assessment of Waste from Olive Oil Production and Design of Biodegradable Packaging. *Foods* **2022**, *11*, 3776. https://doi.org/10.3390/foods11233776

Academic Editors: Stanisław Kowalski and Dorota Gumul

Received: 28 October 2022
Accepted: 21 November 2022
Published: 23 November 2022

1. Introduction

Olive oil is mainly produced in Europe. World olive production was about 2.86 million tons from 2005 to 2018. An important fact is that the preparation of such an amount of oil requires the use of 14.3 to 17.6 million tons of olives, of which about 30% is waste in the form of pomace [1,2]. The latter is an unstable material with a high-water content [3,4]. However, it is a rich source of many valuable components, including omega-3 and -6 fatty acids, mineral compounds, vitamins, carbohydrates, polyphenols and fiber [5–7]. Waste generated in oil production is already successfully used for animal feed or biomass as a source for biofuels [1,8]. Nevertheless, such solutions still have not brought the expected effects. The insufficient management of olive waste is not only a local problem but also a global one. The large mass of pomace produced in a relatively short period of time is an important element for processing plants [8,9].

Moreover, it is also well known that the amount of used disposable plastic tableware continues to grow rapidly, this growth consequently leading to environmental and economic problems that affect not only most developed countries but also the entire ecosystem of the whole planet due to the improper management of such waste [10,11]. To date, disposable tableware was always made of plastic, i.e., polypropylene, polyethylene, or polyethylene terephthalate [12–14]. The usable advantage of such materials is their durability and stability, but on the other hand they have a very long decomposition time. Ironically, the durability of such plastic packaging and dishes, which is a desirable feature, results in their lingering in the oceans and rather than undergoing mineralization they are consumed by marine organisms [15–19]. Starch and bioplastics, in the Life Cycle Assessments (LCAs) process, display a reduced generation of carbon dioxide emissions into the atmosphere [20–25]. The bioplastics based on biomass sources derived from industry by-products or waste reduce the emissions of GHG, since no additional resources are implied in the raw materials production [26,27]. To reduce the content of carbon dioxide emissions, packaging production time can be shortened, but plant waste can also be used, the deposition of which, in landfills, increases the emission of carbon dioxide [28]. Therefore, new packaging materials are being extensively explored and other agricultural products such as bran and corn have been recently used [29,30]. However, these are products of daily necessity in the production of food, e.g., bread and there is increasingly insufficient food in the world, to which 1 in 9 people do not have adequate access [31]. Climate change, which causes floods or droughts, also affects the growth deficit of agricultural products [32]. For this reason, the production of new biodegradable products is highly profitable.

The aim of this research is to develop a mixture of disposable tableware made of olive oil waste and to qualitatively assess the nutritional value.

2. Materials and Methods

2.1. Chemicals and Reagents

2,2-diphenyl-1-(2,4,6-trinitrophenyl) hydrazyl (DPPH), 6-hydroxy-2,5,7,8-tetramethylchroman-2-carboxylic acid (Trolox), methanol, water and galic acid was purchased from Sigma-Aldrich (St. Louis, MO, USA). Folin–Ciocalteu's phenol reagent, sodium carbonate (Na_2CO_3) and sodium nitrate were purchased from Chempur (Piekary Slaskie, Poland). Ultrapure water (resistivity 18.2 MΩ cm), obtained from a Milli-Q purification system (Millipore, Bedford, MA, USA), was used for all analyses. Teff flour, sorghum groats, lecithin, olives and beeswax were purchased at local stores.

2.2. Material and Preparation

Biodegradable disposable dishes (in the shape of cup, plate or bowl) were prepared from by-products of olive oil production (69.5–79.5%) and flour and/or groats and lecithin in various combinations and concentrations (Table 1). The olive pomace is made from green olives of the Manzanilla variety from Spain, cold-pressed. Olive pomace is the residue of pulp and skin, the initial moisture content of which was 51% by weight. The flour/groats weight ratio of 1:1, at 30–20% by weight and lecithin in liquid form of rapeseed origin at 0.5% by weight and flour and groats with a grain size of less than 0.5 mm were used. All ingredients were mixed to form a firm homogeneous mass which was filled into a baking tray. The stuffed disposable dishes were baked in a hot air oven with top and bottom heating for 1.5 h at 180 °C. After baking, the dishes were allowed to cool down (24 °C). Examples of biodegradable disposable utensils are shown in Figure 1.

Table 1. Composition of disposable tableware mixtures.

Mixture	Olive Pomace	Teff Flour	Sorghum Groats	Lecithin
OTSL	+	+	+	+
OTL	+	+	–	+
OSL	+	–	+	+
OTS	+	+	+	–

OTSL: utensils from the mixture of olive pomace, teff flour, sorghum and lecithin; OTL: utensils from the mixture of olive pomace, teff flour and lecithin; OSL: utensils from the mixture of olive pomace, sorghum and lecithin; OTS: utensils from the mixture of olive pomace, teff flour and sorghum. "+"—added ingredient; "–"—no ingredient.

In order to use the cups for liquid dishes, the cups were covered with beeswax. The beeswax was dissolved at a temperature of 70 °C, then the beeswax was evenly applied with a brush on the disposable dishes and heated for 2 min at a temperature of 60 °C. The dishes were then cooled to room temperature.

All packages were stored at room temperature in a dark place (in a cabinet) without additional packages.

The mixture and the method of production can be found in the National patent application 437771 (Section 5).

Figure 1. (**A**)—formed disposable utensils before baking; (**B**)—baked disposable utensils; (**C**)—finished disposable dishes in use (own photos).

2.3. Physical and Mechanical Properties

2.3.1. Flexural Strength

The flexural strength of the disposable utensils was measured instrumentally by 3-point bending test jig flexural test with the EZ Test texturometer (Shimandzu, Kyoto, Japan) using Trapezium software. The procedure was performed according to Olt et al. [29] with minor modifications. The flexural strength test was carried out by crushing a 4 × 4 × 0.3 cm piece of the studied materials, which was placed on two shaped supports with their centres positioned L = 34 mm apart and with a load subsequently being applied to the central part of the test body from the top to the bottom using a knife-shaped element. The distance between fulcrums was 10 mm, test speed 0.75 mm/min and the load was applied until the test body broke, the force necessary to achieve such a break being measured. Five measurements were made. The force sensor was automatically calibrated before the tests using the appropriate calibration cable (Calibration certificate number L1.436.3114.2022). The analyzes were performed on day 1 after baking and after one week, 2 weeks, 4 weeks and 8 weeks of storage. Five repetitions were made.

2.3.2. Drop Test

Samples of disposable utensils with no cracks were dropped from a height of 0.7 m and were thereafter analyzed for cracks or splits after impacting on a level floor. The disposable utensils were then subjected to a stress test [33]. The analyzes were performed on day 1 after baking, after 4 weeks and 8 weeks of storage. Five repetitions were made.

2.3.3. Measurement of Color Parameters

Color parameters were measured in the CIE L*a*b* system, using a Konica Minolta CR-400 (Tokyo, Japan) colorimeter, with SpectraMagicTM NX software, equipped with a CR-A33a protective glass measuring head. In the L*a*b* system, the a* value indicates the proportion of red, the positive b* value the proportion of yellow and L* the brightness of the sample. Measurements were made using the D65 illuminant (PN-65-N-01252), which is the average distribution of daytime radiation power at different times of the day (with UV), at different latitudes of European countries with varying degrees of cloud cover. The analyzes were performed on day 1 after baking, after one week, 2 weeks, 4 weeks and 8 weeks of storage.

2.3.4. Differential Scanning Calorimetry (DSC)

Thermal properties of the composite disposable utensils were analyzed using DSC 2 METTLER TOLEDO (Columbus, OH, USA), at a heating rate of 5 °C/min. Approximately 4 mg of sample was placed in a Tzero® (Columbus, OH, USA) aluminum pin hole hermetic pan, which was heated from −25 °C to 100 °C. The DSC was calibrated with indium. A high purity nitrogen was used as the purge gas with flow rate of 50 mL/min. Samples were analyzed before and after baking.

2.3.5. Water Activity

The water activity in the tested samples was measured with the HP23-AW-A Rotronic (ROTRONIC Instrument Corp, Hauppauge, NY, USA) water activity meter in combination with the HC2-AW Rotronic salt probe, which enables the measurement of the water activity of a whole product. Before the measurement, 2 g of biodegradable dish was ground in a mortar and placed in a desiccator to equilibrate the humidity. The sample was then transferred to a cup and placed in the probe. The measurement was taken. The analyzes were performed on day 1 after baking, after one week, 2 weeks, 4 weeks and 8 weeks of storage.

2.3.6. Water Absorption

Cups and bowls with/without beeswax were weighed empty. Then 40 mL of cold or hot water was added. The dishes were weighed again. They were left for 6 h. The analysis was stopped when the first drop appeared (water leakage) under the package [33]. The analyzes were performed on day 1 after baking, after one week, 2 weeks, 4 weeks and 8 weeks of storage.

2.3.7. Biodegradability Test

The biodegradation was tested on five pieces of each disposable utensil, with 6 × 6 cm samples, of known weight, placed in ground for growing flowers. The samples were retrieved every 7 days from where they were buried under soil, dried at room temperature and placed in a desiccator. Before weighing, each sample was thoroughly cleaned of soil [34]. The analyzes were performed on day 1 after baking, after one week, 2 weeks, 4 weeks and 8 weeks of storage. Three repetitions were made.

2.4. *Nutritional Properties*

2.4.1. Antioxidant Activity of the Extracts by DPPH Assay

The antioxidant activity of the extracts was measured by the DPPH assay and determined according to Zielińska et al. [35] with small modifications. The disposable utensils were ground in a mortar. Next, 1 g was placed into a 50 mL flask and 20 mL of 70% methanol was added. The samples were shaken for 1 h, then filtered. A 0.50–0.25 mL volume of the sample was mixed with 1.95–1.75 mL of a 0.50 mL solution of DPPH in 70% methanol. The samples were homogenized and incubated at room temperature for 30 min in darkness. The absorbance was measured at 517 nm and the control used was 70% methanol. Control samples were prepared in the absence of extracts, following the same procedure. The IC50 value denotes the concentration of the sample required to scavenge 50% of the DPPH free radicals. The scavenging effect was calculated according to Equation: Antioxidant activity (%) = ((AControl − AExtract)/AControl) × 100, based on the calibration curve of Trolox, whose linearity range varied between 20 and 500 μmol/L. The analyzes were performed on day 1 after baking, after one week, 2 weeks, 4 weeks and 8 weeks of storage.

2.4.2. Total Phenolics Content (TPC)

The content of total phenolic compounds (TPC) of the extracts by the Folin–Ciocalteu assay was determined according to a modified procedure from Mattia et al. [36]. The disposable utensils were ground in a mortar. Next, 0.5 g was taken into a 50 mL flask and 20 mL of deionized water was added. The samples were shaken for 1 h, then filtered.

An aliquot of the extracts (0.1 mL) and 4.9 mL of deionized water were mixed with 500 µL of Folin–Ciocalteu reagent and allowed to stand for 3 min. Next, 1.5 mL of a 25% Na_2CO_3 solution was added and then deionized water up to 10 mL final volume. After incubation in darkness for 1 h, the absorbance at 765 nm was measured. The results were expressed as gallic acid equivalents (GAE) (g/100 g) of extract using a standard curve for gallic acid, ranging from 12.5 to 1000 µg/mL. The analyzes were performed on day 1 after baking, after one week, 2 weeks, 4 weeks and 8 weeks of storage.

2.5. Statistical Analysis

Results were expressed as the mean value ± standard deviation (SD). Statistical tests were evaluated by using the Statistica 13.0 software (StatSoft, Inc., Tulsa, OK, USA). Analysis of variance (ANOVA) and the Tukey post-hoc tests were applied to determine differences between means. Differences were considered to be significant at $p < 0.05$.

3. Results

3.1. Mechanical and Thermal Properties

Determining the mechanical properties of food packaging such as cups, plates, or disposable bowls is essential for the achievement of high-quality products. They are mainly analyzed by compression/crush testing to determine the elasticity and the deformation and plasticity limit. Thus, the compressive strength of the product can be determined [37]. The results of the analysis of the mechanical properties of disposable utensils are presented in Table 2.

Table 2. The flexural strength of the disposable utensils.

Mixture	Flexural Strength [N/mm^2]				
	After Baking	1 Week	2 Weeks	4 Weeks	8 Weeks
OTSL	10.25 ± 0.01 ^{a}	10.28 ± 0.02 ^{a}	10.29 ± 0.01 ^{a}	10.31 ± 0.02 ^{a}	11.45 ± 0.02 ^{d}
OTL	10.02 ± 0.02 ^{a}	10.03 ± 0.01 ^{a}	10.31 ± 0.02 ^{a}	10.55 ± 0.01 ^{a}	12.01 ± 0.03 ^{e}
OSL	10.33 ± 0.01 ^{a}	10.35 ± 0.02 ^{a}	10.39 ± 0.02 ^{a}	10.98 ± 0.01 ^{a}	12.22 ± 0.02 ^{e}
OTS	7.45 ± 0.01 ^{b}	8.49 ± 0.02 ^{c}	9.01± 0.01 ^{c}	10.82 ± 0.02 ^{a}	19.21 ± 0.04 ^{f}
OTSL + B	10.28 ± 0.02 ^{a}	10.31 ± 0.02 ^{a}	10.33 ± 0.02 ^{a}	10.55 ± 0.02 ^{a}	11.59 ± 0.03 ^{d}
OTL + B	10.04 ± 0.01 ^{a}	10.12 ± 0.02 ^{a}	10.48 ± 0.02 ^{a}	10.64 ± 0.02 ^{a}	12.25 ± 0.01 ^{e}
OSL + B	10.36 ± 0.02 ^{a}	10.39 ± 0.01 ^{a}	10.95 ± 0.01 ^{a}	11.01 ± 0.01 ^{d}	12.18 ± 0.02 ^{e}
OTS + B	7.49 ± 0.01 ^{b}	8.55 ± 0.01 ^{c}	9.88 ± 0.03 ^{c}	11.05 ± 0.01 ^{d}	19.59 ± 0.03 ^{f}

$^{a–f}$ Same superscript letter in one row indicates no statistically significant differences between mixture ($p < 0.05$). OTSL + B: OTSL covered with beeswax; OTL + B: OTL covered with beeswax; OSL + B: OSL covered with beeswax; OTS + B: OTS covered with beeswax.

Disposable utensils produced from the mixture of olive pomace, teff flour and sorghum groats (OTS) showed the lowest crushing force of 7.45 N/mm^2. The remaining mixtures showed a similar crushing force of about 10 N/mm^2 due to the addition of lecithin. However, during the storage period, disposable tableware needs higher flexural strength. Disposable utensils obtained from the mixture of olive pomace, teff flour, sorghum and lecithin (OTSL) after 8 weeks had the lowest flexural strength 11.45 N/mm^2, compared to those prepared from the mixture of olive pomace, teff flour and lecithin (OTL) and a mixture of olive pomace, sorghum and lecithin (OSL). OTS was the hardest after 8 weeks. Interestingly, the protection of disposable tableware with shin wax did not significantly affect the crushing force. Similar results were obtained by Olt et al. [29], who showed that the flexural strength of rye and wheat bran plates was 14 N/mm^2 and that of tableware made of hulls and buckwheat bran was less than 9 N/mm^2. In view of the above, it can be concluded that the food packaging developed in this study has adequate crushing strength, which in the case of disposable tableware should not be lower than 10 N/mm^2 [29]. Liu et al. [37]

developed cellulose-based cups using sugarcane pomace and bamboo fibers with three times greater crush strength than the products analyzed. The Liu research team also tested the crushing force of a plastic cup, which was in the polystyrene range of 15.6 N/mm^2 [37]. Our material is less crush-resistant compared to the polystyrene cup [37]. Considering that the disposable dishes from the mixtures developed in this study need to be suitable for consumption, the lower squeezing force is an advantage because the material will be much easier to be consumed (crush in the mouth).

In the next stage of the research, the produced disposable packaging was subjected to color analysis. As shown in Table 3, there are differences in the packaging color without lecithin compared to the utensils from the mixture with lecithin. OSL, OTL and OTS packaging was brighter compared to OTSL. The a* value of OTS indicated a significant difference ($p < 0.05$) between the OTSL, OTL and OSL samples and contained a greater degree of redness. The L* value indicates the brightness of the products, which have a gray-brown color. The color did not change significantly, only slightly darkening with storage time, and not noticeable to an ordinary consumer. The addition of beeswax did not significantly change the color of the packaging, and only in large amounts. As the tested samples were immersed in the wax and then the wax surface was evened out with a brush, its thickness was so small that it did not brighten the final product. Disposable dishes were dark after baking, which also resulted in a lack of visual changes in the product. Beeswax does not contain a strong, permanent dye [38,39].

The pomace generated during the production of olive oil contains fatty acids and may oxidize over time. A common knowledge is that the enrichment of pomace with other ingredients including nutrients, can change unfavorable properties, enabling commercial utilization. For example, the addition of antioxidants can reduce the oxidation of sensitive components, while the addition of flour can change the thermal parameters of thermally unstable raw materials. The present study was conducted for packaging with the addition of lecithin, which in this case was used as an antioxidant. Differential scanning calorimetry (DSC) was used to demonstrate changes in the thermal parameters of the tested samples. The results of DSC analysis of the samples taken before and after baking were analyzed (Figure 2). The DSC curve of the unbaked samples is characterized by a single endothermic peak observed in the temperature range of 75–82 °C. The melting temperature peaks (Tm) of the unbaked OTSL, OTL and OSL samples were observed at temperatures of 77 °C, (heat capacity, 105 J/g), 79 °C (112 J/g) and 78 °C (109 J/g), respectively. This temperature distribution may be due to the diversity of components in the matrix including not only non-oxidizing fatty acids, but also starch retrogradation, gelatinization of starch in sorghum groats, protein denaturalization in teff flour and melting of lecithin [40–44]. All mixtures showed a statistically insignificant difference in melting curves. However, the packages showed different thermal properties after baking, in each of the three samples 4–6 (Figure 2). The endothermic peak disappeared, which may indicate a lower water content and greater thermal resistance, as the absence of an endothermic file in the 75–82 °C range may mean that there is no non-surface-bound water in the baked packages and may also indicate a lack of retrograde starch [45–47]. According to Rolandella's research, the glass transition temperature [45] decrease along with increase in the water content in the product. This effect is attributed to enthalpy relaxations or structural relaxations found in biopolymer matrix. The baked product may be more stable under certain storage conditions, and should be stored at temperatures below their Tg.

Table 3. Parameters for measuring the color of disposable packaging.

Mixture	Without the Addition of Beeswax			With the Addition of Beeswax		
	a*	b*	L*	a*	b*	L*
After baking						
OTSL	3.31 ± 0.01 ^{a}	7.56 ± 0.02 ^{c}	38.51 ± 0.06 ^{e}	3.30 ± 0.01 ^{a}	7.57 ± 0.01 ^{c}	38.59 ± 0.05 ^{e}
OTL	3.65 ± 0.02 ^{a}	10.00 ± 0.02 ^{d}	41.56 ± 0.04 ^{f}	3.62 ± 0.01 ^{a}	10.02 ± 0.02 ^{d}	41.62 ± 0.03 ^{f}
OSL	3.77 ± 0.01 ^{a}	10.28 ± 0.01 ^{d}	42.44 ± 0.05 ^{f}	3.74 ± 0.01 ^{a}	10.29 ± 0.02 ^{d}	42.49 ± 0.03 ^{f}
OTS	4.03± 0.01 ^{b}	9.05 ± 0.01 ^{c,d}	39.40 ± 0.06 ^{e}	4.00± 0.01 ^{b}	9.06 ± 0.01 ^{c,d}	39.45 ± 0.05 ^{e}
1 week						
OTSL	3.32 ± 0.01 ^{a}	7.55 ± 0.02 ^{c}	37.50 ± 0.03 ^{e}	3.33 ± 0.01 ^{a}	7.56 ± 0.02 ^{c}	37.68 ± 0.03 ^{e}
OTL	3.72 ± 0.01 ^{a}	10.45 ± 0.01 ^{d}	42.31 ± 0.03 ^{f}	3.73 ± 0.01 ^{a}	10.47 ± 0.02 ^{d}	42.87 ± 0.04 ^{f}
OSL	3.81 ± 0.01 ^{a}	10.32 ± 0.01 ^{d}	43.18 ± 0.04 ^{f}	3.84 ± 0.02 ^{a}	10.33 ± 0.03 ^{d}	43.65 ± 0.03 ^{f}
OTS	4.03 ± 0.01 ^{b}	9.18 ± 0.02 ^{c,d}	39.17 ± 0.04 ^{e}	4.05 ± 0.01 ^{b}	9.20 ± 0.02 ^{c,d}	39.28 ± 0.06 ^{e}
2 weeks						
OTSL	3.32 ± 0.01 ^{a}	7.56 ± 0.01 ^{c}	37.55 ± 0.02 ^{e}	3.34 ± 0.01 ^{a}	7.59 ± 0.01 ^{c}	37.71 ± 0.04 ^{e}
OTL	3.72 ± 0.01 ^{a}	10.45 ± 0.01 ^{d}	42.31 ± 0.03 ^{f}	3.74 ± 0.02 ^{a}	10.48 ± 0.02 ^{d}	42.92 ± 0.03 ^{f}
OSL	3.81 ± 0.01 ^{a}	10.32 ± 0.01 ^{d}	43.18 ± 0.04 ^{f}	3.86 ± 0.01 ^{a}	10.34 ± 0.02 ^{d}	43.99 ± 0.04 ^{f}
OTS	4.06 ± 0.02 ^{b}	7.40 ± 0.01 ^{c}	37.10 ± 0.03 ^{e}	4.08 ± 0.02 ^{b}	7.42 ± 0.02 ^{c}	37.45 ± 0.02 ^{e}
3 weeks						
OTSL	3.33 ± 0.01 ^{a}	7.58 ± 0.01 ^{c}	37.59 ± 0.02 ^{e}	3.34 ± 0.02 ^{a}	7.62 ± 0.02 ^{c}	37.72 ± 0.04 ^{e}
OTL	3.83 ± 0.02 ^{a}	10.62 ± 0.03 ^{d}	42.67 ± 0.03 ^{f}	3.85 ± 0.02 ^{a}	10.68 ± 0.03 ^{d}	43.01 ± 0.02 ^{f}
OSL	3.95 ± 0.02 ^{a}	10.77 ± 0.02 ^{d}	42.97 ± 0.02 ^{f}	3.99 ± 0.02 ^{a}	10.81 ± 0.03 ^{d}	44.11 ± 0.06 ^{f}
OTS	4.11 ± 0.03 ^{b}	10.98 ± 0.05 ^{d}	40.22 ± 0.05 ^{f}	4.15 ± 0.01 ^{b}	11.01 ± 0.02 ^{d}	44.01 ± 0.03 ^{f}
4 weeks						
OTSL	3.36 ± 0.01 ^{a}	7.71 ± 0.02 ^{c}	37.84 ± 0.01 ^{e}	3.38 ± 0.02 ^{a}	7.71 ± 0.03 ^{c}	37.89 ± 0.04 ^{e}
OTL	3.92 ± 0.01 ^{a}	10.78 ± 0.01 ^{d}	43.75 ± 0.02 ^{f}	3.99 ± 0.02 ^{a}	10.78 ± 0.02 ^{d}	43.22 ± 0.02 ^{f}
OSL	4.01 ± 0.01 ^{b}	11.02 ± 0.03 ^{d}	43.12 ± 0.03 ^{f}	4.18 ± 0.02 ^{b}	11.02 ± 0.03 ^{d}	44.18 ± 0.05 ^{f}
OTS	4.18 ± 0.02 ^{b}	11.18 ± 0.02 ^{d}	42.28 ± 0.03 ^{f}	4.21 ± 0.02 ^{b}	11.18 ± 0.02 ^{d}	44.22 ± 0.02 ^{f}
8 weeks						
OTSL	3.39 ± 0.02 ^{a}	7.70 ± 0.01 ^{c}	37.87 ± 0.02 ^{e}	3.41 ± 0.03 ^{a}	7.75 ± 0.02 ^{c}	37.93 ± 0.03 ^{e}
OTL	4.02 ± 0.03 ^{b}	11.03 ± 0.03 ^{d}	42.95 ± 0.03 ^{f}	4.06 ± 0.02 ^{b}	11.09 ± 0.03 ^{d}	43.25 ± 0.05 ^{f}
OSL	4.12 ± 0.02 ^{b}	11.45 ± 0.02 ^{d}	43.98 ± 0.04 ^{f}	4.19 ± 0.01 ^{b}	11.52 ± 0.01 ^{d}	44.25± 0.04 ^{f}
OTS	4.24 ± 0.02 ^{b}	8.90 ± 0.01 ^{c,d}	39.64 ± 0.04 ^{e}	4.31 ± 0.02 ^{b}	8.91 ± 0.02 ^{c,d}	43.61± 0.06 ^{e}

$^{a–f}$ Same superscript letter in one row indicates no statistically significant differences between mixtures ($p < 0.05$).

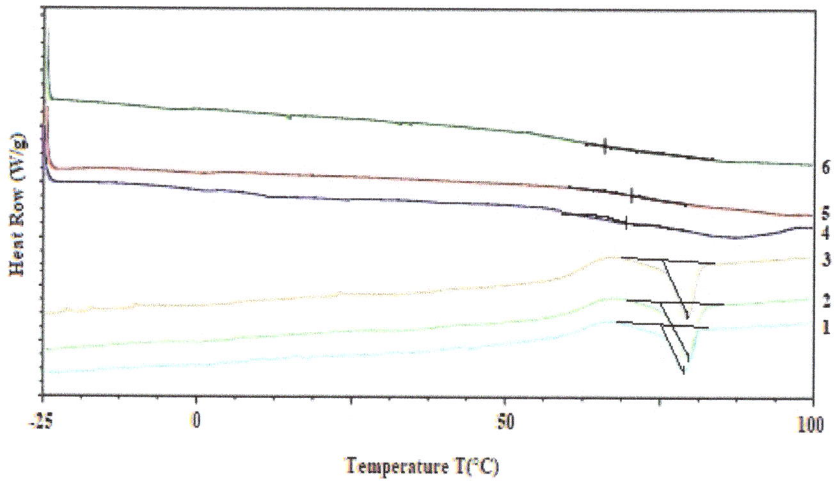

Figure 2. DSC curves of samples before (1—OTSL; 2—OTL; 3—OSL) and after (4—OTSL; 5—OTL; 6—OSL) baking.

3.2. Absorption Properties and Stability of Packaging

An important criterion in determining whether a material can be used as disposable food packaging is its durability, including its ability to resist dropping and low water absorption. For this reason, this research was focused on exploring the packaging durability by drop test. In addition, water absorption capacity was examined. The results of this study showed that the addition of beeswax did not improve nor worsen the structural characteristics of the disposable dishes. Interestingly, the packages made of the OTSL/OTSL + B mixture were not damaged throughout the storage period (Table 4), while OTL and OSL without/with the addition of beeswax were not damaged until 2 weeks after storage. After 2 weeks, disposable dishes dropped on the floor cracked and split. The OTS mixture was the worst and the vessel was destroyed immediately. It can be thus assumed that lecithin improved the plasticity of the flour and increased the durability of the single-dose dishes. The performed stress test correlates with the flexural strength analysis. Buxoo and Jeetah, studied the resistance to dropping of disposable biodegradable paper cups made of pineapple peel, orange peel and Mauritian hemp leaves, coated with a beeswax coating [33]. The use of orange peel and pineapple in the flash test showed that most of the molded cups remained intact, with no cracks. The exception were composites with a higher proportion of orange peel. Buxoo and Jeetah's research showed that fruit waste is a promising material for disposable tableware [33].

Moreover, the water activity of the prepared packages was checked and the results are presented in Table 4. All packages were characterized by low water activity. Low water activity value suggests that microbial growth is inhibited. The low water activity value and resistance to water absorption indicate that finished and baked packages can be stored under normal conditions in places that are not excessively moist. The packaging showed low water activity, i.e., less than 0.17, and the lowest water activity was found in the samples made of the OTSL/OTSL + B mixture. The low water activity of disposable tableware used as direct food packaging preserves food quality and protects against the development of undesirable microorganisms' growth [48]. Populations of yeasts such as Candida, Pichia and fungi can be present in fresh olive pomace. However, this is dependent on the variety of the olives as well as the way the olive oil is produced, as it is important to use high-quality raw materials [49]. The pomace is additionally subjected to high temperature, which degrades the bacterial microflora. Additionally, the low activity of water in the final product has a positive effect on the shelf-life of the product.

Table 4. The results of crack analysis of disposable utensils after drop test and water activity.

Mixture	Crack Analysis				
	After Baking	1 Week	2 Weeks	4 Weeks	8 Weeks
OTSL	remained intact	remained intact	remained intact	remained intact	remained intact
OTL	remained intact	remained intact	remained intact	cracked/split	cracked/split
OSL	remained intact	remained intact	remained intact	cracked/split	cracked/split
OTS	cracked/split	cracked/split	cracked/split	cracked/split	cracked/split
OTSL + B	remained intact	remained intact	remained intact	remained intact	remained intact
OTL + B	remained intact	remained intact	remained intact	cracked/split	cracked/split
OSL + B	remained intact	remained intact	remained intact	cracked/split	cracked/split
OTS + B	cracked/split	cracked/split	cracked/split	cracked/split	cracked/split
			Water Activity		
OTSL	0.059 ± 0.01^{a}	0.059 ± 0.01^{a}	0.060 ± 0.01^{a}	0.059 ± 0.02^{a}	0.052 ± 0.2^{a}
OTL	0.069 ± 0.02^{a}	0.070 ± 0.01^{b}	0.072 ± 0.02^{b}	0.082 ± 0.02^{b}	0.089 ± 0.02^{b}
OSL	0.055 ± 0.01^{a}	0.056 ± 0.01^{a}	0.057 ± 0.02^{a}	0.056 ± 0.01^{a}	0.052 ± 0.02^{a}
OTS	0.098 ± 0.01^{b}	0.170 ± 0.01^{c}	0.175 ± 0.02^{c}	0.198 ± 0.03^{c}	0.115 ± 0.05^{c}
OTSL + B	0.057 ± 0.02^{a}	0.057 ± 0.01^{a}	0.060 ± 0.02^{a}	0.061 ± 0.01^{a}	0.062 ± 0.02^{a}
OTL + B	0.068 ± 0.02^{a}	0.068 ± 0.02^{b}	0.069 ± 0.02^{b}	0.077 ± 0.01^{b}	0.081 ± 0.01^{b}
OSL + B	0.054 ± 0.01^{a}	0.061 ± 0.01^{a}	0.061 ± 0.02^{a}	0.065 ± 0.01^{a}	0.066 ± 0.01^{a}
OTS + B	0.097 ± 0.02^{b}	0.125 ± 0.03^{c}	0.159 ± 0.01^{c}	0.160 ± 0.03^{c}	0.162 ± 0.03^{c}

[a–c] Same superscript letter in one row indicates no statistically significant differences between mixtures ($p < 0.05$).

The disposable food packaging designed in the study the was constructed in the shape of a bowl, a plate and a cup. The cups and bowls have been additionally secured with beeswax to increase their resistance when in contact with water. The cups and bowls were analyzed with and without beeswax protection and tested for their reaction to cold and hot water contact. The results with cold and hot water gave a comparable effect. Therefore, the results are presented in Table 5 in the meantime.

The water resistance test was carried out until the first drop appeared, leaking through the cup/bowl. It was observed that the addition of beeswax increased the water resistance of the cups from 0.5 h to 1 h. The results of the water resistance test correlate with the decreased water absorption into the package. Disposable utensils prepared from the OTS/OTS + B mix were statistically significantly ($p < 0.05$) less resistant to contact with water and the weight of the packaging increased by half of its initial weight. The OSL packaging was characterized by slightly better parameters, which without the addition of beeswax leaked after about 2.5 h with an addition during about 3 h; however, the packaging after 6 h still increased by 1 g. The mix of OTS and OSL with and without the addition of beeswax is more suitable for dressings than for drinks or soups. It was observed that the best mixes for handling liquid meals or hot drinks were OTL and OTSL (4–4.5 h) with the addition of beeswax. However, without this they also fulfilled their role (3–3.5 h until the first drop appeared). An increase in the weight of the cups/bowls was observed only after 6 h, supporting a possible use for serving hot meals and drinks, especially as 3 h would be sufficient time to consume a hot/cold meal or a coffee/tea drink. Noteworthy is that water resistance tests were carried out with both cold and hot water and gave comparable results, so the packs could be used for storing and reheating meals. The water absorption factor is very important; as a general rule, once the temperature increases, the water is being more quickly [50]. Teff flour is characterized by higher water absorbability and gelling properties, which correlates with the results of DSC before baking packaging [51]. The gelling properties allowed plasticization of the mass and made it easier to form the vessels. Teff also promotes oil absorption, therefore it can be assumed that, when combining the mixture with olive waste, it can absorb fat residues and consequently emulsify more strongly [52]. On the other hand, sorghum groats are characterized by the content of caffeine proteins, which is highly hydrophobic. The heat treatment of sorghum grits reduces its water activity, while increasing water absorption [53,54]. This suggests that the combination of olive pomace and teff flour and/or sorghum, although initially increasing water absorption and viscosity of the product, reduced the water absorption properties of both cold and hot water after baking. The addition of a beeswax coating additionally lowers the value of moisture absorption and gives it hydrophobic properties, which allows extension of the time for keeping the liquid in the cups, even when using hot water [55].

Table 5. Occurrence of first drop and water absorption by the package.

Mixture		**The First Drop (h)**		
		After Baking	**4 Weeks**	**8 Weeks**
OTSL		3.5 ± 0.01^{a}	3.5 ± 0.01^{a}	3.5 ± 0.01^{a}
OTL		3 ± 0.01^{a}	3.2 ± 0.01^{a}	3.2 ± 0.01^{a}
OSL		2.5 ± 0.01^{b}	2.5 ± 0.01^{b}	2.4 ± 0.01^{b}
OTS		0.5 ± 0.01^{c}	0.5 ± 0.01^{c}	0.2 ± 0.01^{c}
OTSL + B		4.5 ± 0.01^{d}	4.5 ± 0.01^{d}	4.5 ± 0.01^{d}
OTL + B		4.1 ± 0.01^{d}	4.1 ± 0.01^{d}	4.1 ± 0.01^{d}
OSL + B		3 ± 0.01^{a}	3 ± 0.01^{a}	3 ± 0.01^{a}
OTS + B		1 ± 0.01^{c}	1 ± 0.01^{c}	1 ± 0.01^{c}
	Weight before the test [g]	**Weight after 6 h [g]**		
OTSL	2.21 ± 0.02^{a}	2.26 ± 0.01^{a}	2.32 ± 0.02^{b}	2.36 ± 0.01^{b}
OTL	2.22 ± 0.01^{a}	2.51 ± 0.04^{b}	2.53 ± 0.05^{b}	2.55 ± 0.03^{b}
OSL	2.36 ± 0.01^{a}	3.62 ± 0.03^{c}	3.59 ± 0.04^{c}	3.64 ± 0.02^{c}
OTS	2.44 ± 0.01^{a}	6.49 ± 0.04^{d}	6.48 ± 0.03^{d}	6.51 ± 0.04^{d}
OTSL + B	2.22 ± 0.02^{a}	2.24 ± 0.01^{a}	2.24 ± 0.02^{a}	2.24 ± 0.01^{a}
OTL + B	2.23 ± 0.02^{b}	2.44 ± 0.01^{b}	2.46 ± 0.02^{b}	2.48 ± 0.02^{b}
OSL + B	2.37 ± 0.01^{a}	3.01 ± 0.02^{c}	3.05 ± 0.03^{c}	3.02 ± 0.01^{c}
OTS + B	2.45 ± 0.02^{a}	4.66 ± 0.03^{d}	4.29 ± 0.01^{d}	4.78 ± 0.02^{d}

$^{a–d}$ Same superscript letter in one row indicates no statistically significant differences between mixtures ($p < 0.05$).

Iewkittayakorn et al. showed that paper cups made of pineapple leaf pulp coated with beeswax showed a longer water-uptake time than uncoated ones. The use of this coating increased the emulsifiability [34]. According to Buxoo and Jeetah, the thickness of the beeswax coating determines its water-repellent properties. A coating area of 0.7 mm resulted in resistance of the cups to leakage for up to 30 min [33]. In our case, the times were longer, which may indicate a more even distribution of the beeswax layer, and the use of olive pomace emulsified better with lipids than the use of pineapple/orange peels.

3.3. Nutritional Properties

The developed packaging is potentially edible. Therefore, it was verified whether the antioxidant properties were retained after the production of packaging from the by-products of olive oil production. There are still few studies on consumer biodegradable packaging and its antioxidant properties [33,48,56]. A description of the production of films with a high content of polyphenol compounds can be found in the literature [57,58]. On the other hand, research on disposable tableware is mainly related to mechanical and structural properties [33,48,56]. Still little research can be found on the antioxidant content of disposable tableware [29,37]. For this purpose, the content of polyphenols and the properties of scavenging free radicals were examined. The results are presented in Table 6.

Table 6. Total polyphenol content and antioxidant activity.

Mixture	DPPH (mmol TE/100 g)				
	After Baking	1 Week	2 Weeks	4 Weeks	8 Weeks
OTSL	1.11 ± 0.01 [a]	1.25 ± 0.02 [a]	1.66 ± 0.03 [a]	1.89 ± 0.04 [a]	2.92 ± 0.01 [a]
OTL	3.09 ± 0.02 [b]	3.19 ± 0.01 [b]	3.72 ± 0.02 [b]	3.88 ± 0.03 [b]	5.09 ± 0.03 [c]
OSL	4.11 ± 0.01 [a]	4.36 ± 0.05 [a]	4.57 ± 0.02 [a]	4.85 ± 0.03 [a]	7.50 ± 0.03 [b]
OTS	3.15 ± 0.03 [b]	3.23 ± 0.01 [b]	3.75 ± 0.02 [b]	3.91 ± 0.02 [b]	6.98 ± 0.03 [c]
OTSL + B	0.95 ± 0.01 [a]	0.98 ± 0.03 [a]	1.21 ± 0.02 [a]	1.33 ± 0.02 [a]	1.85 ± 0.03 [a]
OTL + B	2.89 ± 0.02 [b]	2.95 ± 0.01 [b]	3.16 ± 0.03 [b]	3.44 ± 0.04 [b]	4.36 ± 0.05 [c]
OSL + B	3.00 ± 0.06 [a]	3.15 ± 0.03 [a]	3.22 ± 0.04 [a]	3.49 ± 0.03 [a]	4.37 ± 0.03 [b]
OTS + B	3.03 ± 0.02 [b]	3.11 ± 0.03 [b]	3.37 ± 0.03 [b]	3.69 ± 0.04 [b]	5.87 ± 0.04 [c]
	Total polyphenol content [mg GAE/100 g]				
OTSL	10.95 ± 0.01 [a]	10.91 ± 0.02 [a]	10.89 ± 0.01 [a]	10.86 ± 0.01 [a]	10.82 ± 0.02 [a]
OTL	10.44 ± 0.01 [b]	10.38 ± 0.01 [b]	10.35 ± 0.01 [b]	10.30 ± 0.01 [b]	10.25 ± 0.01 [b]
OSL	8.36 ± 0.01 [b]	8.31 ± 0.02 [b]	6.28 ± 0.01 [b]	6.24 ± 0.01 [b]	6.18 ± 0.01 [c]
OTS	10.95 ± 0.01 [a]	10.90 ± 0.01 [a]	10.84 ± 0.01 [a]	10.66 ± 0.01 [a]	10.39 ± 0.01 [a]
OTSL + B	11.16 ± 0.03 [a]	11.12 ± 0.02 [a]	11.04 ± 0.02 [a]	10.59 ± 0.02 [a]	10.54 ± 0.01 [a]
OTL + B	10.66 ± 0.02 [b]	10.59 ± 0.01 [b]	10.52 ± 0.01 [b]	10.49 ± 0.01 [b]	10.47 ± 0.01 [b]
OSL + B	9.55 ± 0.03 [b]	9.52 ± 0.01 [b]	9.48 ± 0.01 [b]	9.31 ± 0.01 [b]	9.25 ± 0.01 [c]
OTS + B	10.03 ± 0.02 [a]	10.58 ± 0.03 [a]	10.57 ± 0.03 [a]	10.56 ± 0.01 [a]	10.49 ± 0.02 [a]

[a–c] Same superscript letter in one row indicates no statistically significant differences between mixtures ($p < 0.05$).

The content of polyphenols ranged from 6.18 to 11.16 mg GAE/100 g, for the 8 weeks stored OSL and the OTSL + B after baking, respectively. The mixtures of OTSL + B (11.16 mg GAE/100 g) and OTSL (10.95 mg GAE/100 g) had the highest content of polyphenols. There is a low decrease in polyphenols with storage time. Noticeably, the addition of teff flour significantly influences the content of polyphenols in the tested products. The addition of polyphenol-rich flour has an enriching effect on the final product, e.g., the addition of grape flour increases the level of polyphenols and antioxidant properties, but the heating process at temperatures above 180 °C reduces polyphenols and antioxidant properties even above 50% [56]. OSL/OSL + B packaging has about half the content of polyphenols but also degrades more quickly during storage. Coating the packages with beeswax had a positive effect on the content of polyphenols in the developed packages and also reduced the degree of polyphenol degradation during storage [59]. The TPC content depends on many factors and is not an exact method, because it also shows other components than just polyphenols, therefore, the TPC content was increased when storing products with a beeswax coating. The increase in TPC could also be influenced by the decomposition of polyphenols contained in beeswax [34,55,59]. The content of phenolic compounds in the olive's pulp is about 2–3%, while olive oil contains about 50–60 mg/100 g of polyphenols [60,61]. Therefore, a conclusion that the use of olive oil pomace preserved the high phenolic content after heat treatment can be drawn. The content of polyphenols correlates with antioxidant activity (Table 6). OTSL + B had the highest antioxidant potential, the dose of 0.95 mmol TE/100 g reducing free radicals by half. The oxidation properties decreased during storage. Dordevic et al. showed that biodegradable teaspoons from various types of flour, i.e., wheat flour, millet smooth flour and grape flour, lose a significant content of polyphenols, and the antioxidant properties of the designed teaspoons decrease [56].

3.4. Biodegradability Properties

All types of the produced packaging samples were subjected to the biodegradation process. The samples were placed in the ground and in the pot and then exposed to the outside to reproduce various weather conditions. Every week, the samples were weighed after prior cleaning and stabilization of the weight in the exicator to drain humidity. After weighing, the test was replaced in the ground and staged outside. Weather conditions were 20–35 °C and moderate rain 6 times per month. The analysis was made for packaging after baking and after 8 weeks of storage. The samples after baking, which were biodegradable within 4 weeks, had reduced their weight by 80%. The products with beeswax coating were the most biodegradable. This may be due to the faster breakdown of beeswax compared to olive pomace. During the further degradation of the samples, those without beeswax degraded by 100%, while those with the beeswax coating by about 90%. The samples with the powder coating degraded completely by week 7. The result of the decrease in masses and the biodegradation process is shown in Figures 3 and 4.

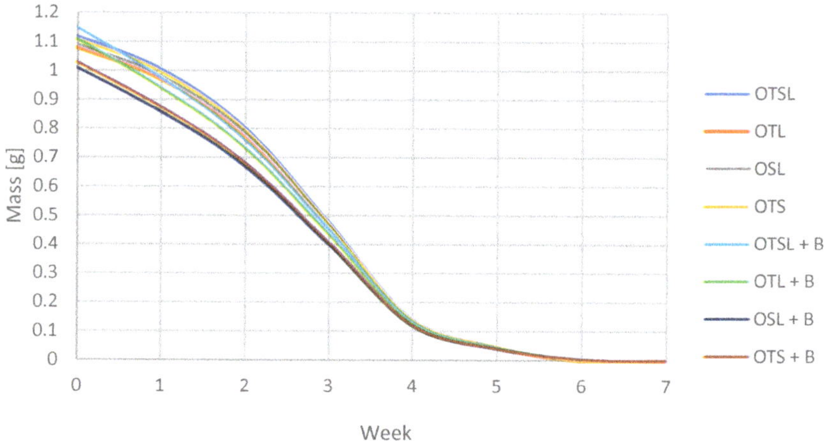

Figure 3. Change in the mass of biodegradable products after baking.

It is worth noting that the storage of the developed disposable tableware does not adversely affect the biodegradation properties. Comparing our product to other biodegradable plant-based products, biodegradation was at a good level. According to research by Iewkittayakorn and others, paper made from pineapple leaf pulp degraded for up to 60 days [34]. In Buxoo and Jeetah biodegradable wax-coated hemp-pineapple composite cup disappeared after 5 weeks in the soil [33]. According to research by Chaabane et al., packaging from sugar cane, palm leaf and paper items is totally degraded, but palm leaf items are degraded by 65% [62].

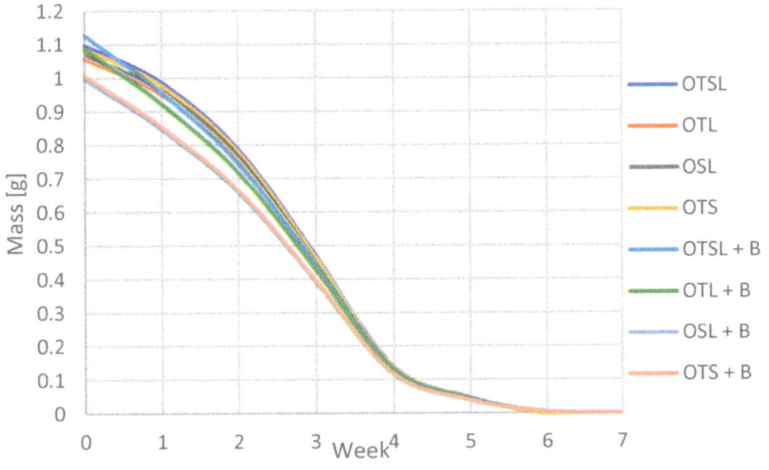

Figure 4. Change in the mass of biodegradable products 8 weeks of storage.

4. Conclusions and Future Perspectives

The study found that disposable packaging such as a plates, cups or bowls can be made from olive oil by-products. The use of olive pomace allows the management and the reduction of landfills. Additively designed disposable tableware can reduce the consumption of plastic products, or products whose manufacturing is not environmentally friendly. The addition of teff flour, sorghum and lecithin enriched the product with valuable polyphenols, increasing the antioxidant potential of the designed products. Disposable dishes can be used for serving cold or hot drinks, soups and for serving dry products. Importantly, the storage of the produced disposable packaging does not affect the characteristics of the product. The developed products exhibit a potential implementation in the following articles and will be explored in further research on the properties of packaging as well as the potential human consumption of the designed packaging. The mixture will also be prepared using the thermos-pressing method, which is less energy-consuming.

5. Patents

National patent application 437771.

Author Contributions: Conceptualization, J.G.; methodology, J.G., J.O. and I.G.-C.; formal analysis, J.G., J.O. and I.G.-C.; writing—original draft preparation, J.G.; writing—review and editing, J.O. and I.G.-C.; project administration, J.G.; funding acquisition, J.G. All authors have read and agreed to the published version of the manuscript.

Funding: This research was funded by "FU2N—Fundusz Udoskonalania Umiejętności Młodych Naukowców" supporting scientific excellence of Lodz University of Technology, grant number W5/FU2N/02/2022.

Institutional Review Board Statement: Not applicable.

Informed Consent Statement: Not applicable.

Data Availability Statement: Data is contained within the article.

Conflicts of Interest: The authors declare no conflict of interest.

References

1. Fernández-Lobato, L.; López-Sánchez, Y.; Blejman, G.; Jurado, F.; Moyano-Fuentes, J.; Vera, D. Life cycle assessment of the Spanish virgin olive oil production: A case study for Andalusian region. *J. Clean. Prod.* **2021**, *290*, 125677. [CrossRef]
2. Hazreen-Nita, M.K.; Kari, Z.A.; Mat, K.; Rusli, N.D.; Sukri, S.A.M.; Harun, H.C.; Lee, S.W.; Rahman, M.M.; Norazmi-Lokman, N.H.; Nur-Nazifah, M.; et al. Olive oil by-products in aquafeeds: Opportunities and challenges. *Aquac. Rep.* **2022**, *22*, 100998. [CrossRef]
3. Chen, C.; Chaudhary, A.; Mathys, A. Nutritional and environmental losses embedded in global food waste. *RCR Adv.* **2020**, *160*, 104912. [CrossRef]
4. Caldeira, C.; Vlysidis, A.; Fiore, G.; De Laurentiis, V.; Vignali, G.; Sala, S. Sustainability of food waste biorefinery: A review on valorisation pathways, techno-economic constraints, and environmental assessment. *Bioresour. Technol.* **2020**, *312*, 123575. [CrossRef] [PubMed]
5. Pattnaik, M.; Pandey, P.; Martin, G.J.O.; Mishra, H.N.; Ashokkumar, M. Innovative Technologies for Extraction and Microencapsulation of Bioactives from Plant-Based Food Waste and their Applications in Functional Food Development. *Foods* **2021**, *10*, 279. [CrossRef] [PubMed]
6. Melini, V.; Melini, F.; Luziatelli, F.; Ruzzi, M. Functional Ingredients from Agri-Food Waste: Effect of Inclusion Thereof on Phenolic Compound Content and Bioaccessibility in Bakery Products. *Antioxidants* **2020**, *9*, 1216. [CrossRef]
7. Guclu, G.; Kelebek, H.; Selli, S. Antioxidant activity in olive oils. In *Olives and Olive Oil in Health and Disease Prevention*; Academic Press: Cambridge, MA, USA, 2021; pp. 313–325. [CrossRef]
8. Arulprakasajothi, M.; Beemkumar, N.; Parthipan, J. Investigating the Physio-chemical Properties of Densified Biomass Pellet Fuels from Fruit and Vegetable Market Waste. *Arab. J. Sci. Eng.* **2020**, *45*, 563–574. [CrossRef]
9. Meena, R.A.A.; Banu, J.R.; Kannah, R.Y.; Yogalakshmi, K.N.; Kumar, G. Biohythane production from food processing wastes—Challenges and perspectives. *Bioresour. Technol.* **2020**, *298*, 122449. [CrossRef] [PubMed]
10. Van der Werf, P.; Gilliland, J.A. A systematic review of food losses and food waste generation in developed countries. *Waste Resour. Manag.* **2017**, *170*, 1747–6534. [CrossRef]
11. Girotto, F.; Alibardi, L.; Cossu, R. Food waste generation and industrial uses: A review. *Waste Manag.* **2015**, *45*, 32–41. [CrossRef] [PubMed]
12. Chen, Y.; Awasthi, A.K.; Wei, F.; Tan, Q.; Li, J. Single-use plastics: Production, usage, disposal, and adverse impacts. *Sci. Total Environ.* **2021**, *752*, 141772. [CrossRef] [PubMed]
13. Yang, Z.; Lu, F.; Zhang, H.; Wang, W.; Shao, L.; Ye, J.; He, P. Is incineration the terminator of plastics and microplastics. *J. Hazard. Mater.* **2021**, *401*, 123429. [CrossRef] [PubMed]
14. Skrypko, T.; Popadynets, N.; Yakhno, T.; Shulla, R.; Vlasenko, T.; Irtyshcheva, I.; Boiko, Y. Optimizing the polymer waste supply chains based on circular economy. *Uncertain Supply Chain Manag.* **2021**, *9*, 343–350. [CrossRef]
15. Egessa, R.; Nankabirwa, A.; Basooma, R.; Nabwire, R. Occurrence, distribution and size relationships of plastic debris along shores and sediment of northern Lake Victoria. *Environ. Pollut.* **2020**, *257*, 113442. [CrossRef] [PubMed]
16. Liu, G.; Wang, J.; Wang, M.; Ying, R.; Li, X.; Hu, Z.; Zhang, Y. Disposable plastic materials release microplastics and harmful substances in hot water. *Sci. Total Environ.* **2021**, *818*, 151685. [CrossRef] [PubMed]
17. Eriksen, M.; Lebreton, L.C.M.; Carson, H.S.; Thiel, M.; Moore, C.J.; Borerro, J.C.; Galgani, F.; Ryan, P.G.; Reisser, J. Plastic Pollution in the World's Oceans: More than 5 Trillion Plastic Pieces Weighing over 250,000 Tons Afloat at Sea. *PLoS ONE* **2014**, *9*, e111913. [CrossRef]
18. Song, J.H.; Murphy, R.J.; Narayan, R.; Davies, G.B. Biodegradable and compostable alternatives to conventional plastics. Philosophical transactions of the Royal Society of London. *Ser. B Biol. Sci.* **2009**, *364*, 2127–2139. [CrossRef]
19. Dybka-Stępień, K.; Antolak, H.; Kmiotek, M.; Piechota, D.; Koziróg, A. Disposable Food Packaging and Serving Materials—Trends and Biodegradability. *Polymers* **2021**, *13*, 3606. [CrossRef]
20. Vroman, I.; Tighzert, L. Biodegradable Polymers. *Materials* **2009**, *2*, 307–344. [CrossRef]
21. George, N.; Debroy, A.; Bhat, S.; Bindal, S.; Singh, S. Biowaste to Bioplastics: An Ecofriendly Approach for a Sustainable Future. *Appl. Biotechnol. Rep.* **2021**, *3*, 221–233. [CrossRef]
22. Razza, F.; Innocenti, F.D.; Dobon, A.; Aliaga, C.; Sanchez, C.; Hortal, M. Environmental profle of a bio-based and biodegradable foamed packaging prototype in comparison with the current benchmark. *J. Clean. Prod.* **2015**, *102*, 493–500. [CrossRef]
23. Spierling, S.; Knüpfer, E.; Behnsen, H.; Mudersbach, M.; Krieg, H.; Springer, S.; Albrecht, S.; Herrmann, C.; Endres, H.-J. Bio-based plastics—A review of environmental, social and economic impact assessments. *J. Clean. Prod.* **2018**, *185*, 476–491. [CrossRef]
24. Shogren, R.; Wood, D.; Orts, W.; Glenn, G. Plant-based materials and transitioning to a circular economy. *Sustain. Prod. Consum.* **2019**, *19*, 194–215. [CrossRef]
25. Espadas-Aldana, G.; Guaygua-Amaguaña, P.; Vialle, C.; Belaud, J.P.; Evon, P.; Sablayrolles, C. Life Cycle Assessment of Olive Pomace as a Reinforcement in Polypropylene and Polyethylene Biocomposite Materials: A New Perspective for the Valorization of This Agricultural By-Product. *Coatings* **2021**, *11*, 525. [CrossRef]
26. Folino, A.; Karageorgiou, A.; Calabrò, P.S.; Komilis, D. Biodegradation of Wasted Bioplastics in Natural and Industrial Environments: A Review. *Sustainability* **2020**, *12*, 6030. [CrossRef]
27. Chamas, A.; Moon, H.; Zheng, J.; Qiu, Y.; Tabassum, T.; Jang, J.H.; Abu-Omar, M.; Scott, S.L.; Suh, S. Degradation Rates of Plastics in the Environment. *ACS Sustain. Chem. Eng.* **2020**, *8*, 3494–3511. [CrossRef]

28. Coelho, A.; de Brito, J. Environmental analysis of a construction and demolition waste recycling plant in Portugal—Part I: Energy consumption and CO_2 emissions. *J. Waste Manag.* **2013**, *33*, 1258–1267. [CrossRef]
29. Olt, J.; Maksarov, V.; Soots, K.; Leemet, T. Technology for the Production of Environment Friendly Tableware. *Environ. Clim. Technol.* **2020**, *24*, 57–66. [CrossRef]
30. Dong, L.; Zhou, Y.; Liu, Y.; Lu, B.; Ji, J.; Ding, Y. High performance and water-degradable poly(neopentyl terephthalate-co-neopentyl succinate) copolymers: Synthesis, properties, and hydrolysis in different aquatic bodies. *J. Appl. Polym. Sci.* **2022**, e53316. [CrossRef]
31. Kılıç, R. The Problem of Hunger In The World and A New Model Proposal To Solve This Problem. *Balk. Sos. Bilim. Derg.* **2022**, *11*, 63–68. [CrossRef]
32. Schmidhuber, J.; Tubiello, F.N. Global food security under climate change. *Proc. Natl. Acad. Sci. USA* **2007**, *104*, 19703–19708. [CrossRef] [PubMed]
33. Buxoo, S.; Jeetah, P. Feasibility of producing biodegradable disposable paper cup from pineapple peels, orange peels and Mauritian hemp leaves with beeswax coating. *SN Appl. Sci.* **2020**, *2*, 1359. [CrossRef]
34. Iewkittayakorn, J.; Khunthongkaew, P.; Wongnoipla, Y.; Kaewtatip, K.; Suybangdum, P.; Sopajarn, A. Biodegradable plates made of pineapple leaf pulp with biocoatings to improve water resistance. *J. Mater. Res. Technol.* **2020**, *9*, 5056–5066. [CrossRef]
35. Di Mattia, C.; Battista, N.; Sacchetti, G.; Serafini, M. Antioxidant Activities in vitro of Water and Liposoluble Extracts Obtained by Different Species of Edible Insects and Invertebrates. *Front. Nutr.* **2019**, *6*, 106. [CrossRef]
36. Zielińska, E.; Karaś, M.; Jakubczyk, A. Antioxidant activity of predigested protein obtained from a range of farmed edible insects. *Food Sci. Technol. Res.* **2017**, *52*, 306–312. [CrossRef]
37. Liu, C.; Luan, P.; Li, Q.; Cheng, Z.; Sun, X.; Cao, D.; Zhu, H. Biodegradable, Hygienic, and Compostable Tableware from Hybrid Sugarcane and Bamboo Fibers as Plastic Alternative. *Matter* **2020**, *3*, 2066–2079. [CrossRef]
38. Szulc, J.; Machnowski, W.; Kowalska, S.; Jachowicz, A.; Ruman, T.; Steglińska, A.; Gutarowska, B. Beeswax-Modified Textiles: Method of Preparation and Assessment of Antimicrobial Properties. *Polymers* **2020**, *12*, 344. [CrossRef]
39. BaratianGhorghi, Z.; Faezian, A.; Yeganehzad, S.; Hesarinejad, M.A. Changes in Thermal, Textural, Color and Microstructure Properties of Oleogel Made from Beeswax with Grape Seed Oil under the Effect of Cooling Rate and Oleogelator Concentration. *Res. Innov. Food Sci. Technol.* **2022**, *11*, 43–54. [CrossRef]
40. Wang, D.; Bean, S.; McLaren, K.; Seib, P.; Madl, R.; Tuinstra, M.; Shi, Y.; Lenz, M.; Wu, X.; Zhao, R. Grain sorghum is a viable feedstock for ethanol production. *J. Ind. Microbiol. Biotechnol.* **2008**, *35*, 313–320. [CrossRef]
41. Adebowale, A.R.A.; Naushad Emmambux, M.; Beukes, M.; Taylor, J.R.N. Fractionation and characterization of teff proteins. *J. Cereal Sci.* **2011**, *54*, 380–386. [CrossRef]
42. Mészáros Szécsény, K.; Esztelecki, I.; Pokol, G. Adventages and limits on usage of thermal methods in complex systems: Bread and bread additives analyses. *J. Therm. Anal. Calorim.* **2007**, *89*, 829–833. [CrossRef]
43. Arvanitoyannis, I. Totally and Partially Biodegradable Polymer Blends Based on Natural and Synthetic Macromolecules: Preparation, Physical Properties, and Potential as Food Packaging Materials. *J. Macromol. Sci. Part. C Polym. Rev.* **1999**, *39*, 205–271. [CrossRef]
44. Droussi, Z.; D'orazio, V.; Provenzano, M.R.; Hafidi, M.; Ouatmane, A. Study of the biodegradation and transformation of olive-mill residues during composting using FTIR spectroscopy and differential scanning calorimetry. *J. Hazard. Mater.* **2009**, *164*, 1281–1285. [CrossRef]
45. Rolandelli, G.; Farroni, A.E.; Buera, M.d.P. Analysis of molecular mobility in corn and quinoa flours through ^{1}H NMR and its relationship with water distribution, glass transition and enthalpy relaxation. *Food Chem.* **2022**, *373*, 131422. [CrossRef]
46. Bogdanova, E.; Kocherbitov, V. Assessment of activation energy of enthalpy relaxation in sucrose-water system: Effects of DSC cycle type and sample thermal history. *J. Therm. Anal. Calorim* **2022**, *147*, 9695–9709. [CrossRef]
47. Garske, R.P.; Mercali, G.D.; Thys, R.C.S.; Cladera-Olivera, F. Cassava starch and chickpea flour pre-treated by microwave as a substitute for gluten-free bread additives. *J. Food Sci. Technol.* **2022**. [CrossRef]
48. Tapia, M.S.; Alzamora, S.M.; Chirife, J. *Effects of Water Activity (a w) on Microbial Stability as a Hurdle in Food Preservation. Water Activity in Foods: Fundamentals and Applications*; John Wiley & Sons: Hoboken, NJ, USA, 2020; Volume 14. [CrossRef]
49. Ntougias, S.; Bourtzis, K.; Tsiamis, G. The microbiology of olive mill wastes. *Biomed. Res. Int.* **2013**, *2013*, 784591. [CrossRef]
50. Songok, J.; Salminen, P.; Toivakka, M. Temperature effects on dynamic water absorption into paper. *J. Colloid Interface Sci.* **2014**, *418*, 373–377. [CrossRef]
51. Pulivarthi, M.K.; Selladurai, M.; Nkurikiye, E.; Li, Y.; Siliveru, K. Significance of milling methods on brown teff flour, dough, and bread properties. *J. Texture Stud.* **2022**, *53*, 478–489. [CrossRef]
52. Alemneh, S.T.; Emire, S.A.; Hitzmann, B. Comparative Study of Chemical Composition, Pasting, Thermal and Functional properties of Teff (Eragrostis tef) Flours Grown in Ethiopia and South Africa. *Int. J. Food Prop.* **2022**, *25*, 144–158. [CrossRef]
53. Batariuc, A.; Ungureanu-Iuga, M.; Mironeasa, S. Effects of Dry Heat Treatment and Milling on Sorghum Chemical Composition, Functional and Molecular Characteristics. *Appl. Sci.* **2021**, *11*, 11881. [CrossRef]
54. Badau, M.H.; Nkama, I.; Jideani, I.A. Water-absorption characteristics of various pearl millet cultivars and sorghum grown in northern Nigeria. *J. Food Process. Eng.* **2005**, *28*, 282–298. [CrossRef]
55. Diyana, Z.N.; Jumaidin, R.; Selamat, M.Z.; Suan, M.S.M. Thermoplastic starch/beeswax blend: Characterization on thermal mechanical and moisture absorption properties. *Int. J. Biol. Macromol.* **2021**, *190*, 224–232. [CrossRef]

56. Gorzynik-Debicka, M.; Przychodzen, P.; Cappello, F.; Kuban-Jankowska, A.; Marino Gammazza, A.; Knap, N.; Wozniak, M.; Gorska-Ponikowska, M. Potential Health Benefits of Olive Oil and Plant Polyphenols. *Int. J. Mol. Sci.* **2018**, *19*, 686. [CrossRef]
57. Chen, T.; Shen, Y.; Wu, D.; Wu, R.; Sheng, J.; Feng, X.; Tang, X. Biodegradable films of chitosan and tea polyphenols catalyzed by laccase and their physical and antioxidant activities. *Food Biosci.* **2021**, *46*, 101513. [CrossRef]
58. De Carli, C.; Aylanc, V.; Mouffok, K.M.; Santamaria-Echart, A.; Barreiro, F.; Tomás, A.; Pereira, C.; Rodrigues, P.; Vilas-Boas, M.; Falcão, S.I. Production of chitosan-based biodegradable active films using bio-waste enriched with polyphenol propolis extract envisaging food packaging applications. *Int. J. Biol. Macromol.* **2022**, *213*, 486–497. [CrossRef]
59. Sawicki, T.; Starowicz, M.; Kłębukowska, L.; Hanus, P. The Profile of Polyphenolic Compounds, Contents of Total Phenolics and Flavonoids, and Antioxidant and Antimicrobial Properties of Bee Products. *Molecules* **2022**, *27*, 1301. [CrossRef]
60. Bucciantini, M.; Leri, M.; Nardiello, P.; Casamenti, F.; Stefani, M. Olive Polyphenols: Antioxidant and Anti-Inflammatory Properties. *Antioxidants* **2021**, *10*, 1044. [CrossRef]
61. Dordevic, D.; Necasova, L.; Antonic, B.; Jancikova, S.; Tremlová, B. Plastic Cutlery Alternative: Case Study with Biodegradable Spoons. *Foods* **2021**, *10*, 1612. [CrossRef]
62. Chaabane, A.B.; Robbe, E.; Schernewski, G.; Schubert, H. Behavior of Biodegradable and Single-Use Tableware Items in the Warnow Estuary (Baltic Sea). *Sustainability* **2022**, *14*, 2544. [CrossRef]

foods

MDPI

Article

Production of Xylooligosaccharides from *Jiuzao* by Autohydrolysis Coupled with Enzymatic Hydrolysis Using a Thermostable Xylanase

Liqin Qin [1], Jinghao Ma [1], Huafeng Tian [2], Yanli Ma [3], Qiuhua Wu [1], Shuang Cheng [3] and Guangsen Fan [1,4,*]

1 School of Food and Health, Beijing Technology and Business University (BTBU), Beijing 100048, China
2 College of Chemistry and Materials Engineering, Beijing Technology & Business University (BTBU), Beijing 100048, China
3 Henan Key Laboratory of Industrial Microbial Resources and Fermentation Technology, Nanyang Institute of Technology, Nanyang 473004, China
4 Beijing Engineering and Technology Research Center of Food Additives, Beijing Technology & Business University (BTBU), Beijing 100048, China
* Correspondence: fanguangsen@btbu.edu.cn

Abstract: The production of xylooligosaccharides (XOS) from *Jiuzao* was studied using a two-stage process based on autohydrolysis pretreatment followed by enzymatic hydrolysis. *Jiuzao* was autohydrolyzed under conditions where temperature, time, particle size, and solid-liquid ratio were varied experimentally. Optimal XOS production was obtained from *Jiuzao* with a >20 mesh particle size treated at 181.5 °C for 20 min with a 1:13.6 solid-liquid ratio. Subsequently, optimal enzymatic hydrolysis conditions for xylanase XynAR were identified as 60 °C, pH 5, and xylanase XynAR loading of 15 U/mL. Using these conditions, a yield of 34.2% XOS was obtained from *Jiuzao* within 2 h. The process developed in the present study could enable effective and ecofriendly industrial production of XOS from *Jiuzao*.

Keywords: *Jiuzao*; autohydrolysis; thermostable xylanase; xylooligosaccharides; enzymatic hydrolysis

Citation: Qin, L.; Ma, J.; Tian, H.; Ma, Y.; Wu, Q.; Cheng, S.; Fan, G. Production of Xylooligosaccharides from *Jiuzao* by Autohydrolysis Coupled with Enzymatic Hydrolysis Using a Thermostable Xylanase. *Foods* **2022**, *11*, 2663. https://doi.org/10.3390/foods11172663

Academic Editors: Stanisław Kowalski and Dorota Gumul

Received: 10 July 2022
Accepted: 26 August 2022
Published: 1 September 2022

1. Introduction

As one of the world's most famous distilled spirits, *Baijiu* is known for its unique flavor [1]. China is the major producer of *Baijiu* and its production amounted to 7.4 million kiloliters in 2020 [2]. *Baijiu* is made with solid-state fermentation and distillation using sorghum, corn, other grains, and rice hulls [3,4]. A mixture of grain and rice hull residues, known as *Jiuzao*, is left after the solid-state fermentation and solid-state distillation processes are complete. Because eight units of *Jiuzao* are generated during the production of one unit of *Baijiu*, nearly 100 million tons of solid waste are generated each year from *Baijiu* production, creating an environmental concern since *Jiuzao* decay produces a foul odor [2]. Because the main components of dry *Jiuzao* are cellulose, hemicellulose, and lignin, it represents a valuable resource for the production of value-added products such as xylooligosaccharides (XOS). The use of *Jiuzao* to produce XOS would be economically and environmentally beneficial.

Currently, XOS derived from xylan-containing lignocellulosic biomass is widely used in food, chemical, pharmaceutical, feed, and nutraceutical industries [5–7]. Although, many different approaches have been reported for XOS production from various sources of biomass, the most effective method employs xylanase to hydrolyze xylan, the main component of hemicellulose [7,8]. This method is not only environmentally friendly, but also produces the main functional components, xylobiose (X2) and xylotriose (X3) [7]. Thus, the enzymatic degradation of hemicellulose in *Jiuzao* to produce XOS should be possible. However, any such new production process must overcome two problems that limit the accessibility of components to hydrolytic enzymes: the strong crystalline structure of the cellulose, and the complex structure crosslinking lignin and hemicellulose with cellulose in *Jiuzao* [7,9]. Effective pretreatments are needed to disrupt the crystalline structure of cellulose and increase hemicellulose exposure to hydrolytic enzymes thereby increasing the efficiency of enzymatic hydrolysis and XOS yield [10,11]. Among many pretreatment methods, autohydrolysis or hot-compressed water pretreatment uses water as the only reagent and is preferred for its economic and environmental benefits [11–14].

Autohydrolysis pretreatment, in which biomass is exposed to pressurized hot water (100–240 °C) extract hemicelluloses into the water phase, is one of the most promising pretreatment methods [15–17]. Water under pressure penetrates the cell walls of the plant biomass, hydrates cellulose, and dissolves hemicellulose and lignin. In addition, the lignocellulosic structure is disrupted by the acidity of the water at high temperature (around 200 °C) and the organic acids released from the hemicellulose [12,18,19]. Autohydrolysis does not require the addition of any chemical, generates reactive cellulose fiber, enables recovery of most of the pentosans, and generates degradation products that do not significantly inhibit subsequent hydrolysis and fermentation [12,16,20]. The fundamental characteristics of the autohydrolysis of lignocellulosic biomass for sugar production have been described in general [20,21]. However, specific features of the process need further clarification to optimize this promising technology, including the details related to both sugar production and inhibitor formation during the hot-water extraction process. Different biomass types have different structure and composition, which would likely require different pretreatment conditions. Thus, the pretreatment process must be tailored to the unique compositional and structural features of the lignocellulosic biomass being used [11].

For production of XOS in the present study, *Jiuzao* was pretreated with autohydrolysis and then hydrolyzed with XynAR, a thermostable xylanase obtained by molecular modification in our laboratory. A method of production of XOS from *Jiuzao* by autohydrolysis with xylanase was established, which can provide a scheme for generating high value products from *Jiuzao*.

2. Materials and Methods

2.1. Materials

Jiuzao was provided in 2019 by the Heibei Bancheng Liquor Group, which is an enterprise producing strong-flavor *Baijiu*, dried at 60 °C for 3 d, and stored in plastic containers at room temperature. Beechwood xylan from Sigma Chemical Company (St. Louis, MO, USA) was used. XOS standards, including X2 (≥95%, HPLC), X3 (≥95%, HPLC), xylotetraose (X4) (≥95%, HPLC), and xylopentaose (X5) (≥95%, HPLC) were obtained from Megazyme (Ireland). All other chemicals and laboratory reagents, unless stated otherwise, were reagent grade and obtained from commercial sources.

2.2. *Jiuzao* Composition

Jiuzao moisture content was calculated using the weight before and after drying at 105 °C. Ash content was measured by igniting in a muffle furnace according to AOAC 923.03. Starch content was determined with the α-amylase method as described in AOAC 996.11. *Jiuzao* that had been completely hydrolyzed with α-amylase to remove starch as described previously [22] was used to assess carbohydrate and lignin content with the method of the National Renewable Energy Laboratory (NREL) [23]. Briefly, 300 mg of *Jiuzao* that had been pretreated with α-amylase was hydrolyzed with 3 mL of 72% (*w*/*w*) sulfuric acid for 1 h at 30 °C, diluted to 4% (*w*/*w*) acid solution by adding 84 mL of double distilled water, and autoclaved at 121 °C for 1 h. The autoclaved samples were filtered through 0.45 μm membranes to separate solid and liquid phases. After thoroughly rinsing with water, the residual biomass was dried at 60 °C to a constant weight. The total lignin content was estimated as acid-soluble and insoluble lignin. The acid-soluble lignin was determined from absorbance at 205 nm with a UV-Vis spectrophotometer (TU-1901, Pgeneral, UV Spectrophotometer). The insoluble lignin in the residual dry solid biomass was estimated gravimetrically by subtracting the residual ash weight obtained after heating at 575 °C for

4 h in a muffle furnace from the total dry weight obtained by drying the original sample at 105 °C to constant weight. Monomeric sugar content in the liquid hydrolysate was used to quantify cellulose and hemicellulose. After neutralizing with $CaCO_3$ and filtering through a 0.22 μm membrane, glucose, xylose (X1), and arabinose (A) in the hydrolysate were analyzed by high-performance liquid chromatography (HPLC, Shimadzu. Co., Ltd., LC-10AD, Kyoto, Japan) with a refractive index detector (RID).

2.3. Autohydrolysis Processing

The autohydrolysis reactor was a stainless-steel autoclave with a working volume of 600 mL (BR-300, JULABO, Seelbach, Germany) and a temperature controller. Thirty-six grams of *Jiuzao* were mixed with 360 mL deionized water and heated with an electric mantle heater while being continuously stirred at 300 rpm with a paddle agitator. When a target temperature was reached, the pretreatment temperature was maintained for 20 min at 180 °C with a PID controller. Upon completion of the set time, the reactor was cooled with cold water. After pretreatment, the solid fraction (SF) and the liquid fraction (LF) were separated by filtration and the solid residue washed several times with deionized water. The SF was then lyophilized, and the LF including wash water was stored at −20 °C for enzymatic hydrolysis. The weight of the SF and the final volume of the LF were measured.

The parameters of temperature, time, *Jiuzao* particle size, and solid-liquid ratio were optimized first by varying single factors as follows: temperatures of 160, 170, 180, 190, and 200 °C, pretreatment times of 0, 20, 40, 60, and 80 min, particle derived from sieving through meshes of >20, 20–40, 40–65, and 65–80, and the solid-liquid ratios of 1:8, 1:10, 1:12, 1:14 and 1:16. According to the single factor experimental results, three factors including temperature, time, and solid-liquid ratio were selected for response surface experiments using the Box–Behnken experimental design (BBD, Design-Expert Software 11.0, StatEase Inc., Minneapolis, MN, USA). The response value was XOS yield after enzymatic hydrolysis of the sample pretreatment by autohydrolysis.

To evaluate the combined effect of temperature (T) and time (t) on thermal treatment efficiency, logarithmic values of the severity factor (log R_o) were calculated for autohydrolysis treatments as expressed in Equation (1) [24].

$$R_0 = \int_0^t \exp \frac{T - T_{ref}}{w} dt \quad (1)$$

Here, t is the holding time (min) at the treatment temperature T, and T_{ref} is the reference temperature where little or no reaction occurs in the system being studied. A fitted value of 14.75, which is an empirical value related to activation energy of the reaction, is used for the arbitrary constant w. A value of 100 °C was used as the reference temperature.

2.4. XynAR Production and Enzyme Assay

Xylanase XynAR, modified from XynA was expressed in *Escherichia coli* (*E. coli*). XynAR production and activity were assessed as described previously [25]. Briefly, the transformants were cultured at 37 °C for 12 h in Luria-broth medium including 40 μg/mL kanamycin. Isopropyl β- D-1-thiogalactopyranoside at a final concentration of 0.5 mmol/L was added into the culture to induce XynAR expression. *E. coli* cells were obtained by centrifugation and disrupted by ultrasonication to get crude XynAR. The activity of XynAR was assayed at 60 °C with 1.0% (*w*/*v*) beechwood xylan (50 mmol/L citrate buffer, pH 6.0) for 10 min as in previous study [25]. The amount of reducing sugar liberated was determined by the 3,5-dinitrosalicylic acid method using xylose as the standard. One unit (U) of enzyme activity was defined as the amount of enzyme required to produce 1 μmoL of reducing sugar per minute.

2.5. Enzymatic Hydrolysis

Preliminary experiments showed that a mixture of the SF and LF yielded more XOS after enzymatic treatment than did the SF alone. Therefore, to obtain maximum XOS yield and improve experimental repeatability, both the SF and LF were used in subsequent work. The enzymatic hydrolysis with XynAR was performed as described previously [26]. In detail, enzymatic hydrolysis was performed on 15 mL samples. The amount of SF required for each 15 mL of LF was calculated to match the proportion in the original autohydrolysis product. After mixing each portion of SF with 15 mL of LF (SF-LF), the 15 mL mixtures were used for enzymatic hydrolysis reactions at 60 °C. After adjusting pH to 6.0, XynAR with a final concentration of 5 U/mL was added to the LF-SF mixture for 2 h. After hydrolysis, the samples were boiled at 100 °C to inactivate the enzyme. The formation of XOS was monitored using HPLC. The influence of different conditions such as temperature (55, 60, 65, 70, and 75 °C), pH (3, 4, 5, 6, and 7), enzyme concentration (5, 15, 45, 135, and 405 U/mL), and hydrolysis period (30, 60, 120, 240, and 480 min) on XOS yield was assessed in single factor experiments. For further optimization a four factor, three-level orthogonal experiment ($L_9(3^4)$) was done using Design-Expert Software 11.0. Based on the results of the single factor tests, nine enzymatic experiments were carried out at temperatures of 60, 65, and 70 °C, pH 3, 4, and 5, enzyme concentrations of 5, 15, and 25 U/mL, and hydrolysis periods of 2, 4, and 6 h.

2.6. Analytical Methods

The concentration of XOS was analyzed using a LC-10AD HPLC (Shimadzu, Japan) equipped with a Shodex Sugar KS-802 packed column (8 mm ID ×300 mm, F6378020) and a RID-10A according to Wu et al. [26]. The column was maintained at 65 °C and eluted with deionized water at a flow rate of 0.8 mL/min. The concentration of glucose, xylose, and arabinose was determined by HPLC system equipped with Aminex HPX-87H column (Bio-Rad, Hercules, USA). The column temperature was kept at 65 °C and the mobile phase was 5 mmol/L H_2SO_4 at a flow rate of 0.6 mL/min.

$$P(X)\ (\%) = C(X)\ (g/L) \times V\ (L) \times 100\%/W\ (g) \quad (2)$$

where P(X) is the XOS, xylose, glucose, or arabinose yield, C(X) is the XOS (DP = 2–5), xylose, glucose, or arabinose concentration in the reaction mixture, V is the volume of LF (L), and W is the hemicellulose or total of cellulose and starch weight in the test sample (g).

2.7. Fourier Transform Infrared Spectroscopy (FTIR)

FTIR experiments (Nicolet™ iS™50, Thermo Scientific™, Waltham, MA, USA) were done using a Fourier transform infrared spectrophotometer equipped with a detector (DTGS) operating at 4000–400 cm^{-1} spectral range at resolution 4 cm^{-1} and 32 scans per sample. About 5 mg of finely milled raw materials and solid residues were used for FTIR analysis.

2.8. Scanning Electron Microscopy (SEM)

Micrographs of samples were observed by scanning electron microscopy SEM (EVO 18, ZEISS, Jena, Germany) at 10 kV. Before observation, the samples were fixed to a 70 mm diameter specimen platform with double-sided adhesive carbon tape and coated with sputter palladium. Each micrograph was taken at different magnification.

2.9. Data Analysis

Each treatment was conducted in triplicate and the results were expressed as the mean. All statistical analyses were performed with Design-Expert Software 11.0 and Excel 2019 (Microsoft Corporattion, Washington, DC, USA).

3. Results and Discussion

3.1. Composition of Jiuzao

The primary components of *Jiuzao* were cellulose at 25.14% followed by lignin at 18.72%. The content of hemicellulose, a polysaccharide composed of xylan, was of special interest because XOS are obtained by its autohydrolysis and enzymatic hydrolysis. *Jiuzao* contained 13.08% hemicellulose, which is within the range of values previously reported [27,28]. In addition, *Jiuzao* also contained 10.08% starch, 11.54% ash, and 11.18% of other components. The various content values reported in the literature may be due to differences in feedstocks, *Baijiu* production process, and/or methods of measurement. Although the hemicellulose content of *Jiuzao* at the present study was lower than that reported for most other biomass sources, the yearly output of *Jiuzao* is large and it is the main pollutant generated during *Baijiu* production [29–31]. Using various *Jiuzao* components to produce value-added products will help mitigate this environmental problem. Hemicellulose would likely be the first component to be used commercially, with the gradual utilization of other components established through a step-by-step approach. The hemicellulose components in *Jiuzao* from various *Baijiu* production facilities could differ and thus the *Jiuzao* from some plants might be better for XOS production [27,28].

3.2. *Optimum Conditions for Autohydrolysis*

3.2.1. Effect of Autohydrolysis Temperature on the Production of XOS and Monosaccharides

Previous reports showed that the yield of XOS depends on the autohydrolysis conditions, especially temperature and reaction time [12,17]. In the present study, the yield of XOS increased first and then decreased with increasing temperature during autohydrolysis and the highest yield was obtained at 190 °C (Figure 1). During the process of autohydrolysis, some treatment conditions produced an oligosaccharide, X2*, with a degree of polymerization (DP) between that of X2 and X3. X2* was obtained with HPLC according to the XOS analytical method. After treatment with 4% (w/w) sulfuric acid at 121 °C for 1 h, the monosaccharide composition of X2* determined by HPLC to be xylose and arabinose, which confirmed that it was an XOS containing arabinose. The production of this oligosaccharide was measured according to the standard curve for X2, and it was likely due to the complexity and structure of the fibrous materials and random interruption of the fibrous structure during autohydrolysis. Therefore, because of the distribution of arabinose residues in *Jiuzao* and their sensitivity to high temperature, XOS containing arabinose can only be produced under appropriate autohydrolysis conditions. In futures studies the preparation conditions, purification, the effects of X2* on probiotics and antioxidant function will be studied. A similar trend for xylose yield with XOS yield was reported previously [32]. As reported in other studies [12,33], with increasing autohydrolysis treatment temperature the hemicellulose components degrade into soluble oligosaccharides and monosaccharides, but as the temperature increases further the oligosaccharides and monosaccharides are completely degraded. In contrast, arabinose yield decreased with increasing temperature. Compared to xylose, arabinose degrades more easily because of the higher susceptibility of arabinosyl linkages to hydrolytic reactions [14,17,32,34]. An increased yield of arabinose was observed at 200 °C, possibly due to the presence of some hemicellulose components that are difficult to degrade as previously reported [14]. The change in glucose yield was complex, showing a decreasing trend followed by an increase and then another decrease. This result may be due to the complex composition of *Jiuzao*. Besides the components described previously, it contains some residual sorghum starch that was not used up during *Baiju* production. This starch is easily degraded by autohydrolysis, producing a relatively high yield of glucose at lower temperature (i.e., 160 °C). The glucose content then decreased as the temperature increased because it was degraded into by-products such as furfural. Finally, at higher temperatures the autohydrolysis of cellulose generated another increase followed by a decrease in glucose content. The proportion of XOS and xylose increased with increasing temperature then decreased, indicating that the degradation of hemicellulose by autohydrolysis may be a disordered process [35]. When the temperature is low, large soluble polymers and xylose are mostly produced as content of XOS with a DP of 2–5 is lower. With increasing temperature, the XOS production rate is higher than that of xylose, thus increasing the XOS-to-xylose ratio. However, with a further

Although the changed behavior of the XOS yield with and without enzymolysis was similar, increasing at first and then decreasing with greater hold time, the yield of XOS after xylanase hydrolysis was higher than that without enzymatic hydrolysis. Moreover, the hold time for the highest yield of XOS was shorter in the autohydrolysis plus enzymatic hydrolysis treatment (at a hold time of 40 min) than that in the autohydrolysis treatment (at a hold time of 60 min). This relative result is like that obtained for the temperature optimization experiment in that the autohydrolysis plus enzymatic hydrolysis treatment peaked at a lower temperature. These results confirm the benefit of enzymatic hydrolysis for XOS preparation from fibrous materials, i.e., lower production cost from both improved yield and reduced autohydrolysis intensity. The peak and decline of XOS yield with different autohydrolysis treatment times after enzymatic hydrolysis parallels that for the different temperature conditions (Figure 2), and the possible mechanisms are the same. In addition, xylose content increased only slightly confirming the specific effect of enzyme hydrolysis. After enzymatic hydrolysis, the ratio of XOS-to-xylose at various hold times all increased compared to treatments lacking enzymatic hydrolysis, and the trend peaked then declined as the XOS and xylose yields changed. The ratio was the highest when the treatment holding time was 20 min, followed by 40 min. Considering both the yield of XOS and its ratio to xylose, the optimum hold time was 40 min.

3.2.3. Effect of Particle Size on the Production of XOS and Monosaccharides

During autohydrolysis, materials with differing particle size having slight differences in structural compactness and/or surface to volume ratio undergo different degrees of action by electrolyzed hydrogen ions and other substances, such as hydronium ions (H_3O^+) and acetic acid [39]. These differences affect the yield of XOS. In the present study, XOS yield after autohydrolysis fluctuated as particle size decreased (Figure 3), likely due to the complex fibrous structure in *Jiuzao*. Fibrous materials may contain xylan components that are easily degraded by autohydrolysis [14], producing XOS and other components easily even from large particles. As particle size decreases, a greater proportion of easily degraded components are available for conversion into XOS due to increased surface to volume ratio compared to the material with larger particle size (such as particle size >20 mesh). Thus, there would be sufficient energy available for the easily degradable parts to be converted in the material with smaller particle size (such as particle size 20–40 mesh), but the excess energy after degrading the easily degradable part is insufficient for production of new XOS by further degradation of the denser parts of the fibrous material and the complex and refractory xylan components. Nevertheless, the excess energy is sufficient to degrade the XOS produced by the easily degradable xylan components to produce other by-products. As a result, the yield of XOS was less when the particle size was 20–40 mesh compared to >20 mesh. As the particle size decreases further, partial components that are not easily degraded will also be destroyed by autohydrolysis to produce XOS thus improving the yield. With further reductions in particle size, the cycle will repeat. This process results in a layer-by-layer degradation of xylan in material as the particle size decreases, which may be due to the presence of relatively complex *Jiuzao* components, including a small amount of wheat and sorghum bran and a large amount of rice hulls, and the differences in fiber structure of these components [40]. In addition, the particle size for high XOS yield from *Jiuzao* treated by autohydrolysis was larger than that for materials used in other studies [33,41,42]. Perhaps the structure of *Jiuzao* was damaged to some extent because of microbial pretreatment and repeated cycles of heating during *Baijiu* production, rendering it more susceptible to autohydrolysis [43]. Xylose yield first decreased and then increased. The mechanisms producing variations in XOS yield across particle sizes are likely those for variations in xylose yield with particle size. The relative yield differences between the two are likely due to differences in the degree of degradation from xylan to XOS versus the degradation from xylan or XOS to xylose, resulting in a slight difference in the changed behavior of xylose yield. The ratio of XOS-to-xylose decreases with decreasing *Jiuzao* particle size, indicating the level of net increment of XOS increase (the total amount of xylan to XOS minus the amount of XOS converted to xylose or other by-products) is less than that of xylose increase (the total amount of xylan or XOS converted to xylose minus the amount of xylose converted to other by-products). With decreasing *Jiuzao* particle size, the changed trend of arabinose yield is like that of XOS yield, which may be related to its ease of degradation and the possible reasons for the XOS yield change trend [39]. The yield of glucose increased first and then decreased with decreasing particle size, which may be the result of relatively slow degradation of cellulose and some residual starch.

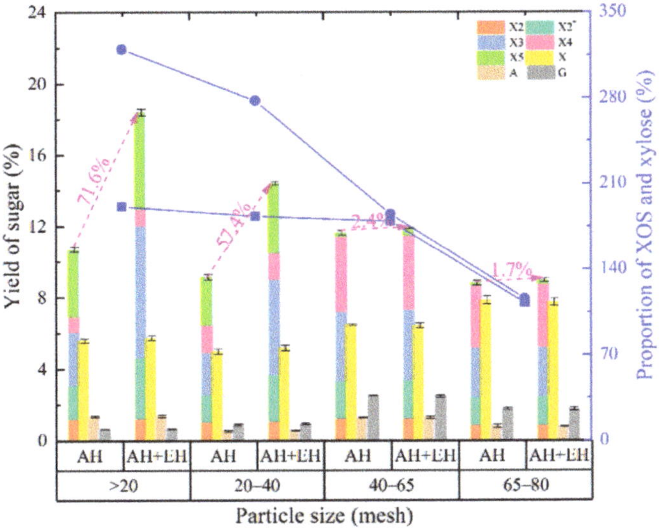

Figure 3. Effects of autohydrolysis pretreatments on sugar yield and the proportion of XOS and xylose at different particle sizes. AH, autohydrolysis-only treatment, no enzymatic hydrolysis treatment; AH + EH, autohydrolysis plus enzymatic hydrolysis treatment. Squares and solid line, the proportion of XOS and xylose for the autohydrolysis only treatment; circles and solid line, the proportion of XOS and xylose for the autohydrolysis plus enzymatic hydrolysis treatment. The magenta text is the percentage increase in the yield of XOS after autohydrolysis plus enzymatic hydrolysis treatment compared with that of the autohydrolysis-only treatment at different particle sizes. X2, xylobiose; X2*, an oligosaccharide with a degree of polymerization between X2 and X3; X3, xylotriose; X4, xylotetraose; X5, xylopentaose; X, xylose; A, arabinose; G, glucose.

After xylanase hydrolysis, the yield of XOS increased to varying degrees for different particle sizes samples. This increase weakened as *Jiuzao* particle size decreased, perhaps due to the inhibition of xylanase hydrolysis by many by-products produced with small particle size material which supports our hypothesis regarding particle size effects for XOS yield. The ratio of XOS-to-xylose in the product after enzymatic hydrolysis was higher than in the absence of enzymatic hydrolysis. The ratio also decreased with the decrease of *Jiuzao* particle size, where greater by-products production caused a decrease in enzyme hydrolysis. Considering the yield and cost of XOS, particles of *Jiuzao* from the 20-mesh screen were selected as the optimum size for further experiments.

3.2.4. Effect of Solid-Liquid Ratio on the Production of XOS and Monosaccharides

The solid-liquid ratio determines the concentration of hydrogen ions, hydronium ions, and acetic acid produced in autohydrolysis required to disrupt the fibrous structure of *Jiuzao*, which in turn affects the yield of XOS [39]. As the solid-liquid ratio increased, the yield of XOS fluctuated (Figure 4). The changed trend of xylose yield paralleled that of XOS, increasing initially, then decreasing, and then increasing again. Although the changed trends of XOS and xylose were the same, the ratio of both shows the opposite trend. This result may be due to different effects of hydrogen ions, hydronium ions, and acetic acid on xylan, XOS, and xylose, resulting in different degrees of net increase for XOS and xylose. The change of glucose yield was like that of XOS, but the inflection point was different due to the different composition and structure of starch and cellulose and the different sensitivity to hydrogen ions, hydronium ions, and/or acetic acid. In addition, the yield of arabinose shows the opposite trend to that of XOS, xylose, and glucose likely due to the ease of arabinose degradation and the structural characteristics of arabinose residues in fibrous materials.

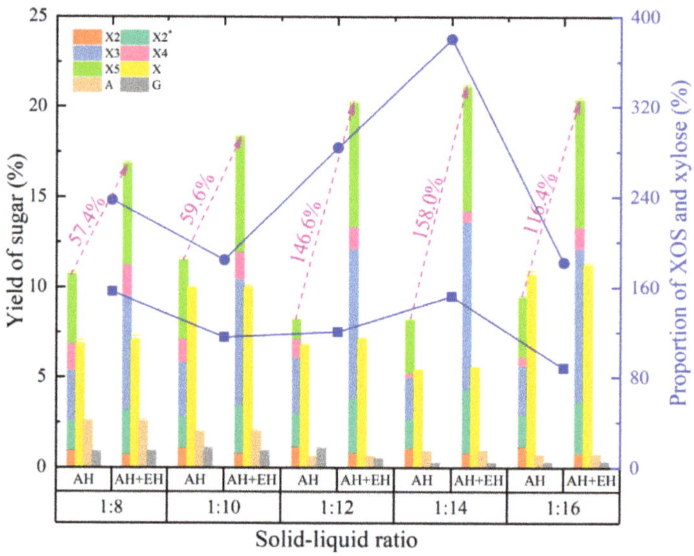

Figure 4. Effects of autohydrolysis pretreatments on sugar yield and the proportion of XOS and xylose at different solid-liquid ratio conditions. AH, autohydrolysis-only treatment, no enzymatic hydrolysis treatment; AH + EH, autohydrolysis plus enzymatic hydrolysis treatment. Squares and solid line, the proportion of XOS and xylose for the autohydrolysis only treatment; circles and solid line, the proportion of XOS and xylose for the autohydrolysis plus enzymatic hydrolysis treatment. The magenta text is the percentage increase in the yield of XOS after autohydrolysis plus enzymatic hydrolysis treatment compared with that of the autohydrolysis-only treatment at different solid-liquid ratios. X2, xylobiose; X2*, an oligosaccharide with a degree of polymerization between X2 and X3; X3, xylotriose; X4, xylotetraose; X5, xylopentaose; X, xylose; A, arabinose; G, glucose.

After coupled enzymatic hydrolysis, the yield of XOS increased and showed an overall upward trend, perhaps due to the relatively low concentration of autohydrolysis products and the decline of product inhibition with the resulting increase in the solid-liquid ratio. At the same time, the concentration of other by-products and their inhibitory effects decreased. In addition, with the solid-liquid ratio increase, the total amount of effective hydrogen ions, hydronium ions, and/or acetic acid in autohydrolysis increased, thus degrading the fibrous material structure. These changes are conducive to bonding of xylanase molecules to xylan fragments thus enhancing hydrolysis. The changed trend of the XOS-to-xylose ratio before and after enzymatic hydrolysis was similar, but the degree of the change differed. This effect is due to the joint action of autohydrolysis and enzymatic hydrolysis. At a low solid-liquid ratio, the effect of autohydrolysis on the XOS-to-xylose ratio was greater than enzymatic hydrolysis, whereas at a high solid-liquid ratio the effect of enzymatic hydrolysis was greater. Unlike the effect of previous factors, the XOS proportion after enzymatic hydrolysis increases and then decreases with an increase in the solid-liquid ratio. This result shows that in post-autohydrolysis samples with a high solid-liquid ratio is conducive to the hydrolytic effect of xylanase and the content of XOS increased after enzymatic hydrolysis. However, when the solid-liquid ratio was too high, the content of hemicellulose after autohydrolysis decreased in lower XOS production by enzymatic hydrolysis. After autohydrolysis plus enzymatic hydrolysis treatment, the yield of XOS at a solid-liquid ratio of 1:14 was not significantly higher than that at 1:12, but the ratio of XOS-to-xylose increased by 34%. Consequently, the solid-liquid ratio of 1:14 was selected as the optimum.

3.2.5. Response Surface Experimental (RSE) Optimized Autohydrolysis Conditions on XOS Yield

Based on the influence of various factors on the yield of XOS in the single factor experimental results, the response surface experimental design was used to analyze the interaction among temperature, time, and solid-liquid ratio to further optimize autohydrolysis conditions for XOS yield (Table 1). Multiple quadratic regression fitting was carried out on the test result data using Design-Expert software. The regression equation is as follows:

$$Y = 20.33 - 3.21A - 1.36B + 0.05C - 5.17AB + 0.35AC + 1.10BC - 6.13A^2 - 0.6292B^2 - 2.40C^2 \quad (3)$$

where Y is the predicted response (yield of XOS), and A, B, and C represent temperature, time, and solid-liquid ratio, respectively.

Table 1. The Box–Behnken experimental design and the responses of the dependent variables.

Test Number	Temperature (°C)		Time (Min)		Solid-Liquid Ratio		Yield of XOS (%)
	A	Code A	B	Code B	C	Code C	Y
1	190	1	40	0	1:16	1	7.8
2	180	0	20	−1	1:16	1	18.1
3	180	0	60	1	1:12	−1	14.3
4	190	1	40	0	1:12	−1	7.5
5	180	0	40	0	1:14	0	19.3
6	180	0	40	0	1:14	0	21.1
7	180	0	40	0	1:14	0	20.6
8	190	1	20	−1	1:14	0	17.6
9	180	0	20	−1	1:12	−1	19.7
10	180	0	60	1	1:16	1	17.1
11	190	1	60	1	1:14	0	5.0
12	170	−1	20	−1	1:14	0	11.8
13	170	−1	40	0	1:16	1	15.4
14	170	−1	40	0	1:12	−1	16.5
15	170	−1	60	1	1:14	0	19.9

Note: XOS, xylooligosaccharides.

The statistical significance of Equation (3) was checked by an *F* test analysis, and the analysis of variance (ANOVA) for the response surface quadratic model was generated (Table 2). The *p* value of the obtained model was significant ($p = 0.0019$), while the mismatch term was not significant ($p > 0.05$), indicating that the regression equation has a good fit and describes well the relationship between the various factors and XOS yield. The determination coefficient R^2 of the regression equation was 0.9739 and the adjustment coefficient R^2_{adj} was 0.9270, which further shows that the second-order regression equation fits the test well. The correlation between the predicted and actual values of the equation was high. Thus, the equation can be used to determine the optimal conditions for XOS yield.

Table 2. Regression coefficients and their significances for XOS production from the results of the Box-Behnken experimental design.

Source	Coefficient Estimate	Sum of Squares	DF	Mean Square	F-Value	p-Value	Significant
Model		362.47	9	40.27	20.74	0.0019	**
Intercept	20.33						
A-Temperature	−3.21	82.56	1	82.56	42.52	0.0013	**
B-Time	−1.36	14.85	1	14.85	7.65	0.0396	*
C-Solid-liquid ratio	0.0500	0.02	1	0.02	0.01	0.9231	
AB	−5.17	107.12	1	107.12	55.17	0.0007	**
AC	0.3500	0.49	1	0.49	0.25	0.6368	
BC	1.10	4.84	1	4.84	2.49	0.1752	
A^2	−6.13	138.71	1	138.71	71.43	0.0004	**
B^2	−0.6292	1.46	1	1.46	0.75	0.4253	
C^2	−2.40	21.34	1	21.34	10.99	0.0211	*
Residual		9.71	5	1.94			
Lack of Fit		7.98	3	2.66	3.08	0.2545	not significant
Pure Error		1.73	2	0.86			
Cor Total		372.18	14				
		$R^2 = 0.9739$	$R^2_{Adj} = 0.9270$	CV = 9.02%			

Note: "*", significant at 5% level ($p < 0.05$); "**", significant at 1% level ($p < 0.01$).

Based on the F value (Table 2), the order of influence of the three factors on the yield of XOS is: temperature > time > solid-liquid ratio, in which the primary terms A and B reach a significant level ($p < 0.05$). Therefore, temperature and time significantly affect XOS yield even when enzymatic hydrolysis is coupled, which is like previous reports on the production of XOS by autohydrolysis only. In addition, A^2 and C^2 have significant effects on their surfaces, and there is a highly significant interaction between temperature and time whereas the interaction between other factors is not significant. According to the regression equation, a response surface plot of the interaction of various factors on the yield of XOS was obtained (Figure 5). When the autohydrolysis temperature is low, the yield of XOS after coupled enzymatic hydrolysis increases with time, while at higher treatment temperature the yield of XOS decreases with time (Figure 5a). Moreover, when time is held constant, the yield of XOS first increases and then decreases with increasing temperature. The temperature optimum for XOS yield decreased with increasing time so that higher XOS yield was either at low temperature for a longer time or high temperature for a shorter time. As in previous studies, XOS yield was affected by the interaction of autohydrolysis temperature and time, which jointly determine the degree of hemicellulose dissolution and degradation in *Jiuzao* [33,39]. The content and accessibility of degradable hemicellulose components in the autohydrolysis samples and the by-products that inhibit the enzymatic hydrolysis efficiency will also affect the activity of xylanase, and then affect XOS yield. Therefore, it is necessary to comprehensively consider the autohydrolysis treatment intensity (including temperature and time), not only to ensure that an appropriate amount of hemicellulose is released and the fibrous structure in *Jiuzao* is sufficiently disrupted after autohydrolysis treatment, but also to control the generation of by-products affecting enzymatic hydrolysis efficiency. The response surface describing the relationship of temperature and solid-liquid ratio is convex, indicated that the yield of XOS increases first and then decreases with increases either in temperature or in the solid-liquid ratio (Figure 5b). Thus, a combination of conditions of treatment temperature and solid-liquid ratio is better for high XOS yield. The degradation of the fibrous structure and the release hemicellulose components are reduced at low temperature and solid-liquid ratio in the autohydrolysis process. However, high temperature and solid-liquid ratio would cause excess damage to the fibrous structure and produce many xylanase inhibitors that affect the subsequent enzymatic hydrolysis for XOS production. Therefore, appropriate autohydrolysis temperature and solid-liquid ratio are required to obtain optimal XOS yield. At a constant treatment time, the XOS yield increases then decreases with increasing the solid-liquid ratio (Figure 5c). When the solid-liquid ratio is held constant, the change of XOS yields with treatment time is complex and depends on treatment temperature. When the temperature is low, the XOS yield increased with treatment time. At higher temperature the yield decreased as treatment time increased, probably a result of greater fiber destruction and an increase in the concentration of by-products that inhibit enzymatic hydrolysis.

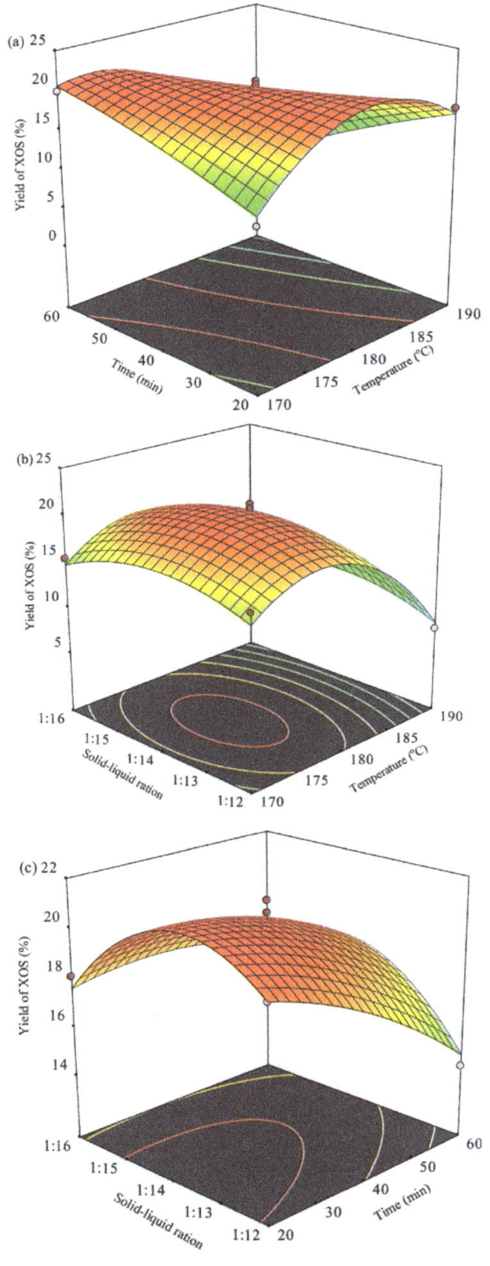

Figure 5. Response surface plot and contour plot for XOS production. (**a**) Effect of interaction of temperature and time on the XOS production when solid-liquid ratio is held at zero level. (**b**) Effect of interaction of temperature and solid-liquid ratio on the XOS production when time is held at zero level. (**c**) Effect of interaction of time and solid-liquid ratio on the XOS production when temperature is held at zero level.

In conclusion, the best treatment conditions are a balance between XOS produced by autohydrolysis and many factors affecting the subsequent production of XOS via enzymatic hydrolysis. The optimal conditions for XOS yield were identified from the simulation equation generated with the Design-Expert software as follows: temperature of 181.5 °C, time of 20 min, and solid-liquid ratio of 1: 13.6. Under these conditions, the XOS yield was predicted at 21.3%. This prediction was verified from the average value of three repeated experiments where the yield of XOS reaches 21.7%, a value nearly identical to the model's predicted value. This result shows that the simulation equation has good fit and that the model is reliable. Under the optimized conditions, the yield of XOS after enzymatic hydrolysis was two-fold higher than that without enzymatic hydrolysis, and the ratio of XOS-to-xylose was 232.9%. In addition, although the yield of XOS after response surface optimization was like that after single factor optimization, the treatment intensity Log R_o of autohydrolysis was reduced from 3.96 to 3.70. Thus, these autohydrolysis conditions were used for subsequent research.

3.3. Optimum Conditions of Enzymatic Hydrolysis

The effect of various physical-chemical parameters on enzymatic hydrolysis was studied to optimize production of XOS. Temperature is a key parameter for enzyme action. Temperature affects the thermal energy of substrate molecules and enzyme activity [44]. In the present study, the effect of various temperatures on XOS production was tested. The production of XOS after 2 h of incubation was significantly higher at 60 or 65 °C than other temperatures (Figure 6a). XynAR was genetically modified for thermal stability and its temperature optimum is 75 °C and its half-life at 80 °C is more than 150 min (unpublished data). These characteristics enable xylanase XynAR to better degrade agricultural waste and generate XOS at relatively high temperature. However, the enzymatic hydrolysis efficiency of XynAR was reduced unexpectedly when the temperature exceeded 65 °C. This yield reduction is likely due the inhibitory effect on xylanase activity of substances produced from autohydrolysis.

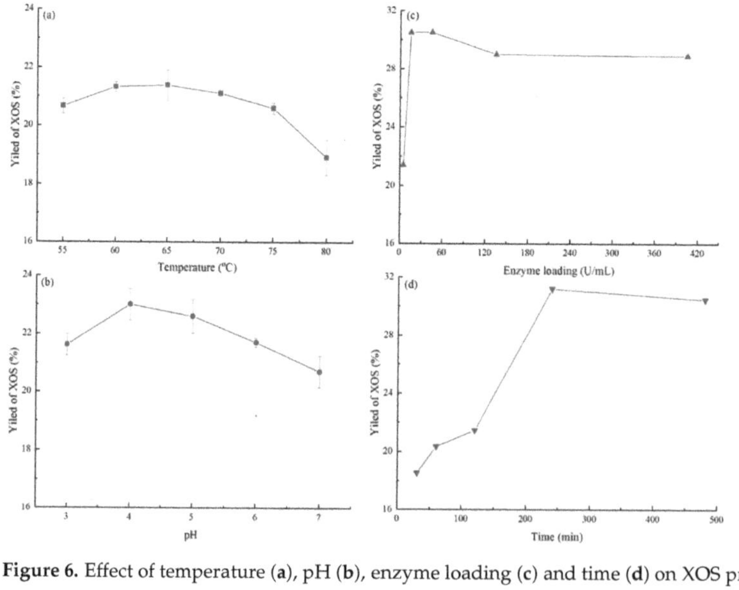

Figure 6. Effect of temperature (**a**), pH (**b**), enzyme loading (**c**) and time (**d**) on XOS production.

The pH is another important factor affecting the enzymatic hydrolysis of xylanase. It affects the activity of xylanase, the ionic state of substances, the adsorption capacity of substrate to xylanase, and the inhibition capacity of inhibitor to xylanase [44]. The hydrolysis reaction was tested at five pH values from 3.0–7.0 at 60 °C for 2 h. The yield of XOS increased initially and later decreased and maximum XOS yield occurred between pH 4.0–5.0 (Figure 6b). The activity of XynAR was stable over the range of 3.0–9.0 and high activity remained at pH range 4.0–6.5 (unpublished data). In addition, the optimum pH of xylanase XynAR for autohydrolysis of *Jiuzao* samples for XOS production differed from its optimum pH. This difference may be related to the effects of substances in the autohydrolysis samples on the xylanase activity of XynAR. The pH can affect these substances and the inhibitory activity of xylanase by these substances was reduced when the pH was relatively acidic.

In enzymatic hydrolysis, the ratio of enzyme to substrate must be considered because of product yield, enzymatic hydrolysis rate, and cost [44]. The effect of varying enzyme loading on the volume basis (5–405 U/mL) was evaluated. Increasing the xylanase concentration from 5 to 15 U/mL produced 23.0% to 30.5% of XOS after 2 h of incubation (Figure 6c). The yield of XOS slightly decreased at higher enzyme concentration probably due to decreased efficiency either because of competition between enzyme molecules for substrates or substrate limitation [32]. Thus, 15.0 U/mL was used for production of XOS in further experiments.

The production of XOS over time increased as the incubation period increased and a hydrolysis period of 4 h achieved the highest production of XOS (Figure 6d). The yield of XOS was 31.2% after 4 h of incubation and the yield then declined to 30.4% after 8 h of incubation. This decline may have been due to either a decreased level of easily accessible hydrolytic sites in the xylan chain or decreased xylanase activity from end-product inhibition. The increased incubation for production of XOS was not required as it did not enhance production of XOS. Compared to other studies, a hydrolysis period of 4 h in the present study was short due to the use of thermostable xylanase [32,44]. These results suggest that the thermostable xylanase is effective in a SF-LF production system and shows great potential for production of XOS.

To obtain high XOS yield with more cost-effective conditions, an orthogonal test was conducted using the single factor test results. The enzymolysis condition of $A_1B_3C_3D_2$ of 60 °C, pH 5, enzyme loading 25 U/mL, and hydrolysis period 4 h, gave the best yield of XOS at 33.0% (Table 3). According to the R values, in the process of xylanase enzymolysis of SF-LF, the influential of the four factors on XOS yield was in the order: B (pH) > C (enzyme loading) > A (temperature) > D (hydrolysis period) (Table 3). The optimized combination is $A_1B_3C_2D_1$ was established from analyzing the K values and a validation test was done in which the yield of XOS was 34.2%, higher than that the highest value listed in Table 3. Therefore, the optimal conditions for XOS production with xylanase XynAR were 60 °C, pH 5, enzyme loading of 15 U/mL, and a hydrolysis period 2 h. The XOS yield from *Jiuzao* was lower than that of previous reports using other substrates [45–47]. Aside from the ingredient differences, perhaps production from *Jiuzao* was lower because some hemicellulose structures from rice hulls were decomposed under the combined action of high temperature treatment and microbial activity during the *Baijiu* production process [12,33]. However, *Jiuzao* has certain advantages in the conversion of XOS compared with some other substrates such as brewers' spent grain, pineapple peel waste, and tobacco stalk (Table 4), and a large amount of *Jiuzao* is produced every year. In addition, *Jiuzao* produced by different *Baijiu* production processes will have different XOS production potential and the same raw materials may produce different XOS due to different sources, pretreatment methods and types of enzymes used. Therefore, different methods may be needed to optimize XOS yields from *Jiuzao* obtained from different *Baijiu* production processes. Primarily due to the hydrolysis characteristics of xylanase XynAR in the present study, X3 and X2 were the main components of the XOS, accounting for 44.5% and 41.0% of the total, respectively.

Table 3. The L9 orthogonal array applied for XOS production.

Test Number	Factors				Yield of XOS (%)
	A (Temperature, °C)	B (pH)	C (Enzyme Loading, U/mL)	D (Time, Min)	
1	1 (60 °C)	1 (3)	1 (5 U/mL)	1 (120 min)	28.6
2	2 (65 °C)	1	2 (15 U/mL)	2 (240 min)	29.0
3	3 (70 °C)	1	3 (25 U/mL)	3 (360 min)	28.0
4	1	2 (4)	2	3	31.3
5	2	2	3	1	31.5
6	3	2	1	2	27.9
7	1	3 (5)	3	2	33.0
8	2	3	1	3	30.4
9	3	3	2	1	32.3
Average K1	31.0	28.5	29.0	30.8	
Average K2	30.3	30.2	30.9	30.0	
Average K3	29.4	31.9	30.9	29.9	
Range (R)	1.6	3.4	1.9	0.9	
Optimal level	1	3	2	1	

Note: XOS, xylooligosaccharides.

Table 4. Summary of representative XOS productions from different lignocellulosic materials.

Type of Biomass	Hemicellulose (%)	Yield of XOS (%)	References
Rice husk	11.2	69 [a]	[48]
		33.8 [b]	[48]
		12.9	[49]
Wheat straw	40	40.3	[50]
	20.1	21	[51]
Rye bran	21	60 [c]	[52]
		40 [d]	[52]
Corncob	31.2	75	[53]
	22	57.6	[54]
	33.4	10.7	[29]
Rice straw	24.14	54.3	[55]
	17	50.5	[54]
Mahogany sawdust	24.3	36.8	[31]
Mango sawdust	26.5	25.5	[31]
Sugarcane bagasse	25.7	55.4	[56]
	26.5	53.2	[57]
	20.75	51.1	[58]
	32.7	36.4	[59]
	30	55.9	[60]
Wheat bran	17.6	22.8 [e]	[61]
		19 [f]	[61]
		18.6 [g]	[61]
Hawthorn kernels	28	66.8	[62]
Moso bamboo	17.3	42.7	[63]
Almond shell	20.2	40.6	[64]
Arecanut husk	24.6	35.1	[30]
Jiuzao	13.08	34.2	This study
Brewers' spent grain	16.5	31.5	[65]
Pineapple peel waste	31.8	25.6	[66]
Tobacco stalk	22	11.4	[67]
Natural grass	28	11	[68]
Poplar	15.8	10.7	[69]
Quinoa stalks	10.9	1.26	[70]

Note: [a] XOS was produced by *Aspergillus nidulans* XynC A773; [b] XOS was produced by *Aspergillus brasiliensis* BLf; [c] XOS was produced by RmXyn10A from GH10; [d] XOS was produced by Pentopan Mono BG from GH11; [e] XOS was produced by Pentopan; [f] XOS was produced by RmXyn10A-CM; [g] XOS was produced by NpXyn11A.

3.4. FTIR Spectroscopy Analysis

FTIR spectroscopy is a common analytical method used to identify the functional groups and molecular conformations of substances. Thus, the effects of the autohydrolysis protocol and enzymatic hydrolysis were determined by FTIR spectroscopy. For comparison purposes, the spectra of untreated rice hulls were also measured. The different treatment samples and raw material had the similar general spectral profile except some characteristic bands (Figure 7). All samples, including rice hulls, *Jiuzao*, *Jiuzao* after autohydrolysis treatment, and *Jiuzao* after autohydrolysis and enzymatic hydrolysis treatments, had a wide band positioned around 3400 cm^{-1}, which indicated the vibrations of the hydrogen bound mainly from cellulose and hemicellulose [71]. Compared with rice hulls, *Jiuzao* had blue shifts and different peak intensity at this frequency band and others, which may be related to the destruction of rice hulls in *Jiuzao* and consequent differences in composition. Aside from main component of rice hull, *Jiuzao* also contains sorghum residue and some fungal mycelium. After autohydrolysis treatment, the absorption intensity near this area decreased significantly due to the disruption of hydrogen bonds, indicating that the cellulose and hemicellulose components in *Jiuzao* were damaged by autohydrolysis. The reduced peak intensity after enzymatic hydrolysis indicated that the hemicellulose was further degraded by xylanase. The band at about 2920 cm^{-1} was related to C-H stretching vibration of methyl and methylene in lignin, cellulose, and hemicellulose [72,73]. Compared with *Jiuzao*, the intensity of the absorption peak decreased and there was a slight blue shift as the treatment procedure progressed (autohydrolysis and enzymatic hydrolysis). This result indicated that the lignin, cellulose, and hemicellulose components in *Jiuzao* were degraded after treatment, especially with the enzymatic hydrolysis treatment. The peak at 1740 cm^{-1} in the rice hull sample is attributed to either the acetyl and uronic ester groups of the hemicelluloses or the ester linkage of carboxylic group of the ferulic and p-coumeric acids of lignin and/or hemicelluloses [72]. After *Baijiu* production was complete, this peak was absent, indicating the cleavage of these bonds. These results show that some hemicellulose in rice hulls was degraded during *Baijiu* production, and this hemicellulose may be the most easily degraded component. This result also explains the formation of ferulic acid and other flavor substances in *Baijiu* and the reasons for the low XOS yield from *Jiuzao*. The C=O stretching bands are attributed to the polysaccharides and hemicellulose or lignin ester groups observed at 1640 cm^{-1} [73]. The change in absorption peak intensity of different treatment samples shows that autohydrolysis and enzymatic hydrolysis are important for the degradation of hemicellulose or lignin in *Jiuzao*. The intensive band at 1514 cm^{-1} is due to aromatic skeletal vibrations in bound lignin [74]. Comparing different treatment samples, this band weakens after enzymatic hydrolysis treatment indicating that some lignin combined with hemicellulose will decompose with the degradation of hemicellulose via enzymatic hydrolysis by xylanase after autohydrolysis. The small bands between -1453 cm^{-1} and -1235 cm^{-1}, such as 1453, 1373, and 1234 cm^{-1}, represent either C-H and C-O or OH bending vibrations in hemicelluloses [74]. There are slight differences between different samples at these range bands, which indicates that the hemicellulose components have been damaged to varying degrees after different treatments. A similar trend was observed at the 1031 cm^{-1} band for the C-OH stretching and β-glycosidic linkages of the cellulose glucose ring, which indicates that some cellulose components in material are more degradation resistant [73]. In conclusion, the cellulose, hemicellulose, and lignin in the components of *Jiuzao* were damaged after autohydrolysis and enzymatic hydrolysis, which shows the value of these two treatments for treating agricultural waste.

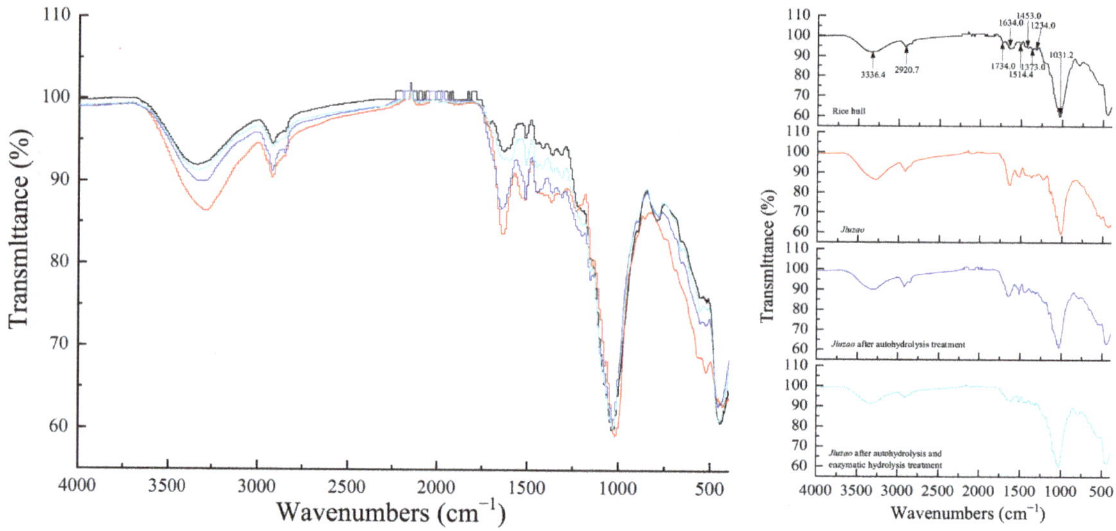

Figure 7. FTIR spectra of rice hulls, *Jiuzao*, *Jiuzao* after autohydrolysis treatment, and *Jiuzao* after autohydrolysis and enzymatic hydrolysis treatments. Right part of chart, combination of FTIR spectra of four samples; left part of chart, FTIR spectra of each of the four samples. Black line, FTIR spectra of rice hulls; red line, FTIR spectra of *Jiuzao*; blue line, FTIR spectra of *Jiuzao* after autohydrolysis treatment; cyan line, FTIR spectra of *Jiuzao* after autohydrolysis and enzymatic hydrolysis treatment.

3.5. Morphological Structure of Rice Hull by Different Pretreatments

Microscopic differences were observed on the surfaces rice hulls, *Jiuzao*, *Jiuzao* after autohydrolysis treatment, and *Jiuzao* after autohydrolysis treatment and enzymatic hydrolysis treatment (Figure 8). The surface of rice hulls, the main component of *Jiuzao*, were flat and smooth, indicating a solid structure (Figure 8a). Large irregular structures are visible, which may epicuticular wax. The rice hulls that had undergone *Baijiu* production showed a marked change, with a corrugated and slightly broken surface likely induced by high temperature and microbial treatment (Figure 8b). The irregular block structure of the raw rice hull was rougher and partly degraded. Some of the surface structure, which may have been wax and partial easily degraded cellulose, hemicellulose, or lignin components, appeared to have been lost during *Baijiu* production and possibly show an indirect reason for the relatively low XOS yield by autohydrolysis coupled with xylanase treatment. Noticeable differences were present in the autohydrolysis treatment sample. The surface structure was further damaged and the epidermis rough, which would enhance contact between xylanase and the internal structure thus increasing accessibility between enzyme and substrate and the surface area available for enzymatic hydrolysis (Figure 8c). Although the surface of the materials appeared to be eroded, with many dents, it still presents a relatively regular structural outline, which could possibly hinder subsequent enzymatic hydrolysis. The surface anatomy of materials changed significantly after enzymatic hydrolysis with debris, holes, and cracks (Figure 8d). The obvious degradation of structural perhaps shows the effects of enzymatic hydrolysis by xylanase.

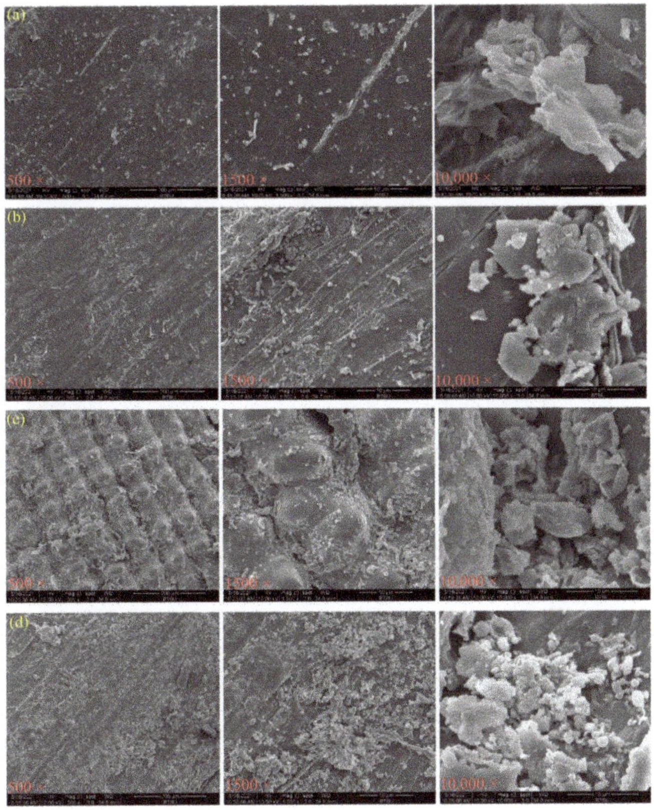

Figure 8. Scanning electron microscope image of rice hulls (**a**), *Jiuzao* (**b**), *Jiuzao* after autohydrolysis treatment (**c**), and *Jiuzao* after autohydrolysis and enzymatic hydrolysis treatment (**d**) at different magnifications.

4. Conclusions

The present study established the potential for producing XOS from *Jiuzao* using autohydrolysis pretreatment and enzymatic hydrolysis with a thermostable xylanase. Treating *Jiuzao* at 181.5 °C for 20 min with solid-liquid ratio of 1:13.6 enabled effective hydrolysis with thermostable xylanase to produce XOS. After optimization of enzymatic hydrolysis conditions, the highest yield of XOS from *Jiuzao* was 34.2% at 60 °C and pH 5 with 15 U/mL of XynAR for 2 h. The process has great promise for practical production of XOS from *Jiuzao*.

Author Contributions: L.Q. carried out the experiments and wrote the manuscript. J.M. and Q.W. carried out the experiments. G.F. conceived of the study, designed the experimental protocol, and wrote the manuscript. Y.M. and S.C. revised the manuscript. H.T. analyzed the data. All authors have read and agreed to the published version of the manuscript.

Funding: This research was supported by the Beijing Natural Science Foundation (6222003, 2202014 and 6164029), the Open Research Fund Program of Henan Key Laboratory of Industrial Microbial Resources and Fermentation Technology (HIMFT20190205), the General Project of Scientific Research Program of Beijing Municipal Education Commission (KM201910011006/PXM2019_014213_000007), the School Level Cultivation Fund of Beijing Technology and Business University for Distinguished and Excellent Young Scholars (BTBUYP2021).

Institutional Review Board Statement: Not applicable.

Informed Consent Statement: Not applicable.

Data Availability Statement: All data and materials have been provided in this manuscript.

Acknowledgments: We thank John Michael Bonman and Katrolia, Priti, for providing language help.

Conflicts of Interest: The authors declare no conflict of interest.

References

1. Song, X.B.; Wang, G.N.; Zhu, L.; Zheng, F.P.; Ji, J.; Sun, J.Y.; Li, H.H.; Huang, M.Q.; Zhao, Q.Z.; Zhao, M.M.; et al. Comparison of two cooked vegetable aroma compounds, dimethyl disulfide and methional, in Chinese Baijiu by a sensory-guided approach and chemometrics. *LWT-Food Sci. Technol.* **2021**, *146*, 111427. [CrossRef]
2. Ke, Q.M.; Zeng, W.; Shuai, Y.T.; Jin, J. Study on ultrasonic-assisted enzymatic hydrolysis of distiller's grains cellulose. *Sci. Technol. Food Ind.* **2021**, *43*, 196–203. [CrossRef]
3. Fan, G.S.; Liu, P.X.; Chang, X.; Yin, H.; Cheng, L.J.; Teng, C.; Gong, Y.; Li, X.T. Isolation and identification of a high-yield ethyl caproate-producing yeast from *Daqu* and optimization of its fermentation. *Front. Microbiol.* **2021**, *12*, 663744. [CrossRef] [PubMed]
4. Zhu, L.; Song, X.B.; Pan, F.; Tuersuntuoheti, T.; Zheng, F.P.; Li, Q.; Hu, S.Q.; Zhao, F.F.; Sun, J.Y.; Sun, B.G. Interaction mechanism of kafirin with ferulic acid and tetramethyl pyrazine: Multiple spectroscopic and molecular modeling studies. *Food Chem.* **2021**, *363*, 130298. [CrossRef] [PubMed]
5. Vázquez, M.J.; Alonso, J.L.; Domínguez, H.; Parajó, J.C. Xylooligosaccharides: Manufacture and applications. *Trends Food Sci. Technol.* **2000**, *11*, 387–393. [CrossRef]
6. Ayyappan, P.; Abirami, A.; Anbuvahini, N.A.; Kumaran, P.S.T.; Naresh, M.; Malathi, D.; Antony, U. Physicochemical properties of cookies enriched with xylooligosaccharides. *Food Sci. Technol. Int.* **2016**, *22*, 420–428. [CrossRef]
7. Palaniappan, A.; Antony, U.; Emmambux, M.N. Current status of xylooligosaccharides: Production, characterization, health benefits and food application. *Trends Food Sci. Technol.* **2021**, *111*, 506–519. [CrossRef]
8. Charalampopoulos, D.; A Rastall, R.A. Prebiotics in foods. *Curr. Opin. Biotechnol.* **2012**, *23*, 187–191. [CrossRef]
9. Zoghlami, A.; Paës, G. Lignocellulosic biomass: Understanding recalcitrance and predicting hydrolysis. *Front. Chem.* **2019**, *7*, 874. [CrossRef]
10. Qing, Q.; Li, H.J.; Kumar, R.; Wyman, C.E. Xylooligosaccharides production, quantification, and characterization in context of lignocellulosic biomass pretreatment. In *Aqueous Pretreatment of Plant Biomass for Biological and Chemical Conversion to Fuels and Chemicals*; Stevens, C.V., Wyman, C.E., Eds.; John Wiley and Sons, Ltd.: Hoboken, NJ, USA, 2013; pp. 391–415. [CrossRef]
11. Galbe, M.; Zacchi, G. Pretreatment: The key to efficient utilization of lignocellulosic materials. *Biomass- Bioenergy* **2012**, *46*, 70–78. [CrossRef]
12. Zhang, W.W.; You, Y.Z.; Lei, F.H.; Li, P.F.; Jiang, J.X. Acetyl-assisted autohydrolysis of sugarcane bagasse for the production of xylo-oligosaccharides without additional chemicals. *Bioresour. Technol.* **2018**, *265*, 387–393. [CrossRef] [PubMed]
13. Garrote, G.; Dominguez, H.; Parajo, J.C. Mild autohydrolysis: An environmentally friendly technology for xylooligosaccharide production from wood. *J. Chem Technol. Biot.* **1999**, *74*, 1101–1109. [CrossRef]
14. El Hage, R.; Chrusciel, L.; Desharnais, L.; Brosse, N. Effect of autohydrolysis of *Miscanthus* × *giganteus* on lignin structure and organosolv delignification. *Bioresour. Technol.* **2010**, *101*, 9321–9329. [CrossRef] [PubMed]
15. Xiao, L.P.; Shi, Z.J.; Xu, F.; Sun, R.C. Characterization of lignins isolated with alkaline ethanol from the hydrothermal pretreated *Tamarix ramosissima*. *BioEnergy Res.* **2013**, *6*, 519–532. [CrossRef]
16. Sonkar, R.M.; Gade, P.S.; Bokade, V.; Mudliar, S.N.; Bhatt, P. Ozone assisted autohydrolysis of wheat bran enhances xylooligosaccharide production with low generation of inhibitor compounds: A comparative study. *Bioresour. Technol.* **2021**, *338*, 125559. [CrossRef]
17. Xiao, X.; Bian, J.; Peng, X.P.; Xu, H.; Xiao, B.; Sun, R.C. Autohydrolysis of bamboo (*Dendrocalamus giganteus* Munro) culm for the production of xylo-oligosaccharides. *Bioresour. Technol.* **2013**, *138*, 63–70. [CrossRef] [PubMed]
18. Gullón, P.; Muñoz, M.J.G.; Parajó, J.C. Manufacture and prebiotic potential of oligosaccharides derived from industrial solid wastes. *Bioresour. Technol.* **2011**, *102*, 6112–6119. [CrossRef] [PubMed]
19. Lachos-Perez, D.; Martinez-Jimenez, F.; Rezende, C.A.; Tompsett, G.; Timko, M.; Forster-Carneiro, T. Subcritical water hydrolysis of sugarcane bagasse: An approach on solid residues characterization. *J. Supercrit. Fluids* **2016**, *108*, 69–78. [CrossRef]
20. Ruiz, H.A.; Rodríguez-Jasso, R.M.; Fernandes, B.D.; Vicente, A.A.; Teixeira, J.A. Hydrothermal processing, as an alternative for upgrading agriculture residues and marine biomass according to the biorefinery concept: A review. *Renew. Sustain. Energy Rev.* **2013**, *21*, 35–51. [CrossRef]
21. Xiao, L.P.; Shi, Z.J.; Xu, F.; Sun, R.C. Hydrothermal treatment and enzymatic hydrolysis of *Tamarix ramosissima*: Evaluation of the process as a conversion method in a biorefinery concept. *Bioresour. Technol.* **2013**, *135*, 73–81. [CrossRef]
22. Fu, Z.L.; Zhu, Y.T.; Teng, C.; Fan, G.S.; Li, X.T. Biochemical characterization of a novel feruloyl esterase from *Burkholderia pyrrocinia* B1213 and its application for hydrolyzing wheat bran. *3 Biotech* **2022**, *12*, 24. [CrossRef] [PubMed]
23. Sluiter, A.; Hames, B.; Ruiz, R.; Scarlata, C.; Sluiter, J.; Templeton, D.; Crocker, D. Determination of structural carbohydrates and lignin in biomass. *Lab. Anal. Proced.* **2008**, *1617*, 1–16.

24. Overend, R.P.; Chornet, E.; Gascoigne, J.A.; Hartley, B.S.; Broda, P.M.A.; Senior, P.J. Fractionation of lignocellulosics by steam-aqueous pretreatments. *Philos. Trans. R. Soc. Lond. Ser. A Math. Phys. Sci.* **1987**, *321*, 523–536. [CrossRef]
25. Fan, G.S.; Wu, Q.H.; Li, Q.; Sun, B.G.; Ma, Y.L.; Wu, K.L.; Wang, C.Y.; Teng, C.; Yang, R.; Li, X.T. Impact of the disulfide bond on hydrolytic characteristics of a xylanase from *Talaromyces thermophiles* F1208. *Int. J. Biol. Macromol.* **2020**, *164*, 1748–1757. [CrossRef] [PubMed]
26. Wu, Q.H.; Fan, G.S.; Yu, T.F.; Sun, B.G.; Tang, H.H.; Teng, C.; Yang, R.; Li, X.T. Biochemical characteristics of the mutant xylanase T-XynC(122)C(166) and production of xylooligosaccharides from corncobs. *Ind. Crop. Prod.* **2019**, *142*, 111848. [CrossRef]
27. Wang, D.D.; Wang, L.Y.; Wei, Y.X.; Cui, Q.; Sun, Z.J. Effects of solid-state fermentation of Nongxiang Baijiu on the recalcitrance of rice husk. *Liquor-Making Sci. Technol.* **2017**, 25–29. [CrossRef]
28. Fan, E.D.; Feng, M.X.; Li, C.Y.; Wu, D.G.; Chen, Y.F.; Xiao, D.G.; Guo, X.W. Study on improving the quality of distiller's grains feed by steam explosion combined with various microorganisms. *J. Agri. Biotechnol.* **2021**, *1*, 194–206. [CrossRef]
29. Seesuriyachan, P.; Kawee-Ai, A.; Chaiyaso, T. Green and chemical-free process of enzymatic xylooligosaccharide production from corncob: Enhancement of the yields using a strategy of lignocellulosic destructuration by ultra-high pressure pretreatment. *Bioresour. Technol.* **2017**, *241*, 537–544. [CrossRef]
30. Singh, R.D.; Banerjee, J.; Sasmal, S.; Muir, J.; Arora, A. High xylan recovery using two stage alkali pre-treatment process from high lignin biomass and its valorisation to xylooligosaccharides of low degree of polymerisation. *Bioresour. Technol.* **2018**, *256*, 110–117. [CrossRef]
31. Rajagopalan, G.; Shanmugavelu, K.; Yang, K.L. Production of prebiotic-xylooligosaccharides from alkali pretreated mahogany and mango wood sawdust by using purified xylanase of *Clostridium strain* BOH3. *Carbohydr. Polym.* **2017**, *167*, 158–166. [CrossRef]
32. Sabiha-Hanim, S.; Noor, M.A.M.; Rosma, A. Effect of autohydrolysis and enzymatic treatment on oil palm (*Elaeis guineensis* Jacq.) frond fibres for xylose and xylooligosaccharides production. *Bioresour. Technol.* **2011**, *102*, 1234–1239. [CrossRef] [PubMed]
33. Surek, E.; Buyukkileci, A.O. Production of xylooligosaccharides by autohydrolysis of hazelnut (*Corylus avellana* L.) shell. *Carbohydr. Polym.* **2017**, *174*, 565–571. [CrossRef]
34. Chen, M.H.; Bowman, M.J.; Dien, B.S.; Rausch, K.D.; Tumbleson, M.E.; Singh, V. Autohydrolysis of *Miscanthus* × *giganteus* for the production of xylooligosaccharides (XOS): Kinetics, characterization and recovery. *Bioresour. Technol.* **2014**, *155*, 359–365. [CrossRef] [PubMed]
35. Parajó, J.C.; Garrote, G.; Cruz, J.M.; Dominguez, H. Production of xylooligosaccharides by autohydrolysis of lignocellulosic materials. *Trends Food Sci. Technol.* **2004**, *15*, 115–120. [CrossRef]
36. Akpinar, O.; Gunay, K.; Yilmaz, Y.; Levent, O.; Bostanci, S. Enzymatic processing and antioxidant activity of agricultural waste autohydrolysis liquors. *Bioresources* **2010**, *5*, 699–711.
37. Surek, E.; Buyukkileci, A.O.; Yegin, S. Processing of hazelnut (*Corylus avellana* L.) shell autohydrolysis liquor for production of low molecular weight xylooligosaccharides by *Aureobasidium pullulans* NRRL Y–2311–1 xylanase. *Ind. Crop. Prod.* **2021**, *161*, 113212. [CrossRef]
38. Álvarez, C.; González, A.; Negro, M.J.; Ballesteros, I.; Oliva, J.M.; Sáez, F. Optimized use of hemicellulose within a biorefinery for processing high value-added xylooligosaccharides. *Ind. Crop. Prod.* **2017**, *99*, 41–48. [CrossRef]
39. Zheng, B.; Zhu, Y.K.; Zheng, S.H.; Mo, Y.E.; Sun, S.L.; Ren, J.L.; Li, Y.H.; Wu, A.M.; Li, H.L. Upgrade the torrefaction process of bamboo based on autohydrolysis pretreatment. *Ind. Crop. Prod.* **2021**, *166*, 113470. [CrossRef]
40. Jiang, Y.S.; Xing, M.Z.; Kang, Q.; Sun, J.Y.; Zeng, X.A.; Gao, W.H.; Li, H.H.; Gao, Y.H.; Li, A.J. Pulse electric field assisted process for extraction of Jiuzao glutelin extract and its physicochemical properties and biological activities investigation. *Food Chem.* **2022**, *383*, 132304. [CrossRef]
41. Gütsch, J.S.; Nousiainen, T.; Sixta, H. Comparative evaluation of autohydrolysis and acid-catalyzed hydrolysis of *Eucalyptus globulus* wood. *Bioresour. Technol.* **2012**, *109*, 77–85. [CrossRef]
42. Ho, A.L.; Carvalheiro, F.; Duarte, L.C.; Roseiro, L.B.; Charalampopoulos, D.; Rastall, R.A. Production and purification of xylooligosaccharides from oil palm empty fruit bunch fibre by a non-isothermal process. *Bioresour. Technol.* **2014**, *152*, 526–529. [CrossRef] [PubMed]
43. Jiang, Y.S.; Sun, J.Y.; Yin, Z.T.; Li, H.H.; Sun, X.T.; Zheng, F.P. Evaluation of antioxidant peptides generated from Jiuzao (residue after Baijiu distillation) protein hydrolysates and their effect of enhancing healthy value of Chinese Baijiu. *J. Sci. Food Agric.* **2020**, *100*, 59–73. [CrossRef]
44. Wang, Y.S.; Zhu, X.; Jin, L.Q.; Zheng, Y.; Liao, C.J.; Zheng, Y.G. Post-hydrolysis of the prehydrolysate from eucalyptus pulping with xylanase. *J. Clean. Prod.* **2017**, *142*, 2865–2871. [CrossRef]
45. Vázquez, M.J.; Alonso, J.L.; Domínguez, H.; Parajó, J.C. Enhancing the potential of oligosaccharides from corncob autohydrolysis as prebiotic food ingredients. *Ind. Crop. Prod.* **2006**, *24*, 152–159. [CrossRef]
46. Dias, L.M.; Neto, F.S.P.P.; Brienzo, M.; de Oliveira, S.C.; Masarin, F. Experimental design, modeling, and optimization of production of xylooligosaccharides by hydrothermal pretreatment of sugarcane bagasse and straw. *Biomass-Convers. Biorefinery* **2022**. [CrossRef]
47. Lian, Z.N.; Wang, Y.N.; Luo, J.; Lai, C.H.; Yong, Q.; Yu, S.Y. An integrated process to produce prebiotic xylooligosaccharides by autohydrolysis, nanofiltration and endo-xylanase from alkali-extracted xylan. *Bioresour. Technol.* **2020**, *314*, 123685. [CrossRef]
48. Menezes, B.D.S.; Rossi, D.M.; Squina, F.; Ayub, M.A.Z. Xylooligosaccharides production by fungi cultivations in rice husk and their application as substrate for lactic acid bacteria growth. *Bioresour. Technol. Rep.* **2018**, *2*, 100–106. [CrossRef]

49. da Silva Menezes, B.; Rossi, D.M.; Ayub, M.A.Z. Screening of filamentous fungi to produce xylanase and xylooligosaccharides in submerged and solid-state cultivations on rice husk, soybean hull, and spent malt as substrates. *World J. Microbiol. Biotechnol.* **2017**, *33*, 58. [CrossRef]
50. Liu, M.Q.; Huo, W.K.; Xu, X.; Weng, X.Y. Recombinant *Bacillus amyloliquefaciens* xylanase A expressed in *Pichia pastoris* and generation of xylooligosaccharides from xylans and wheat bran. *Int. J. Biol. Macromol.* **2017**, *105*, 656–663. [CrossRef]
51. Faryar, R.; Linares-Pastén, J.A.; Immerzeel, P.; Mamo, G.; Andersson, M.; Stålbrand, H.; Mattiasson, B.; Karlsson, E.N. Production of prebiotic xylooligosaccharides from alkaline extracted wheat straw using the K80R-variant of a thermostable alkali-tolerant xylanase. *Food Bioprod. Process.* **2015**, *93*, 1–10. [CrossRef]
52. Falck, P.; Aronsson, A.; Grey, C.; Stålbrand, H.; Karlsson, E.N.; Adlercreutz, P. Production of arabinoxylan-oligosaccharide mixtures of varying composition from rye bran by a combination of process conditions and type of xylanase. *Bioresour. Technol.* **2014**, *174*, 118–125. [CrossRef]
53. Liu, X.Q.; Liu, Y.; Jiang, Z.Q.; Liu, H.J.; Yang, S.Q.; Yan, Q.J. Biochemical characterization of a novel xylanase from *Paenibacillus barengoltzii* and its application in xylooligosaccharides production from corncobs. *Food Chem.* **2018**, *264*, 310–318. [CrossRef]
54. Purohit, A.; Rai, S.K.; Chownk, M.; Sangwan, R.S.; Yadav, S.K. Xylanase from *Acinetobacter pittii* MASK 25 and developed magnetic cross-linked xylanase aggregate produce predominantly xylopentose and xylohexose from agro biomass. *Bioresour. Technol.* **2017**, *244*, 793–799. [CrossRef]
55. Le, B.; Yang, S.H. Production of prebiotic xylooligosaccharide from aqueous ammonia-pretreated rice straw by β-xylosidase of *Weissella cibaria*. *J. Appl. Microbiol.* **2019**, *126*, 1861–1868. [CrossRef]
56. Goldbeck, R.; Gonçalves, T.A.; Damasio, A.R.L.; Brenelli, L.B.; Wolf, L.D.; Paixão, D.A.A.; Rocha, G.J.M.; Squina, F.M. Effect of hemicellulolytic enzymes to improve sugarcane bagasse saccharification and xylooligosaccharides production. *J. Mol. Catal. B Enzym.* **2016**, *131*, 36–46. [CrossRef]
57. Zhou, X.; Zhao, J.; Zhang, X.; Xu, Y. An eco-friendly biorefinery strategy for xylooligosaccharides production from sugarcane bagasse using cellulosic derived gluconic acid as efficient catalyst. *Bioresour. Technol.* **2019**, *289*, 121755. [CrossRef]
58. Reddy, S.S.; Krishnan, C. Production of high-pure xylooligosaccharides from sugarcane bagasse using crude β-xylosidase-free xylanase of *Bacillus subtilis* KCX006 and their bifidogenic function. *LWT—Food Sci. Technol.* **2016**, *65*, 237–245. [CrossRef]
59. Bragatto, J.; Segato, F.; Squina, F.M. Production of xylooligosaccharides (XOS) from delignified sugarcane bagasse by peroxide-HAc process using recombinant xylanase from *Bacillus subtilis*. *Ind. Crop. Prod.* **2013**, *51*, 123–129. [CrossRef]
60. Ruthes, A.C.; Martínez-Abad, A.; Tan, H.T.; Bulone, V.; Vilaplana, F. Sequential fractionation of feruloylated hemicelluloses and oligosaccharides from wheat bran using subcritical water and xylanolytic enzymes. *Green Chem.* **2017**, *19*, 1919–1931. [CrossRef]
61. Mathew, S.; Aronsson, A.; Karlsson, E.N.; Adlercreutz, P. Xylo- and arabinoxylooligosaccharides from wheat bran by endoxylanases, utilisation by probiotic bacteria, and structural studies of the enzymes. *Appl. Microbiol. Biotechnol.* **2018**, *102*, 3105–3120. [CrossRef]
62. Liu, X.Q.; Yang, S.Q.; Ma, J.W.; Yu, J.; Yan, Q.J.; Jiang, Z.Q. Efficient production of acetylated xylooligosaccharides from Hawthorn kernels by a xylanase from *Paecilomyces aerugineus*. *Ind. Crop. Prod.* **2020**, *158*, 112962. [CrossRef]
63. Huang, C.X.; Wang, X.C.; Liang, C.; Jiang, X.; Yang, G.; Xu, J.; Yong, Q. A sustainable process for procuring biologically active fractions of high-purity xylooligosaccharides and water-soluble lignin from *Moso* bamboo prehydrolyzate. *Biotechnol. Biofuels* **2019**, *12*, 189. [CrossRef]
64. Singh, R.D.; Nadar, C.G.; Muir, J.; Arora, A. Green and clean process to obtain low degree of polymerisation xylooligosaccharides from almond shell. *J. Clean. Prod.* **2019**, *241*, 118237. [CrossRef]
65. Amorim, C.; Silvério, S.C.; Rodrigues, L.R. One-step process for producing prebiotic arabino-xylooligosaccharides from brewer's spent grain employing Trichoderma species. *Food Chem.* **2019**, *270*, 86–94. [CrossRef]
66. Banerjee, S.; Patti, A.F.; Ranganathan, V.; Arora, A. Hemicellulose based biorefinery from pineapple peel waste: Xylan extraction and its conversion into xylooligosaccharides. *Food Bioprod. Process.* **2019**, *117*, 38–50. [CrossRef]
67. Akpinar, O.; Erdogan, K.; Bakir, U.; Yilmaz, L. Comparison of acid and enzymatic hydrolysis of tobacco stalk xylan for preparation of xylooligosaccharides. *LWT—Food Sci. Technol.* **2010**, *43*, 119–125. [CrossRef]
68. Samanta, A.K.; Jayapal, N.; Kolte, A.P.; Senani, S.; Sridhar, M.; Suresh, K.P.; Sampath, K.T. Enzymatic production of xylooligosaccharides from alkali solubilized xylan of natural grass (*Sehima nervosum*). *Bioresour. Technol.* **2012**, *112*, 199–205. [CrossRef]
69. Hao, X.X.; Wen, P.Y.; Wang, J.; Wang, J.Y.; You, J.X.; Zhang, J.H. Production of xylooligosaccharides and monosaccharides from hydrogen peroxide-acetic acid-pretreated poplar by two-step enzymatic hydrolysis. *Bioresour. Technol.* **2019**, *297*, 122349. [CrossRef]
70. Salas-Veizaga, D.M.; Villagomez, R.; Linares-Pastén, J.A.; Carrasco, C.; Álvarez, M.T.; Adlercreutz, P.; Karlsson, E.N. Extraction of glucuronoarabinoxylan from quinoa stalks (*Chenopodium quinoa* Willd.) and evaluation of xylooligosaccharides produced by GH10 and GH11 xylanases. *J. Agric. Food Chem.* **2017**, *65*, 8663–8673. [CrossRef]
71. Zhang, M.Y.; Liao, A.M.; Thakur, K.; Huang, J.H.; Zhang, J.G.; Wei, Z.J. Modification of wheat bran insoluble dietary fiber with carboxymethylation, complex enzymatic hydrolysis and ultrafine comminution. *Food Chem.* **2019**, *297*, 124983. [CrossRef]
72. Kaushik, A.; Singh, M. Isolation and characterization of cellulose nanofibrils from wheat straw using steam explosion coupled with high shear homogenization. *Carbohydr. Res.* **2011**, *346*, 76–85. [CrossRef]

73. Gil-López, D.L.; Lois-Correa, J.A.; Sánchez-Pardo, M.E.; Domínguez-Crespo, M.A.; Torres-Huerta, A.M.; Rodríguez-Salazar, A.E.; Orta-Guzmán, V.N. Production of dietary fibers from sugarcane bagasse and sugarcane tops using microwave-assisted alkaline treatments. *Ind. Crop. Prod.* **2019**, *135*, 159–169. [CrossRef]
74. Peng, F.; Ren, J.L.; Xu, F.; Bian, J.; Peng, P.; Sun, R.C. Comparative study of hemicelluloses obtained by graded ethanol precipitation from sugarcane bagasse. *J. Agric. Food Chem.* **2009**, *57*, 6305–6317. [CrossRef]

MDPI AG
Grosspeteranlage 5
4052 Basel
Switzerland
Tel.: +41 61 683 77 34

Foods Editorial Office
E-mail: foods@mdpi.com
www.mdpi.com/journal/foods

www.ingramcontent.com/pod-product-compliance
Lightning Source LLC
Chambersburg PA
CBHW061516210926
51850CB00008B/1940
9783725814831